The Audible Past

Jonathan Sterne

DUKE UNIVERSITY PRESS Durham & London   2003

The Audible Past

CULTURAL ORIGINS OF SOUND REPRODUCTION

©2003 Duke University Press

All rights reserved

Printed in the United States of America on acid-free paper ∞

Designed by Amy Ruth Buchanan

Typeset in Garamond 3 by G&S Typesetters, Inc.

Library of Congress Cataloging-in-Publication Data appear on the last printed page of this book.

For Carrie

Contents

List of Figures ix

List of Abbreviations for Archival and Other Historical Materials Cited xi

Acknowledgments xiii

Hello! 1

1. Machines to Hear for Them 31
2. Techniques of Listening 87
3. Audile Technique and Media 137
4. Plastic Aurality: Technologies into Media 179
5. The Social Genesis of Sound Fidelity 215
6. A Resonant Tomb 287

Conclusion: Audible Futures 335

Notes 353

Bibliography 415

Index 437

List of Figures

1. Bell and Blake's ear phonautograph 32
2. Leon Scott's phonautograph 36
3. Drawing of sound refraction from S. Morland, *Tuba Stentoro-Phonica* 43
4. Edison's tinfoil phonograph 47
5. Berliner's gramophone 47
6. Early gramophone record 48
7. Early Reis telephone 78
8. The human telephone, 1897 82
9. "You *Need* a Headset"—1925 advertisement for Brandes headphones 88
10. "How Far Can I Hear with the MR-6?"—DeForest advertisement from the early 1920s 89
11. Diagram of Laennec's stethoscope 104
12. Monaural stethoscopes 105
13. Cammann's stethoscope 112
14. Binaural stethoscopes 113
15a-b. Telegraphy—ancient and modern 139
16. Diagram of printing telegraph 145
17. Train telegraph operator, 1890 159
18. Hearing tubes in Edison catalog, ca. 1902 162
19. Advertisement for the Berliner gramophone 164
20. Students at the Marconi Wireless School 165
21. Cartoon of frustrated housewife, 1923 166

22. "The Instantaneous Answer"—an N. W. Ayer advertisement for AT&T's phone service, 1910 169
23. "The Man in the Multitude"—another N. W. Ayer advertisement for AT&T's phone service, 1915 170
24. "Her Voice Alluring Draws Him On"—cover of the July 1913 *Telephone Review* (New York) 171
25. "Tuning In"—cartoon from the 6 September 1923 *Syracuse Telegram* 173
26. "Ecstatic Interference"—artist's drawing, 1922 175
27. Victorian woman with headset and radio 175
28. Cartoons from the April 1923 *Wireless Age* 176
29. "Which Is Which?"—Victor Talking Machine advertisement, 1908 217
30. "The Human Voice *Is* Human"—Victor Victrola advertisement, 1927 224
31. "How Radio Broadcasting Travels"—RCA diagram, 1920s 226
32a-c. Advertising cards that depict telephone networks, ca. 1881 227
33. Artist's illustration of Elisha Gray's telephone, 1890 228
34. Drawing from the home notebook of Charles Sumner Tainter, 3 April 1881 229
35. Drawing from the home notebook of Charles Sumner Tainter, 19 November 1882 229
36. Dictating and listening to the graphophone, 1888 230
37. Patent drawing for photophonic receiver, 1881 231
38. Another patent drawing for photophonic receiver, 1881 232
39. Speaking to the photophone, 1884 233
40. Listening to the photophone, 1884 233
41. Telephone drawings from Alexander Graham Bell's notebooks, 1876 234
42a-b. Posters for telephone concerts 252–53
43. "Dr. Jekyll and Mr. Hyde at the Telephone"—AT&T advertisement 265
44a-b. "Ye Telephonists of 1877"—cartoon 271
45. "Winnie Winkle, Breadwinner"—1924 cartoon 273
46. "The Indestructible Records"—ca. 1908 advertisement for the Indestructible Phonographic Record Co. 300
47. "Indestructible Phonographic Records—Do Not Wear Out"—1908 advertisement 301
48. *His Master's Voice* 302

List of Abbreviations

for Archival and Other Historical Materials Cited

AGB	Alexander Graham Bell Collection, Manuscripts Division, Library of Congress
CST	Charles Sumner Tainter Collection, Archives Center, National Museum of American History
EB	Emile Berliner Collection, Recorded Sound Reference Center, Library of Congress
EBM	Emile Berliner Collection, Manuscripts Division, Library of Congress
EG	Elisha Gray Collection, Archives Center, National Museum of American History
GHC	George H. Clark Radioana Collection, Archives Center, National Museum of American History
NWA	N. W. Ayer Collection, Archives Center, National Museum of American History
WBA	Warshaw Business Americana Collection, Archives Center, National Museum of American History
Phonogram I	Periodical published by the National Phonograph Company, 1891–93
Phonogram II	Periodical published by the Edison Phonograph Company, 1900–1902

Acknowledgments

Although a single name appears on the cover, no one ever writes a book alone. There are many people I want to thank for their support and camaraderie as I have journeyed through this project. There was always someone around to share in my feelings of enchantment, confusion, exhilaration, and occasional despondence. This long book requires many acknowledgments: I have a lot of pent-up gratitude.

I would like to start by thanking *all* my teachers, although there are a few who warrant special distinction. Several teachers at the University of Illinois made the intellectual and institutional space for me to do this kind of work. Many of my teachers have also gone on to become trusted friends and confidants. Larry Grossberg helped me develop an interest in cultural theory and cultural studies into a facility, and he encouraged my intellectual eclecticism and scholarly imagination in a world that too often values disciplinarity over creativity. John Nerone almost single-handedly taught me the craft of historiography and has been a model of intellectual generosity and pluralism. James Hay taught me to see themes in my writing that I would not have otherwise seen and always pushed me to take the next step in interpreting my material. Tom Turino helped me rethink my approach to the study of sound and music from the ground up and helped me integrate my experiential knowledge as a musician into my academic research.

Many other teachers have made essential contributions to my development as a scholar. John Archer, Cliff Christians, Richard Leppert, John Lie, Steve Macek, Lauren Marsh, Cameron McCarthy, Roger Miller, Cary

Nelson, Gary Thomas, and Paula Treichler have all nurtured my writing and thinking. Mark Rubel helped me remember why I undertook this project in the first place. Andrea Press and Bruce Williams have offered some indispensable professional mentorship.

I am grateful to the University of Pittsburgh, the Richard D. and Mary Jane Edwards Endowed Publication Fund, the Smithsonian Institution, and the Graduate College of the University of Illinois for financial support while researching and writing this book. Some of the material in chapter 6 appears in different form as "Sound Out of Time/Modernity's Echo," in *Turning the Century: Essays in Cultural Studies,* ed. Carol Stabile (Boulder, Colo.: Westview, 2000), 9–32. Part of an earlier version of chapter 2 appears as "Mediate Auscultation, the Stethoscope, and the 'Autopsy of the Living': Medicine's Acoustic Culture," *Journal of Medical Humanities* 22, no. 2 (summer 2001): 115–36. An earlier version of chapter 1 appears as "A Machine to Hear for Them: On the Very Possibility of Sound's Reproduction," *Cultural Studies* 15, no. 2 (spring 2001): 259–94.

While I was a fellow at the Smithsonian, my intern, Keith Bryson, helped me track down interesting leads in the archives and in periodicals. Steve Lubar and Charles McGovern provided ample intellectual and practical support as my advisers and guides while I was there. Katherine Ott, Carlene Stephens, and Elliott Sivowich helped me sort through materials and pointed me in promising directions. Other fellows provided crucial conversation and support: Atsushi Akera, Adrienne Berney, John Hartigan, Rebecca Lyle, Scott Sandage, and Scott Trafton are all great company if you get the chance. I would also like to thank the many librarians, archivists, and reference staff who have helped me throughout the research process. Sam Brylawski at the Library of Congress and Deborra Richardson at the Smithsonian have been particularly helpful.

I have been blessed with a remarkable group of colleagues, friends, and students in Pittsburgh. Carol Stabile defies categorization: she is a remarkable teacher, a wise mentor, a trusted colleague, an intellectual coconspirator, a confidante, and most of all, a wonderful friend. More than anyone else, she has taught me the value of bringing humor and joy to scholarship and to politics. Melissa Butler, Danae Clark, Zack Furness, Bill Fusfield, Mark Harrison, Bridget Kilroy, Gordon Mitchell, Jessica Mudry, Lester Olson, John Poulakos, Michelle Silva, Pete Simonson, and Vanda Rakova Thorne have also been cherished conversation partners, colleagues, and friends. They challenge me to stay true to my intellectual commitments, but they never let me take them for granted. John Lyne has been a supportive chair. John Erlen has been a helpful and graceful colleague in making Pitt's history-of-medicine collections available to me.

My graduate students have been incredibly supportive and (when needed) understanding as well. I want to thank *all* of them, especially the many students in my seminars these past three years for their engagement and camaraderie: you have taught me well. Additional thanks to Maxwell Schnurer and Elena Cattaneo for adopting me shortly after my arrival and to Regina Renk and Janet McCarthy (and Diane Tipps at the University of Illinois) for their help in the many mundane tasks involved in finishing a book manuscript.

Other friends and colleagues have offered essential help along the way. John Durham Peters and Will Straw provided careful, challenging, and immensely stimulating comments on earlier drafts of the manuscript. Without their help, this would be a different book. Jennifer Daryl Slack, M. Medhi Semati, and Patty Sotirin helped me rethink some crucial ideas about nature and culture over a long weekend in the Upper Peninsula. Kevin Carollo, Greg Dimitriadis, Jayson Harsin, Anahid Kassabian, Pete Simonson, and Carol Stabile gave me careful readings when I needed them most.

My editor at Duke University Press, Ken Wissoker, is outstanding: he pushed me at just the right times, and he encouraged me all the time. It has been a *great* experience working with everyone at the press—the entire staff has been immensely helpful and accommodating.

Many friends have been a source of immense joy and personal and intellectual support; their friendship helps make my life what it is: Michael Bérubé, Kevin Carollo, Janet Lyon, Tom Robbins, and Mike Witmore have been great friends, great intellectual companions, and great bandmates over the years. In addition to my bandmates, many friends, colleagues (near and far), and interlocutors helped shape my life during the writing of this book: Steve Bailey, Shannon and Craig Bierbaum, Jack Bratich, Dave Breeden, Andy Cantrell, Julie Davis, Melissa Deem, Greg Dimitriadis, Ariel Ducey, Greg Elmer, Lisa Friedman, Loretta Gaffney, Paula Gardner, Kelly Gates, Ron Greene, Melanie Harrison, Jayson Harsin, Rob Henn, Toby Higbie, Amy Hribar, Nan Hyland, Steve Jahn, Steve Jones, Lisa King, Sammi King, Elizabeth Majerus, Craig Matarrese, Dan McGee, Matt Mitchell, Radhika Mongia, Negar Mottahedeh, Dave Noon, Jeremy Packer, Craig Robertson, Gil Rodman, Wayne Schneider, Heather and Rob Sloane, Gretchen Soderlund, Mark Unger, Mary Vavrus, Dan Vukovich, Fred Wasser, Greg Wise, and many others whom I cannot name here deserve much more lauding and gratitude than one can offer in the space of an acknowledgments section.

The members of the *Bad Subjects* Production Team have been constant interlocutors for the past nine years now. My work as an editor of the 'zine and my almost-daily interactions with listserv members has shaped my thinking in countless ways. Special thanks to Charlie Bertsch and Joel Schalit for bringing me in, filling me in, and being there to bail me out on more than one occasion.

My family, Muriel Sterne, Philip Griffin, David, Lori, Abby, and Adam Sterne, and Helen and Mario Avati, have always been there for me—even when they were not quite sure what I was up to. Myron Weinstein was a great uncle: he put me up for a summer while I researched in Washington, D.C.; I learned what I know about research from one of the true masters. I think of him every time I visit Washington.

The final space in these acknowledgments belongs to my life companion, Carrie Rentschler, who is everything to me. Carrie has read parts of this book too many times to count. Over years of conversation and reading, she has offered essential comments at every stage of the process, from conception to completion. I am unspeakably grateful for her generosity, wit, humor, intellect, and affection every day that we are together. But, even more, I am grateful for her company as we go through life. I dedicate this book to her.

 # Hello!

Here are the tales currently told: Alexander Graham Bell and Thomas Watson had their first telephone conversation in 1876. "Mr. Watson— Come here— I want to see you!" yelled Bell to Watson, and the world shook. Thomas Edison first heard his words—"Mary had a little lamb"— returned to him from the cylinder of a phonograph built by his assistants in 1878, and suddenly the human voice gained a measure of immortality. Guglielmo Marconi's wireless telegraph conquered the English channel in 1899. Unsuspecting navy personnel first heard voices coming over their radios in 1906. Each event has been claimed as a turning point in human history. Before the invention of sound-reproduction technologies, we are told, sound withered away. It existed only as it went out of existence. Once telephones, phonographs, and radios populated our world, sound had lost a little of its ephemeral character. The voice became a little more unmoored from the body, and people's ears could take them into the past or across vast distances.

These are powerful stories because they tell us that something happened to the nature, meaning, and practices of sound in the late nineteenth century. But they are incomplete.[1] If sound-reproduction technologies changed the way we hear, where did they come from? Many of the practices, ideas, and constructs associated with sound-reproduction technologies predated the machines themselves. The basic technology to make phonographs (and, by extension, telephones) existed for some time prior to their actual invention.[2] So why did sound-reproduction technologies emerge when they did and not at some other time? What preceded them that made them possible, desirable, effective, and meaningful? In what milieu did they dwell? How and why did sound-reproduction technologies take on the particular technological and cultural forms and functions that they did? To answer these questions, we move from considering simple mechanical possibility out into the social and cultural worlds from which the technologies emerged.

The Audible Past offers a history of the *possibility* of sound reproduction—the telephone, the phonograph, radio, and other related technologies. It examines the social and cultural conditions that gave rise to sound reproduction and, in turn, how those technologies crystallized and combined larger cultural currents. Sound-reproduction technologies are artifacts of vast transformations in the fundamental nature of sound, the human ear, the faculty of hearing, and practices of listening that occurred over the long nineteenth century. Capitalism, rationalism, science, colonialism, and a host of other factors—the "maelstrom" of modernity, to use Marshall Berman's phrase—all affected constructs and practices of sound, hearing, and listening.[3]

As there was an Enlightenment, so too was there an "Ensoniment." A series of conjunctures among ideas, institutions, and practices rendered the world audible in new ways and valorized new constructs of hearing and listening. Between about 1750 and 1925, sound itself became an object and a domain of thought and practice, where it had previously been conceptualized in terms of particular idealized instances like voice or music. Hearing was reconstructed as a physiological process, a kind of receptivity and capacity based on physics, biology, and mechanics. Through techniques of listening, people harnessed, modified, and shaped their powers of auditory perception in the service of rationality. In the modern age, sound and hearing were reconceptualized, objectified, imitated, transformed, reproduced, commodified, mass-produced, and industrialized. To be sure, the transformation of sound and hearing took well over a century. It is not that people woke up one day and found everything suddenly different. Changes in sound, listening, and hearing happened bit by bit, place by place, practice by practice, over a long period of time.

"The golden age of the ear never ended," writes Alan Burdick. "It continues, occluded by the visual hegemony."[4] *The Audible Past* tells a story where sound, hearing, and listening are central to the cultural life of modernity, where sound, hearing, and listening are foundational to modern modes of knowledge, culture, and social organization. It provides an alternative to the pervasive narrative that says that, in becoming modern, Western culture moved away from a culture of hearing to a culture of seeing. There is no doubt that the philosophical literature of the Enlightenment—as well as many people's everyday language—is littered with light and sight metaphors for truth and understanding.[5] But, even if sight is in some ways the privileged sense in European philosophical discourse since the Enlightenment, it is fallacious to think that sight alone or in its supposed difference from hearing explains modernity.

There has always been a heady audacity to the claim that vision is the social chart of modernity. While I do not claim that listening is *the* social chart of modernity, it certainly charts a significant field of modern practice. There is always more than one map for a territory, and sound provides a particular path through history. In some cases—as this book will demonstrate—modern ways of hearing prefigured modern ways of seeing. During the Enlightenment and afterward, the sense of hearing became an object of contemplation. It was measured, objectified, isolated, and simulated. Techniques of audition developed by doctors and telegraphers were constitutive characteristics of scientific medicine and early versions of modern bureaucracy. Sound was commodified; it became something that can be bought and sold. These facts trouble the cliché that modern science and rationality were outgrowths of visual culture and visual thinking. They urge us to rethink exactly what we mean by the *privilege* of vision and images.[6] To take seriously the role of sound and hearing in modern life is to trouble the visualist definition of *modernity*.

Today, it is understood across the human sciences that vision and visual culture are important matters. Many contemporary writers interested in various aspects of visual culture (or, more properly, visual aspects of various cultural domains)—the arts, design, landscape, media, fashion—understand their work as contributing to a core set of theoretical, cultural, and historical questions about vision and images. While writers interested in visual media have for some time gestured toward a conceptualization of *visual culture,* no such parallel construct—*sound culture* or, simply, *sound studies*—has broadly informed work on hearing or the other senses.[7] While sound is considered as a unified intellectual problem in some science and engineering fields, it is less developed as an integrated problem in the social and cultural disciplines.

Similarly, visual concerns populate many strains of cultural theory. The question of *the gaze* haunts several schools of feminism, critical race theory, psychoanalysis, and poststructuralism. The cultural status of *the image* and seeing occupies great minds in semiotics, film studies, several schools of literary and art-historical interpretation, architecture, and communication. While sound may interest individual scholars in these areas, it is still too often considered a parochial or specialized concern. While there are many scholars of sound active in communication, film studies, music, and other human sciences, sound is not usually a central theoretical problem for major schools of cultural theory, apart from the privilege of the voice in phenomenology and psychoanalysis and its negation in deconstruction.[8]

It would be possible to write a different book, one that explains and criticizes scholars' preference for visual objects and vision as an object of study. For now, it is enough to note that the fault lies with both cultural theorists and scholars of sound. Cultural theorists too easily accept pieties about the dominance of vision and, as a result, have elided differences between the privilege of vision and the totality of vision. Meanwhile, studies of sound tend to shy away from questions of sound culture as such (with a few notable exceptions) and prefer instead to work within other disciplinary or interdisciplinary intellectual domains. By *not* gesturing back toward a more general level of questioning, these works offer an implicitly cumulativist epistemology of the history of sound. The promise of cumulativist approaches is that one day we will have enough historical information to begin generalizing about society. The problem with this perspective is that such a remarkable day is always just over the horizon.[9] If sound and hearing are indeed significant theoretical problems, then now is as good a time as any to begin dealing with them as broad intellectual matters.

Many authors have claimed that hearing is the neglected sense in modernity, a novel sense for analysis.[10] It would perhaps be polemically acceptable at this point to lament the relative lack of scholarly work on sound as compared with images and vision, chart the pioneers, and then claim that this book will fill the gap. But the reality is somewhat different. There is a vast literature on the history and philosophy of sound; yet it remains conceptually fragmented. For the interested reader, there is a wealth of books and articles available on different aspects of sound written by scholars of communication, music, art, and culture.[11] But, without some kind of overarching, shared sensibility about what constitutes *the history of sound, sound culture,* or *sound studies,* piecing together a history of sound from the bewildering array of stories about speech, music, technology, and other sound-related practices has all the promise and appeal of piecing together a pane of shattered glass. We know that the parts line up somehow, we know that they can connect, but we are unsure of how they actually link together. We have histories of concert audiences, telephones, speeches, sound films, soundscapes, and theories of hearing. But only rarely do the writers of histories of sound suggest how their work connects with other, related work or with larger intellectual domains. Because scholarship on sound has not consistently gestured toward more fundamental and synthetic theoretical, cultural, and historical questions, it has not been able to bring broader philosophical questions to bear on the various intellectual fields that it inhabits. The challenge, then, is to imagine sound as a problem that moves beyond its immediate empirical context. The history of sound is already connected to the larger projects of the human sciences; it is up to us to flesh out the connections.

In positing a history of sound, *The Audible Past* extends a long tradition of interpretive and critical social thought. Some authors have quoted the young Marx on the importance of sensory history: "The forming of the five senses is a labor of the entire history of the world down to the present." Marx's passage signals that the very capacity to relate to the world through one's senses is organized and learned differently in different social settings. The senses are "cultivated or brought into being." "Man himself becomes the object" to be shaped and oriented through historical and social process.[12] Before the senses are real, palpable, concrete, or available for contemplation, they are already affected and effected through the particular historical conditions that also give rise to the subject who possesses them. We can fully consider the senses as historical only if we consider society, culture, technology, and the body as themselves artifacts of human history. A truly historicist understanding of the senses—or of a particular sense—therefore requires a commitment to the constructivist and contextualist strain of social and cultural thought. Conversely, a vigorous constructivism and a vigorous contextualism require a history of the senses. It is no accident that Marx's discussion of the senses appears in a section on communism in the *Economic and Philosophic Manuscripts of 1844*. Even to begin imagining (another) society, the young Marx had to consider the historical dynamics of sensation itself. As we imagine the possibilities of social, cultural, and historical change—in the past, present, or future—it is also our task to imagine histories of the senses. It is widely accepted that "the individual observer became an object of investigation and a locus of knowledge beginning in the first few decades of the 1800s" and that, during that same period, "the status of the observing subject was transformed."[13] So, too, transformations in sound, hearing, and listening were part of massive shifts in the landscapes of social and cultural life of the last three centuries.

The emergence of sound-reproduction technology in the nineteenth and twentieth centuries provides a particularly good entry into the larger history of sound. It is one of the few extant sites in the human sciences where scholars have acknowledged and contemplated the historicity of hearing. As Theodor Adorno, Walter Benjamin, and countless other writers have argued, the problem of mechanical reproduction is central to understanding the changing shape of communication in the late nineteenth and early twentieth centuries. For them, the compelling problem of sound's reproducibility, like the reproduction of images, was its seeming abstraction from the social world even as it was manifested more dynamically within it.[14] Other writers have offered even stronger claims for sound reproduction: it has been described as a "material foundation" of the changing senses of space and time at the turn of the twentieth century, part of a "perceptual revolution" in the early twentieth century. Sound technologies are said to have amplified and extended sound and our sense of hearing across time and space.[15] We are told that telephony altered "the conditions of daily life"; that sound recording represented a moment when "everything suddenly changed," a "shocking emblem of modernity"; that radio was "the most important electronic invention" of the twentieth century, transforming our perceptual habits and blurring the boundaries of private, public, commercial, and political life.[16]

Taken out of context or with a little hostility, claims for the historical significance of sound reproduction may seem overstated or even grandiose. D. L. LeMahieu writes that sound recording was one of "a score of new technologies thrust upon a population increasingly accustomed to mechanical miracles. In a decade when men learned to fly, the clock-sprung motor of a portable gramophone or the extended playing time of a double-sided disk hardly provoked astonishment. Indeed, what may be most remarkable was the rapidity with which technological innovations became absorbed into everyday, commonplace experience."[17] The same could be said for telephony, radio, and many other technologies. Yet LeMahieu's more sober prose still leaves room for wonder—not at the revolutionary power of sound-reproduction technology, but at its banality. If modernity, in part, names the experience of rapid social and cultural change, then its "shocking emblems" may very well have been taken in stride by some of its people.

Because sound-reproduction technology's role in history is so easily treated as self-evidently decisive, it makes sense to begin rewriting the history of sound by reconsidering the historical significance of sound technologies. A focus on sound-reproduction technology has an added advantage for the historian of sound: during their early years, technologies leave huge paper trails, thus making them especially rich resources for historical research. In early writings about the telephone, phonograph, and radio, we find a rich archive of reflections on the nature and meaning of sound, hearing, and listening. Douglas Kahn writes that, "as a historical object, sound cannot furnish a good story or consistent cast of characters nor can it validate any ersatz notion of progress or generational maturity. The history is scattered, fleeting, and highly mediated—it is as poor an object in any respect as sound itself."[18] Prior to the twentieth century, very little of the sonic past was physically preserved for historical analysis at a later date. So it makes sense to look instead at a particular domain of practice associated with sound. The paper trail left by sound-reproduction technologies provides one useful starting point for a history of sound.

Like an examination of the sense organs themselves, an examination of sound technologies also cuts to the core of the nature/nurture debate in thinking about the causes of and possibilities for historical change. Even the most basic mechanical workings of sound-reproduction technologies are historically shaped. As I will argue, the vibrating diaphragm that allowed telephones and phonographs to function was itself an artifact of changing understandings of human hearing. Sound-reproduction technologies are artifacts of particular practices and relations "all the way down"; they can be considered archaeologically. The history of sound technology offers a route into a field of conjunctures among material, economic, technical, ideational, practical, and environmental changes. Situated as we are amid torrential rains of capitalist development and marketing that pelt us with new digital machinery, it is both easy and tempting to forget the enduring connection between any technology and a larger cultural context. Technologies sometimes enjoy a certain level of deification in social theory and cultural history, where they come to be cast as divine actors. In "impact" narratives, technologies are mysterious beings with obscure origins that come down from the sky to "impact" human relations. Such narratives cast technologies themselves as primary agents of historical change: technological deification is the religion behind claims like "the telephone changed the way we do business," "the phonograph changed the way we listen to music." Impact narratives have been rightly and widely criticized as a form of technological determinism; they spring from an impoverished notion of causality.[19]

At the same time, technologies are interesting precisely because they can play a significant role in people's lives. Technologies are repeatable social, cultural, and material processes crystallized into mechanisms. Often, they perform labor that had previously been done by a person. It is this process of crystallization that makes them historically interesting. Their mechanical character, the ways in which they commingle physics and culture, can tell us a great deal about the people who build and deploy them. Technologies manifest a designed mechanical agency, a set of functions cordoned off from the rest of life and delegated to them, a set of functions developed from and linked to sets of cultural practices. People design and use technologies to enhance or promote certain activities and discourage others. Technologies are associated with habits, sometimes crystallizing them and sometimes enabling them. They embody in physical form particular dispositions and tendencies. The door closer tends to close the door unless I stop it with my hand or a doorstop. The domestic radio set receives but does not broadcast unless I do a little rewiring and add a microphone. The telephone rings while I write the introduction to this book. After years of conditioning to respond to a ringing telephone, it takes some effort to ignore it and finish the sentence or paragraph. To study technologies in any meaningful sense requires a rich sense of their connection with human practice, habitat, and habit. It requires attention to the fields of combined cultural, social, and physical activity—what other authors have called *networks* or *assemblages*—from which technologies emerge and of which they are a part.[20]

The story presented in these pages spirals out from an analysis of the mechanical and physical aspects of the technologies themselves to the techniques, practices, and institutions associated with them. At each juncture in the argument, I show how sound-reproduction technologies are shot through with the tensions, tendencies, and currents of the culture from which they emerged, right on down to their most basic mechanical functions. Our most cherished pieties about sound-reproduction technologies—for instance, that they separated sounds from their sources or that sound recording allows us to hear the voices of the dead—were not and are not innocent empirical descriptions of the technologies' impact. They were wishes that people grafted onto sound-reproduction technologies—wishes that became programs for innovation and use.

For many of their inventors and early users, sound-reproduction technologies encapsulated a whole set of beliefs about the age and place in which they lived. Sound-reproduction technologies represented the promise of science, rationality, and industry and the power of the white man to co-opt and supersede domains of life that were previously considered to be magical. For their early users, sound technologies were—in a word—modern.[21] *Modernity* is of course a cloudy analytic category, fraught with internal contradictions and intellectual conflicts. Its difficulty probably stems from its usefulness as a heuristic term, and my use of it is deliberately heuristic. When I claim that sound-reproduction technology indexes an acoustic modernity, I do not mean quite the same thing as the subjects of my history. *The Audible Past* explores the ways in which the history of sound contributes to and develops from the "maelstrom" of modern life (to return to Berman): capitalism, colonialism, and the rise of industry; the growth and development of the sciences, changing cosmologies, massive population shifts (specifically migration and urbanization), new forms of collective and corporate power, social movements, class struggle and the rise of new middle classes, mass communication, nation-states, bureaucracy; confidence in progress, a universal abstract humanist subject, and the world market; and a reflexive contemplation of the constancy of change.[22] In modern life, sound becomes a problem: an object to be contemplated, reconstructed, and manipulated, something that can be fragmented, industrialized, and bought and sold.

But *The Audible Past* is not a simple modernization narrative for sound and hearing. *Modernization* can too easily suggest a brittle kind of universalism, where the specific historical developments referenced by *modernity* are transmogrified into a set of historical stages through which all cultures must pass. In Johannes Fabian's apt phrase, the idea of modernity as modernization turns relations of space—relations between cultures—into relations of time, where the white man stands at the pinnacle of world evolution.[23] While I am not an exponent of a developmental theory of modernity as "modernization," it is such a central element of some discourses about sound reproduction that we will confront it more than once in the following pages. A long line of inventors, scholars, businesspeople, phonographic anthropologists, and casual users thought of themselves as partaking in a modern way of life, as living at the pinnacle of the world's progress. They believed that their epoch rode the crest of modernization's unstoppable wave. So, in addition to being a useful heuristic for describing the context of the project as a whole, *modernity* and its conjugates are also important categories to be analyzed and carefully taken apart within this history.

The remainder of this introduction provides some conceptual background for the history that follows. The next section is an extended consideration of sound as an object of historical study: what does it mean to write a history of something so apparently natural and physical as sound and hearing? A more detailed map of the book's arguments then follows.

Rethinking Sound's Nature:
Of Forests, Fallen Trees, and Phenomenologies

All this talk of modernity, history, and sound technology conjures an implied opposite: the *nature* of sound and hearing. Insofar as we treat sound as a fact of nature, writing something other than its natural history might seem like an immodest or inappropriate endeavor—at best it could aspire only to partiality. Although film scholars have been using the phrase *history of sound* for some time, it has an uneasy ring to it. After all, scholars of the visible world do not write "histories of light" (although perhaps they should), instead preferring to write histories of "visual culture," "images," "visuality," and the like. Bracketing light in favor of "the visual" may be a defensive maneuver since the various visual terms conveniently bracket questions of the nature of nature. But, besides sounding good, *history of sound* already embodies a hard-to-grasp but necessary paradox of nature and culture central to everything that follows in this book. At its core, the phenomenon of sound and the history of sound rest at the in-between point of culture and nature.

It is impossible to "merely describe" the faculty of hearing in its natural state. Even to try is to pretend that language has no figurative dimension of its own. The language that we use to describe sound and hearing comes weighted down with decades or centuries of cultural baggage. Consider the careers of two adjectives associated with the ear in the English language. The term *aural* began its history in 1847 meaning "of or pertaining to the organ of hearing"; it did not appear in print denoting something "received or perceived by the ear" until 1860. Prior to that period, the term *auricular* was used to describe something "of or pertaining to the ear" or perceived by the ear.[24] This was not a mere semantic difference: *auricular* carried with it connotations of oral tradition and hearsay as well as the external features of the ear visible to the naked eye (the folded mass of skin that is often synecdochally referred to as the ear is technically either the *au-*

ricle, the *pinna,* or the *outer ear*). *Aural,* meanwhile, carried with it no connotations of oral tradition and referred specifically to the middle ear, the inner ear, and the nerves that turn vibrations into what the brain perceives as sound (as in *aural surgery*). The idea of the aural and its decidedly medical inflection is a part of the historical transformation that I describe in the following pages.

Generally, when writers invoke a binary coupling between culture and nature, it is with the idea that culture is that which changes over time and that nature is that which is permanent, timeless, and unchanging. The nature/culture binary offers a thin view of nature, a convenient straw figure for "social construction" arguments.[25] In the case of sound, the appeal to something static is also a trick of the language. We treat sound as a natural phenomenon exterior to people, but its very definition is anthropocentric. The physiologist Johannes Müller wrote over 150 years ago that, "without the organ of hearing with its vital endowments, there would be no such a thing as sound in the world, but merely vibrations."[26] As Müller pointed out, our other senses can also perceive vibration. Sound is a very particular perception of vibrations. You can take the sound out of the human, but you can take the human out of the sound only through an exercise in imagination. Sounds are defined as that class of vibrations perceived—and, in a more exact sense, sympathetically produced—by the functioning ear when they travel through a medium that can convey changes in pressure (such as air). The numbers for the range of human hearing (which absolutely do not matter for the purposes of this study) are twenty to twenty thousand cycles per second, although in practice most adults in industrial society cannot hear either end of that range. We are thus presented with a choice in our definition: we can say either that sound is a class of vibration that *might* be heard or that it is a class of vibration that *is* heard, but, in either case, the hearing of the sound is what makes it. My point is that human beings reside at the center of any meaningful definition of sound. When the hearing of other animals comes up, it is usually contrasted with human hearing (as in "sounds that only a dog could hear"). As part of a larger physical phenomenon of vibration, sound is a product of the human senses and not a thing in the world apart from humans. Sound is a little piece of the vibrating world.

Perhaps this reads like an argument that falling trees in the forest make no sounds if there are no people there to hear them. I am aware that the squirrels would offer another interpretation. Certainly, once we establish an operational definition of sound, there may be those aspects of it that can be identified by physicists and physiologists as universal and unchanging. By our definition of sound, the tree makes a noise whether or not anyone is there to hear it. But, even here, we are dealing in anthropocentric definitions. When a big tree falls, the vibrations extend outside the audible range. The boundary between vibration that is sound and vibration that is not-sound is not derived from any quality of the vibration in itself or the air that conveys the vibrations. Rather, the boundary between sound and not-sound is based on the understood possibilities of the faculty of hearing—whether we are talking about a person or a squirrel. Therefore, as people and squirrels change, so too will sound—by definition. Species have histories.

Sound history indexes changes in human nature and the human body—in life and in death. The very shape and functioning of technologies of sound reproduction reflected, in part, changing understandings of and relations to the nature and function of hearing. For instance, in the final chapter of this book, I discuss how Victorian writers' desire for permanence in sound recording was an extension of changing practices and understandings of preserving bodies and food following the Civil War. The connections among canning, embalming, and sound recording require that we consider practices of sound reproduction in relation to other bodily practices. In a phrase, the history of sound implies a history of the body.

Bodily experience is a product of the particular conditions of social life, not something that is given prior to it. Michel Foucault has shown that, in the eighteenth and nineteenth centuries, the body became "an object and target of power." The modern body is the body that is "is manipulated, shaped, trained," that "obeys, responds, becomes skillful and increases its forces." Like a machine, it is built and rebuilt, operationalized and modified.[27] Beyond and before Foucault, there are scores of authors who reach similar conclusions. Already in 1801, a Dr. Jean-Marc Gaspard Itard concluded, on the basis of his interactions with a young boy found living "wild" in the woods, that audition is learned. Itard named the boy Victor. Being a wild child, Victor did not speak—and his silence led to questions about his ability to hear. Itard slammed doors, jingled keys, and made other sounds to test Victor's hearing. The boy even failed to react when Itard shot off a gun near his head. But Victor was not deaf: the young doctor surmised that the boy's hearing was just fine. Victor simply showed no interest in the same sounds as "civilized" French people.[28]

While the younger Marx argued that the history of the senses was a core component of human history, the older Marx argued that the physical conditions under which laborers "reproduced" themselves would vary from society to society—that their bodies and needs were historically determined.[29] The French anthropologist Marcel Mauss, one of Foucault's many influences, offered that "man's first and most natural technical object, and at the same time technical means, is his body." What Mauss called *body techniques* were "one of the fundamental moments of history itself: education of the vision, education in walking—ascending, descending, running."[30] To Mauss's list we could add the education and shaping of audition. Phenomenology always presupposes culture, power, practice, and epistemology. "Everything is knowledge, and this is the first reason why there is no 'savage experience': there is nothing beneath or prior to knowledge."[31]

The history of sound provides some of the best evidence for a dynamic history of the body because it traverses the nature/culture divide: it demonstrates that the transformation of people's physical attributes is part of cultural history. For example, industrialization and urbanization decrease people's physical capacities to hear. One of the ways in which adults lose the upper range of their hearing is through encounters with loud machinery. A jackhammer here, a siren there, and the top edge of hearing begins to erode. Conflicts over what does and does not constitute environmental noise are themselves battles over what sounds are admissible in the modern landscape.[32] As Nietzsche would have it, modernity is a time and place where it becomes possible for people to be measured.[33] It is also a place where the human-built environment modifies the living body.

If our goal is to describe the historical dynamism of sound or to consider sound from the vantage point of cultural theory, we must move just beyond its shifting borders—just outside sound into the vast world of things that we think of as not being about sound at all. The history of sound is at different moments strangely silent, strangely gory, strangely visual, and always contextual. This is because that elusive inside world of sound—the sonorous, the auditory, the heard, the very density of sonic experience—emerges and becomes perceptible only through its exteriors. If there is no "mere" or innocent description of sound, then there is no "mere" or innocent description of sonic experience. This book turns away from attempts to recover and describe people's interior experience of listening—an auditory past—toward the social and cultural grounds of sonic experience. The "exteriority" of sound is this book's primary object of study. If sound in itself is a variable rather than a constant, then the history of sound is of necessity an externalist and contextualist endeavor. Sound is an artifact of the messy and political human sphere.

To borrow a phrase from Michel Chion, I aim to "disengage sound thinking . . . from its naturalistic rut."[34] Many theorists and historians of sound have privileged the static and transhistorical, that is, the "natural," qualities of sound and hearing as a basis for sound history. A surprisingly large proportion of the books and articles written about sound begin with an argument that sound is in some way a "special case" for social or cultural analysis. The "special case" argument is accomplished through an appeal to the interior nature of sound: it is argued that sound's natural or phenomenological traits require a special sensibility and special vocabulary when we approach it as an object of study. To fully appreciate the strangeness of beginning a history with a transhistorical description of human listening experience, consider how rare it is for histories of newspapers or literature to begin with naturalistic descriptions of light and phenomenologies of reading.

Sound certainly has natural dimensions, but these have been widely misinterpreted. I want to spend the next few pages considering other writers' claims about the supposed natural characteristics of sound in order to explain how and why *The Audible Past* eschews transhistorical constructs of sound and hearing as a basis for a history of sound. Transhistorical explanations of sound's nature can certainly be compelling and powerful, but they tend to carry with them the unacknowledged weight of a two-thousand-year-old Christian theology of listening.

Even if it comes at the beginning of a history, an appeal to the "phenomenological" truth about sound sets up experience as somehow outside the purview of historical analysis. This need not be so—phenomenology and the study of experience are not by definition opposed to historicism. For instance, Maurice Merleau-Ponty's work has a rich sense of the historical dimensions of phenomenological experience.[35] But founding one's analysis on the supposed transhistorical phenomenological characteristics of hearing is an incredibly powerful move in constructing a cultural theory of sound. Certainly, it asserts a universal human subject, but we will see that the problem is less in the universality per se than in the universalization of a set of particular religious prejudices about the role of hearing in salvation. That these religious prejudices are embedded at the very center of Western intellectual history makes them all the more intuitive, obvious, or otherwise persuasive.

To offer a gross generalization, assertions about the difference between hearing and seeing usually appear together in the form of a list.[36] They begin at the level of the individual human being (both physically and psychologically). They move out from there to construct a cultural theory of the senses. These differences between hearing and seeing are often considered as biological, psychological, and physical facts, the implication being that they are a necessary starting point for the cultural analysis of sound. This list strikes me as a litany—and I use that term deliberately because of its theological overtones—so I will present it as a litany here:

- —hearing is spherical, vision is directional;
- —hearing immerses its subject, vision offers a perspective;
- —sounds come to us, but vision travels to its object;
- —hearing is concerned with interiors, vision is concerned with surfaces;
- —hearing involves physical contact with the outside world, vision requires distance from it;
- —hearing places us inside an event, seeing gives us a perspective on the event;
- —hearing tends toward subjectivity, vision tends toward objectivity;
- —hearing brings us into the living world, sight moves us toward atrophy and death;
- —hearing is about affect, vision is about intellect;
- —hearing is a primarily temporal sense, vision is a primarily spatial sense;
- —hearing is a sense that immerses us in the world, vision is a sense that removes us from it.[37]

The audiovisual litany—as I will hereafter call it—idealizes hearing (and, by extension, speech) as manifesting a kind of pure interiority. It alternately denigrates and elevates vision: as a fallen sense, vision takes us out of the world. But it also bathes us in the clear light of reason. One can also see the same kind of thinking at work in Romantic conceptualizations of music. Caryl Flinn writes that nineteenth-century Romanticism promoted the belief that "music's immaterial nature lends it a transcendent, mystical quality, a point that then makes it quite difficult for music to speak to concrete realities. . . . Like all 'great art' so construed, it takes its place outside of history where it is considered timeless, universal, functionless, operating beyond the marketplace and the standard social relations of consumption and production."[38] Outlining the *differences* between sight and hearing begs the prior question of what we mean when we talk about their nature. Some authors refer back to physics; others refer back to transcendental phenomenology or even cognitive psychology. In each case, those citing the litany do so to demarcate the purportedly special capacities of each sense as the starting point for historical analysis. Instead of offering us an entry into the history of the senses, the audiovisual litany posits history as something that happens *between* the senses. As a culture moves from the dominance of one sense to that of another, it changes. The audiovisual litany renders the history of the senses as a zero-sum game, where the dominance of one sense by necessity leads to the decline of another sense. But there is no scientific basis for asserting that the use of one sense atrophies another. In addition to its specious zero-sum reasoning, the audiovisual litany carries with it a good deal of ideological baggage. Even if that were not so, it would still not be a very good empirical account of sensation or perception.

The audiovisual litany is ideological in the oldest sense of the word: it is derived from religious dogma. It is essentially a restatement of the long-standing spirit/letter distinction in Christian spiritualism. The spirit is living and life-giving—it leads to salvation. The letter is dead and inert—it leads to damnation. Spirit and letter have sensory analogues: hearing leads a soul to spirit, sight leads a soul to the letter. A theory of religious communication that posits sound as life-giving spirit can be traced back to the Gospel of John and the writings of Saint Augustine. These Christian ideas about speech and hearing can in turn be traced back to Plato's discussion of speech and writing in the *Phaedrus*.[39] The hearing-spirit/sight-letter framework finds its most coherent contemporary statement in the work of Walter Ong, whose later writing (especially *Orality and Literacy*) is still widely cited as an authoritative description of the phenomenology and psychology of sound. Because Ong's later work is so widely cited (usually in ignorance of the connections between his ideas on sound and his theological writings), and because he makes a positive statement of the audiovisual litany such a central part of his argument about cultural history, Ong's work warrants some consideration here.

To describe the balance sheet of the senses, Ong used the word *sensorium,* a physiological term that denoted a particular region of the brain that was thought to control all perceptual activity. *Sensorium* fell out of favor in the late nineteenth century as physiologists learned that there is no such center in the brain. Ong's use of the term should therefore be considered metaphoric. For him, the sensorium is "the entire sensory apparatus as an organizational complex," the combined balance among a fixed set of sensory capacities.

Although *Orality and Literacy* reads at times like a summary of scientific findings, Ong's earlier writings clearly state that his primary interest in the senses is explicitly driven by theological concerns: "The question of the sensorium in the Christian economy of revelation is particularly fascinating because of the primacy which this economy accords to the word of God and thus in some mysterious way to sound itself, a primacy already suggested in the Old Testament pre-Christian [*sic*] tradition."[40] For Ong, "divine revelation itself . . . is indeed inserted in a particular sensorium, a particular mixture of the sensory activity typical of a given culture." Ong's balance-sheet history of the senses is clearly and urgently linked to the problem of how to hear the word of God in the modern age. The sonic dimension of experience is closest to divinity. Vision suggests distance and disengagement. Ong's history of the move from sound-based oral culture to sight-based literate culture is a history of "a certain silencing of God" in modern life. Ong's assertions about the difference between the world of "oral man" and the "hypertrophy of the visual" that marks the modern age parallel perfectly the spirit/letter distinction in Catholic spiritualism. It is a sophisticated and iconoclastic antimodernist Catholicism. Still, Ong argues that the audiovisual litany transcends theological differences: "Faith or no, we must all deal with the same data."[41]

Of course, parts of the audiovisual litany have come under heavy criticism. The work of Jacques Derrida can be read as an inversion of Ong's value system—Ong himself suggests as much.[42] Derrida uses his well-known phrase *the metaphysics of presence* to criticize and dismantle the connections among speech, sound, voice, and presence in Western thought. Although Derrida's most celebrated critiques of presence find him tarrying with Edmund Husserl's transcendental phenomenology, Ferdinand de Saussure's semiotic theory, and Martin Heidegger's ontology, his criticisms are certainly applicable to Ong's thought as well. Ong argues for exactly the metaphysics of presence that Jacques Derrida attacks as "ontotheological," as a creeping Christian spiritualism that inhabits Western philosophy: "The living act, the life-giving act [hearing oneself speak], the *Lebendigkeit,* which animates the body of the signifier and transforms it into a meaningful expression, the soul of language, seems not to separate itself from itself, from its own self-presence."[43] For Derrida, the elevation of speech as the center of subjectivity and the point of access into the divine is "essential to the history of the West, therefore to metaphysics in its entirety, even when it professes to be atheist."[44] Derrida uses this position to argue for the visual side of the audiovisual litany—an emphasis on vision, writing, difference, and absence. Deconstruction inverts, inhabits, and reanimates the sound/vision binary, privileging writing over speech and refusing both speech-based metaphysics and presence-based positive assertions.

Here, I want to make a slightly different move: the audiovisual litany carries with it the theological weight of the durable association among sound, speech, and divinity, even in its scientific guise. Rather than inverting the audiovisual litany, why not redescribe sound? Since this book is not bound by Christian doctrine, there is no law—divine or otherwise—requiring us to assume the interiority of sound and the connection between sound, subjective self-presence, and intersubjective experience. We do not need to assume that sound draws us into the world while vision separates us from it. We can reopen the question of the sources of rationality and modern ways of knowing. If history exists *within* the senses as well as *between* them, then we need not begin a history of sound with an assertion of the transhistorical dimensions of sound.

My criticism of the audiovisual litany goes far beyond the questions of essentialism or social construction, which usually degenerate into philosophical hygienics. Even if we grant the possibility of a transcendental subject of sensation, the audiovisual litany falls short on its own terms. Despite all the appeals to nature in the name of the litany, the phenomenology implied by the audiovisual litany is highly selective—it stands on shaky empirical (and transcendental) ground. As Rick Alman has argued, claims about the transhistorical and transcultural character of the senses often derive their support from culturally and historically specific evidence—limited evidence at that. In the audiovisual litany, "an apparently ontological claim about the role of sound [or vision] has been allowed to take precedence over actual analysis of sound's functioning."[45] Consider the purportedly unique temporal and spatial characteristics of auditory phenomenology. Ong argues that "sound is more real or existential than other sense objects, despite the fact that it is also more evanescent. Sound itself is related to present actuality rather than to past or future"; sounds exist only as they go out of existence.[46] But, strictly speaking, Ong's claim is true for any event—any *process* that you can possibly experience—and so it is not a quality special or unique to sound. To say that ephemerality is a special quality of sound, rather than a quality endemic to any form of perceptible motion or event in time, is to engage in a very selective form of nominalism.[47] The same criticism can be made of the litany's attribution of a "surface"-oriented spatiality to vision as opposed to an "interior" orientation to sound: it is a very selective notion of surface. Anyone who has heard fingernails on a chalkboard or footsteps in a concrete hallway (or on a wooden floor) can recognize that listening has the potential to yield a great deal of information about surfaces very quickly. The phenomenologist Don Ihde has shown that writers who take sound as a weakly spatial sense wholly disregard "the contemporary discoveries of very complex spatial attributes to auditory experience."[48] He demonstrates that hearing has many spatial aspects and possibilities to which we do not normally attend. We can learn a great deal about shape, surface, or texture from listening. Perhaps the biggest error of the audiovisual litany lies in its equation of hearing and listening. Listening is a directed, learned activity: it is a definite cultural practice. Listening requires hearing but is not simply reducible to hearing.

There is no "mere" or innocent description of interior auditory experience. The attempt to describe sound or the act of hearing in itself—as if the sonic dimension of human life inhabited a space prior to or outside history—strives for a false transcendence. Even phenomenologies can change. In this respect, we follow in Dr. Itard's footsteps. Like the studious Itard, who was perplexed by the wild boy who could hear but did not speak, historians of sound must surmise that our subjects' hearing is fine medically. But we can know their sonic world only through their efforts, expressions, and reactions. History is nothing but exteriorities. We make our past out of the artifacts, documents, memories, and other traces left behind. We can listen to recorded traces of past history, but we cannot presume to know exactly what it was like to hear at a particular time or place in the past. In the age of technological reproduction, we can sometimes experience an audible past, but we can do no more than presume the existence of an auditory past.

What Is Sound Reproduction? Plan of the Present Work

I have argued that technologies of sound reproduction provide us with a compelling entry into the history of sound, but sound-reproduction technology is not necessarily a well-bounded historical object. One could argue that ancient uses of animal horns to amplify the voice and aid the hard-of-hearing are, in a certain sense, sound-reproduction technologies. Certainly, musical instruments could have some claim to that status, as could speaking-head or piano-playing automatons and other sound-synthesis technologies from the seventeenth to the nineteenth centuries. So what is different about telephones, phonographs, radios, and other technologies commonly conjured up as "sound reproduction"? A number of writers have offered semiexperiential definitions of modern sound-reproduction technologies based on their power to separate a sound from its "source."

Since the power to split sources and copies is the most common definition of sound-reproduction technology, it warrants some scrutiny. Pierre Schaeffer, the composer who pioneered *musique concrète,* argued that sound-reproduction technologies produced "acousmatic" sounds—sounds that one hears without seeing their source. John Corbett extends the line of thought by using an explicitly psychoanalytic framework to talk about reproduced sound in terms of visual lack: "It is the lack of the visual, endemic to recorded sound, that initiates desire in relation to the popular music object."[49] For Corbett, our inability to see the recording leads us to want it, to attend to it. Barry Truax and R. Murray Schafer have coined the term *schizophonia* to describe the "split between an original sound and its electroacoustic reproduction" enabled by sound-reproduction technologies.[50] The Greek prefix *schizo-* means "split" and also has a convenient connotation of "psychological aberration" for these authors. Truax and Schafer also argue that reproduction removes sound from its original context.

By my own historicization of practices and ideologies of sound, one could hypothesize a particular context where the acousmatic definition of sound reproduction holds explanatory force. Indeed, the concept of acousmatic sound may seem intuitively plausible to many people today. But that does not make it true. Recall, with Stuart Hall, that that which is most obvious is most ideological: "When people say to you 'Of course that's so, isn't it?' that 'of course' is the most ideological moment, because that's the moment at which you're least aware that you are using a particular ideological framework, and that if you used another framework the things that you are talking about would have a different meaning."[51] Acousmatic or schizophonic definitions of sound reproduction carry with them a questionable set of prior assumptions about the fundamental nature of sound, communication, and experience. Most important, they hold human experience and the human body to be categories outside history:

1. They assume that face-to-face communication and bodily presence are the yardsticks by which to measure all communicative activity. They define sound reproduction negatively, as negating or modifying an undamaged interpersonal or face-to-face copresence. For these authors, the difference between sound reproduction and interpersonal interaction is important because the former lacks some of the qualities of the latter.
2. Because they assume the primacy of face-to-face interaction, these authors assume that sound-reproduction technologies will have

a disorienting effect on the senses that are otherwise oriented or grounded in coherent bodily experience. The assumption of prior sensory coherence requires a notion of a human body that exists outside history. For instance, the claim that sound reproduction has "alienated" the voice from the human body implies that the voice and the body existed in some prior holistic, unalienated, and self-present relation. As I have already argued, phenomenological understandings of subjectivity need not privilege self-presence or reject historicism.

3. They assume that, at some time prior to the invention of sound-reproduction technologies, the body was whole, undamaged, and phenomenologically coherent. By extension, this is to argue that all modern life is disorienting, that the only subject that is whole or at peace with itself is one that is not mediated or fragmented by technology. But the idea of the body's phenomenological unity and sanctity gains power precisely at the moment in its history that the body is being taken apart, reconstructed, and problematized—the eighteenth and nineteenth centuries. In contrast, medieval thought and practice often constructed the body as a filthy container for the soul, something to be transcended and overcome in the afterlife.

4. They assume that sound-reproduction technologies can function as neutral conduits, as instrumental rather than substantive parts of social relationships, and that sound-reproduction technologies are ontologically separate from a "source" that exists prior to and outside its affiliation with the technology. Attending to differences between "sources" and "copies" diverts our attention from processes to products; technology vanishes, leaving as its by-product a source and a sound that is separated from it.

Assertions of the primacy of face-to-face communication or interpersonal immediacy have been widely criticized on a variety of theoretical fronts, and I will not rehearse those arguments here.[52] Treating face-to-face communication as primary also predetermines the history of sound reproduction before we even tell the story. If interpersonal interaction is the presumptively primary or "authentic" mode of communication, then sound reproduction is doomed to denigration as inauthentic, disorienting, and possibly even dangerous by virtue of its "decontextualizing" sound from its "proper" interpersonal context. But, to begin a theory and history of sound's reproducibility, we do *not* need final, fundamental, or transhistorical answers to questions about the relations between hearing and seeing, between technological reproduction and sensory orientation, between original and copy, and between presence and absence in communication. We can provide more robust answers to those questions by reconsidering them in the course of studying sound reproduction. This history of sound begins by positing sound, hearing, and listening as historical problems rather than as constants on which to build a history.

So let us take a ride on Ockham's razor and work from a simpler definition of sound-reproduction technology, one that does not require us to posit a transcendental subject of hearing: modern technologies of sound reproduction use devices called *transducers,* which turn sound into something else and that something else back into sound. All sound-reproduction technologies work through the use of transducers. Telephones turn your voice into electricity, sending it down a phone line and turning it back into sound at the other end. Radio works on a similar principle but uses waves instead of wires. The diaphragm and stylus of a cylinder phonograph change sound through a process of inscription in tinfoil, wax, or any number of other surfaces. On playback, the stylus and diaphragm transduce the inscriptions back into sounds. Digital sound-reproduction technologies all use transducers; they simply add another level of transformation, converting electric current into a series of zeros and ones (and back again).

My definition is certainly reductive and incomplete, but it is a very instructive reduction. It offers us a useful starting point for a history of sound reproduction, especially for a history that will proceed analytically rather than chronologically. Even though transducers operate on a very simple set of physical principles, they are also cultural artifacts. This is where *The Audible Past* begins its history of sound.

Chapter 1 takes as its central exhibit the ear phonautograph, a machine for "writing" sound waves. By following around the device, its inventors, and the ideas that it operationalized, the chapter offers a genealogy of new constructs of sound and hearing. The ear phonautograph used an excised human middle ear as a transducer, and the functioning of the tympanic membrane (also known as the diaphragm or the eardrum) in the human ear was the model for the diaphragms in all subsequent sound-reproduction technologies. As a result, I call the mechanical principle behind transducers *tympanic.* The history of the isolation and reproduction of the tympanic function leads us back into the construction of sound and hearing as objects of knowledge and experimentation in the late eighteenth century and the nineteenth. The tympanic function emerged at the intersection of modern acoustics, otology, and physiology and the pedagogy of the deaf.

The ways in which the middle ear conducts vibration may seem like a simple mechanical function, something that we feel is without history. But the tympanic function opens out into changing constructions of sound, hearing, and humanity. Sound reproduction is historical all the way down.[53] In acoustics, physiology, and otology, sound became a waveform whose source was essentially irrelevant; hearing became a mechanical function that could be isolated and abstracted from the other senses and the human body itself. Although these developments may on their own seem minor or merely matters of technical discovery, they mark a larger shift in the history of sound.

Prior to the nineteenth century, philosophies of sound usually considered their object through a particular, idealized instance such as speech or music. Works of grammar and logic distinguished between significant and insignificant sounds by calling all significant sounds *vox*—voice.[54] Other philosophers took music as an idealized theoretical instance of sound, leading to the analysis of pitch and harmony, all the way up to the harmony of the spheres and, for Saint Augustine, God. In contrast, the concept *frequency*—previously developed by Descartes, Mersenne, and Bernoulli—offered a way to think about sound as a form of motion or vibration. As the notion of frequency took hold in nineteenth-century physics, acoustics, otology, and physiology, these fields broke with the older philosophies of sound. Where speech or music had been the general categories through which sound was understood, they were now special cases of the general phenomenon of sound. The emergence of the tympanic function thus coincided with an inversion of the general and the specific in philosophies of sound. Sound itself became the general category, the object of knowledge, research, and practice.[55] Chapter 1 also inverts a historical commonplace: the objectification and abstraction of hearing and sound, their construction as bounded and coherent objects, was a prior condition for the construction of sound-reproduction technologies; the objectification of sound was not a simple "effect" or result of sound-reproduction technology.

While chapter 1 considers the construction of sound and hearing, chapters 2 and 3 offer histories of various practices of listening during the same period. They chronicle the development of *audile technique,* a set of practices of listening that were articulated to science, reason, and instrumentality and that encouraged the coding and rationalization of what was heard. By *articulation,* I mean the process by which different phenomena with no necessary relation to one another (such as hearing and reason) are connected in meaning and/or practice.[56] For a time, hearing surpassed vision as a tool of examination, conception, and understanding in selected regions of medicine and telecommunications. Chapter 2 provides an introduction to the idea of audile technique and explores how, in the first decades of the nineteenth century, doctors moved away from listening to their patients' speech and began listening more closely to patients' bodies to distinguish signs of health and illness. As it became a symbol of the medical profession, the stethoscope signaled both virtuosic and highly technical listening skills. Chapter 3 explores how American telegraph operators from the 1840s to the 1880s and early users of sound-reproduction technologies from the 1880s to the 1920s developed other forms of audile technique. Telegraphers started listening to their machines instead of reading their printouts. In a cacophonous room, they would focus on the noise of their machine alone and take down telegraphic messages at ever-increasing speeds. Listening skill was a mark of professional distinction in sound telegraphy. Physicians' use of stethoscopes and sound telegraphers' virtuosic message taking prefaced a much wider dissemination of audile technique with the telephone, phonograph, and radio. Even today, when listeners in a music library treat the surface noise of an LP record or the hiss of a tape as "exterior" to the music on the recording, they use some of the same techniques of listening that physicians and telegraphers developed over 150 years ago.

A new practical orientation toward acoustic space developed alongside audile technique: listening became more directional and directed, more oriented toward constructs of private space and private property. The construct of acoustic space as private space in turn made it possible for sound to become a commodity. Audile technique did not occur in the collective, communal space of oral discourse and tradition (if such a space ever existed); it happened in a highly segmented, isolated, individuated acoustic space. Listening technologies that promoted the separation of hearing from the other senses and promoted these traits were especially useful. Stethoscopes and headphones allowed for the isolation of listeners in a "world of sounds" where they could focus on the various characteristics of the sounds to which they attended. Thus, as early as 1820, R. T. H. Laennec, the inventor and first popularizer of the stethoscope, could characterize listening to a patient's body without a stethoscope as *immediate,* by which he meant to connote "lacking in the proper mediation." While other techniques of listening likely developed in other contexts, chapters 2 and

3 offer a genealogy of those techniques that were central for constructing sound reproduction as we know it today.

Chapter 4 moves from the subjective to the industrial: it shows how the technologies that came to be organized as the sound media emerged from a small, industrializing field of invention that was in continuous flux from the 1870s through the 1920s. The new sound media were part of an emergent field of mass communication and mass culture that was itself organized by and oriented toward an American middle class shifting from Victorian ideals to consumerism as a way of life. Moreover, the shape of the sound media was not guaranteed at the outset. There is no necessary connection between the technology of radio and that of broadcasting; nor is there an essential connection between the technology of telephony and that of point-to-point communication. At prior moments, the telephone was a broadcast medium, and radio was a point-to-point medium. Social forms did not necessarily follow logically from technologies: those connections had to be made. Technologies had to be articulated to institutions and practices to become media. The sound media thus emerged in the tumultuous context of turn-of-the-century capitalism and colonialism.

Chapter 5 historicizes "acousmatic" understandings of sound-reproduction technologies—the idea that they separate a sound from its "source"—through examining the idea of a reproduced sound's "fidelity" to its source. Acousmatic understandings of sound reproduction (which conceptualized it as splitting copies of sounds from their ontologically separate sources) depended on three prior conditions: (1) the emergence of audile technique as a way of abstracting some reproduced sounds (such as voices or music) as worthy of attention or "interior," and others (such as static or surface noise) as "exterior" and therefore to be treated as if they did not exist; (2) the organization of sound-reproduction technologies into whole social and technical networks; and (3) the representation of these techniques and networks as purely natural, instrumental, or transparent conduits for sound.

The idea that sound-reproduction technologies separated sounds from their sources turns out to have been an elaborate commercial and cultural project. Early auditors of sound-reproduction technologies did not always assume that reproduced sound reflected an "original" at the other end. In response, manufacturers and marketers of sound-reproduction technologies felt that they had to convince audiences that the new sound media belonged to the same class of communication as face-to-face speech. While other rhetorical strategies may have been possible, this rhetoric of equivalence allowed advertisers to render sound-reproduction technologies in familiar terms. Through an examination of the idea of sound fidelity before it denoted a quality that can be physically measured (covering the period 1878–1930), chapter 5 argues that early skeptical listeners essentially had it right: sound-reproduction technologies are inseparable from the "sources" of reproduced sound. To put it another way, the social organization of sound-reproduction technology conditioned the possibility for *both* "original" and "copy" sounds. Performers had to develop whole new performance techniques in order to produce "originals" suitable for reproduction. Even the very grounds on which the ability of sound-reproduction technologies "faithfully" to reproduce sound could be tested in laboratories had to be established. The ever-shifting figure of sound fidelity crystallized a whole set of problems around the experience of reproducibility, the aesthetics of technologically reproduced sound, and the relations between original and copy. Considering sound-reproduction technologies as articulated to particular techniques and as media forces us to trouble the supposed objectivity of acousmatic descriptions; it shows them to be historically motivated.

Chapter 6 offers a history of the audible past itself. It considers the conditions under which recordings came to be understood as historical documents, yielding insight into the past. Although early recordings were far from permanent records, early images of and overtures to sound recording's permanence—and the newfound ability to hear "the voices of the dead"—promoted and gradually propelled technological and institutional innovation. New, innovative recording equipment and media were developed with the specific aim of producing longer-lasting recordings. In this respect, sound recording followed innovations in other major nineteenth-century industries like canning and embalming. Institutions grew that were dedicated to the collection and preservation of sound recordings. Chapter 6 argues that through the historical process of making sound recording more "permanent"—which began as nothing more than a Victorian fantasy about a machine—the historical process was itself altered. As beliefs surrounding death, the preservation of the dead body, transcendence, and temporality shaped or explained sound reproduction, sound reproduction itself became a distinctive way of relating to, understanding, and experiencing death, history, and culture.

Developmental ideas of history and culture were bound up with the political currents of American society at the turn of the twentieth century.

After decades of pursuing genocidal policies toward Native Americans, the U.S. government and other agencies began in the 1890s to employ anthropologists, who would use sound recording to "capture and store" the music and language of their native subjects. Embedded in this anthropological project were loaded conceptions of American culture as embodying a universal tendency toward "progress" that would simply engulf Native American life ways along the way. As Johannes Fabian has argued, the idea of modernity and its doctrine of progress was often taken to imply the historical superiority of "modern" civilization (generally urban, cosmopolitan, largely white, middle-class culture in the United States and Western Europe) over other cultures by casting those different (yet actually contemporaneous) cultures as if they existed in the collective past of the moderns. The military and economic domination of other cultures by the United States and Western Europe—and the larger projects of racism and colonialism—became explainable in the late nineteenth century as the product of a difference between that which is modern and that which is not (yet) modern. Relations of space become relations of time.[57] The drive to build and fill phonographic archives with the sounds of "dying" nations and cultures, the desire to make sound recordings permanent, was inextricably linked to early anthropologists' ambivalent relations to history and their subjects. Phonography's much-touted power to capture the voices of the dead was thus metonymically connected to the drive to dehistoricize and preserve cultures that the U.S. government had actively sought to destroy only a generation earlier. Permanence in sound recording was much more than a mechanical fact; it was a thoroughly cultural and political program. To a great degree, inventing reproducibility was about reconstructing sound and hearing and developing technologies to fit and promote these new constructs. The idea of sound recording's permanence is a striking example of the movement from wish to practice to technological form.

A note on my approach concludes this introduction. Given the scope of my task, I offer no pretense to finality or totality in the account that I offer. *The Audible Past* is a deliberately speculative history. My intent is not to establish once and for all a small set of historical facts, although clearly facts are important to my history. Rather, this book uses history as a kind of philosophical laboratory—to learn to ask new questions about sound, technology, and culture. If all accounts of human action carry with them some concept of human nature, then we would do well to reflect on the choices that we make in describing human nature. *The Audible Past* offers a speculative foray into moments when the many natures of sounding and hearing were objects of practice and reflection. It is not a complete statement on human nature itself, nor is my primary goal the recovery of lived experience, although certainly people's own accounts of their experiences can provide insight into the history of sound.

Like any intellectual product, this book bears the mark of its author's biases. My own distaste for the cult of Edison in phonograph historiography has led me to emphasize Berliner and Bell (who are much less fully treated in the critical historiography). The greater depth of the film and radio historiography has led me to place greater emphasis on the telephone and phonograph. In foregrounding the history of sound, I deemphasize many of the metanarratives of cultural and political history. It would be equally possible to orient a history of sound around points of change or transformation in the history of speech, music, or even industrial and other forms of environmental noise.[58] But the history of sound reproduction provides a uniquely powerful entry into the history of sound precisely because it is a history of attempts to manipulate, transform, and shape sound.

My emphasis on the very early moments of technologies and practices at times leads me to concentrate on a relatively small, elite (white, male, European or American, middle-class, able-bodied, etc.) group of people. My archival material, perhaps limited by some measures of historiography, has a distinctly American and East Coast bias. In the early years of sound-reproduction technologies, their use was heavily scattered and atomized. Each technology took decades to "diffuse" fully throughout American society and elsewhere. The emphasis on sound itself also risks a certain level of audism (a term used by scholars of deaf culture; we might best think of it as an ethnocentrism of those who hear). But these are risks worth taking.

The Audible Past focuses on hearing elites because they provide a wealth of documentation about the meaning of sound and listening—qua sound and listening—on which to build a study. As a result, I have not been very concerned with recovering the experiences of my historical subjects. Alexander Graham Bell does not need *The Audible Past* to save him from historical oblivion—and one does not need to identify with elites in order to study them. But, more important, the history of sound must move beyond recovering experience to interrogating the conditions under which that experience became possible in the first place. Experiences are themselves variables shaped by the contexts through which they then help their subjects navigate.[59]

Of course the question of experience still lingers. While acknowledging the plurality of possible audible pasts, this book outlines some common bases for modern sound culture in the West—especially around practices of sound reproduction. It is doubtful that they are truly universals, but they are sufficiently general to be worth considering. There are certainly other dominant, emergent, or subjugated constructs of sound, listening, and hearing beyond the ones considered in these pages. Histories of sound could contribute to a much wider range of themes in cultural and political history than I cover in this book. As always, there are other histories to be written. We will have to write them in order to know if they fundamentally challenge my conclusions here.

This is not to succumb to the localism, cumulativism, and neopositivism that has ravaged much contemporary cultural historiography. Events or phenomena merely need to exist to carry some intellectual significance; they do not need to pass a test of universality. Sound history, however partial, must continually move between the immediate and the general, the concrete and the abstract. There is a burden of sound history, just as there is a burden of history, to borrow a phrase from Hayden White. To offer a compelling account of humanity, sound history must remain "sensitive to the more general world of thought *and* action from which it proceeds and to which it returns."[60]

1   Machines to Hear for Them

If at some later point, instead of doing a "history of ideas," one were to read the state of the cultural spirit off of the sundial of human technology, then the prehistory of the gramophone could take on an importance that might eclipse that of many a famous composer.—**THEODOR ADORNO**, "The Form of the Phonograph Record"

I would merely direct your attention to the apparatus itself, as it gave me the clue to the present form of the telephone.—**ALEXANDER GRAHAM BELL**

The ancestor of the telephone you are used to using remains the remains of a real human ear.—**AVITAL RONELL**, *The Telephone Book*

In 1874, Alexander Graham Bell and Clarence Blake constructed a most curious machine (figure 1).[1] A direct ancestor of the telephone and the phonograph, it consisted of an excised human ear attached by thumbscrews to a wooden chassis. The ear phonautograph produced tracings of sound on a sheet of smoked glass when sound entered the mouthpiece. One at a time, users would speak into the mouthpiece. The mouthpiece would channel the vibrations of their voices through the ear, and the ear would vibrate a small stylus. After speaking, users could immediately afterward see the tracings of their speech on the smoked glass. This machine, a version of the phonautograph invented by Leon Scott in 1857, used the human ear as a mechanism to *transduce* sound: it turned audible vibrations into something else. In this case, it turned speech into a set of tracings.

But the ear phonautograph did not use the whole ear: that folded mass of flesh on the side of the head—known as the *outer ear, auricle, pinna,* or

often simply *ear*—was loosely modeled in the mouthpiece and thereby rendered unnecessary; the inner ear was superfluous because the machine merely transduced sound for writing. The ear phonautograph was not an attempt to reproduce the actual *perception* of sound. This left only the middle ear, which in a living person ordinarily focuses audible vibrations and conveys them to the inner ear, where the auditory nerve can perceive them as sound. In using the tympanum or eardrum and the small bones to channel and transduce sonic vibrations, the ear phonautograph imitated (or, more accurately, isolated and extracted) this process of transducing sound for the purpose of hearing and thereby applied it to another purpose—tracing. Bell and Blake attached a small piece of straw directly to the small bones to serve as a stylus, producing tracings that were a direct effect of the tympanic vibrations. Inasmuch as we can say that the ear phonautograph embodies the basic principles of other inventions that followed it like the telephone, phonograph, radio, or microphone, we could claim for it a minor technological significance. But, here, I am interested in the ear phonautograph as a cultural artifact in a deeper sense.

How is it that a human ear came to be affixed to a machine at this time, in this way, and in this place? For Bell, the ear phonautograph was the clue

Figure 1. Bell and Blake's ear phonautograph

to the functioning of the telephone. For our purposes, it gives a clue to a more general characteristic of the machines and relations that follow it in time: it places the human ear, *as a mechanism*, as the source and object of sound reproduction. The ear phonautograph is an artifact of a shift from models of sound reproduction based on imitations of the mouth to models based on imitations of the ear. This is more than merely a matter of the choice between two models for imitation; it marks a shift in understandings of sound and practices of sound reproduction. As sound became problematized in physics, acoustics, physiology, and otology, these fields moved toward contemplating and constructing sound as a kind of *effect* in the world. As we will see, prior analyses of sound had been more oriented toward a particular source—theories of sound took the voice and the mouth, or music and a particular instrument (such as the violin), as ideal-typical for the analysis, description, and modeling of sonic phenomena. The mouths and instruments were taken as *general* cases for understanding sound. Sound-reproduction technologies informed by this perspective attempted to synthesize sound by modeling human sonic activities like speech or musical performance. In contrast, the new sciences of sound would in a sense (or, rather, in the sense of hearing) invert the general and the specific in theories of sound. No longer themselves general categories of sound fit for theory construction, the mouth, the voice, music, and musical instruments would become specific contenders for audition in a whole world of sonic phenomena. In this new regime, hearing was understood and modeled as operating uniformly on sounds, regardless of their source. Sound itself, irrespective of its source, became the general category or object for acoustics and the study of hearing. Thus, the ear displaced the mouth in attempts to reproduce sound technologically because it was now possible to treat sound as any phenomenon that excites the sensation of hearing. Under this new regime, the ear's powers to transduce vibrations held the key to sound reproduction.

Using the ear phonautograph as a nodal point in alternately commingled and twisted historical streams, this chapter traces out the tributary currents shaping the very possibility of sound reproduction as we know it. Although I argue that the ear phonautograph represents the maturation of a new sonic regime of sorts (two years before telephony actualized sound reproduction), this is not a strictly Foucauldian tale of a single epistemic or historical "break" between epochs. A multitude of cracks, fissures, tipping points, displacements, and inversions make up this history. So this chapter follows first one tributary current and then another. If you can understand all the prior factors coming together in Bell and Blake's attachment of a human ear to a machine, you can understand the conditions necessary for the reproduction of sound as we now know it. For our purposes, the ear phonautograph will, therefore, serve as a tour guide to nineteenth-century approaches to sound. This chapter can be read as an archaeology of the machine and its construction. Beginning with attempts to construct sound as an object of physics, and following in turn the education of the deaf, the growth of otology ("ear medicine"), physiological studies of hearing as a function, anatomy laws, and elaborate models of human vocal and musical process, this chapter retraces a group of paths that crisscross the history of sound in the nineteenth century. I dispense with a linear chronology in order to map and highlight the different historical processes that made possible the mechanical reproduction of sound.

The ear phonautograph embodies a very simple principle: the ear is a mechanism that can be used — instrumentally — to a variety of ends. So foremost in this history is the notion of the ear as a mechanism for transducing vibrations. Specifically, this mechanism can be called a *tympanic* mechanism. I use the word *tympanic* deliberately: its linguistic evolution reflects the same cultural movements that I describe below. It begins as a description of a specific location in human and animal bodies: the tympanic bone and tympanic membrane make up the eardrum, the tympanum. By 1851, this location becomes an operation. It is possible to speak of a "Tympanic apparatus," the purpose of which is to "receive the sonorous vibrations from the air and to transmit them to the membranous wall of the labyrinth." By the end of the century, *tympanic* also refers to the function of a telephone's diaphragm or anything else resembling a drum.[2] Following the etymology, the word moves from connoting a region, to a functional description of the region, to a pure function. This is the history of hearing itself during the nineteenth century.

To speak of a set of sound-reproduction technologies as *tympanic* is to understand them as all functionally related, as sharing a set of common operational and philosophical principles, and, most important, as embodiments and intensifications of tendencies that were already existent elsewhere in the culture. Even today, every apparatus of sound reproduction has a tympanic function at precisely the point where it turns sound into something else (usually electric current) and when it turns something else into sound. Microphones and speakers are transducers; they turn sound into other things, and they turn other things into sound. It is still impossible to think of a configuration of technologies that makes sense as sound reproduction without either microphones or speakers. What began as a theory of hearing became an operational principle of hearing machines. The very workings of the telephone, phonograph, and related technologies were thus an outgrowth of changes in practical understandings of hearing, scientific understandings of sound, and medical approaches to the human ear in the mid-nineteenth century.[3]

At the most basic level—how they worked—tympanic sound-reproduction technologies are best understood as the *result* of a proliferation of a particular set of practices and practical understandings concerning sound and the ear, not as the *cause*. Of course new sound technologies had an impact on the nature of sound or hearing, but they were part of social and cultural currents that they themselves did not create. The growth of tympanic machines represents—and is an effect of—a reorganization of these cultural constructs of the ear and hearing, rather than a singular point of origin for these new constructs. This can be demonstrated historically: as the ear phonautograph shows, we can see this set of beliefs and practices literally inscribed in some of sound reproduction's technological predecessors. Since physics and mechanics are so often mistaken for transcendent, a priori, causal conditions of technological history, it makes sense to begin with a cultural, intellectual, and social history of the tympanic mechanism. Even the most basic mechanical functions have their histories. Thus, I turn to a history of the function that Bell and Blake sought to render in its purest form through the ear phonautograph.

Delegation, Synesthesia, and the Appearance of Sound

The ear phonautograph was the progeny of a longer line of experimentation. As of 1874, it was the latest innovation of Leon Scott's phonautograph (figure 2). Scott's phonautograph produced a visual representation of sound—called a *phonautogram*—by partially imitating the processes of the human ear. Like the outer ear, this machine channels sounds through a conic funnel to vibrate a small, thin membrane. This membrane, called a *diaphragm,* is attached to a stylus (a needle or some other instrument for writing). The diaphragm vibrates the stylus, which then makes tracings on a cylinder. Different sounds provide different vibrations, resulting in different patterns on the cylinder. In conceiving the phonautograph, Scott experimented with diaphragms made from both synthetic material and animal membrane, although it was known that his own machine was modeled on the action of the membrane and small bones of the human ear. Scott,

2

who was a typesetter, came on the idea for the phonautograph when proofreading drawings of the anatomy of the ear for a physics textbook.[4] As we will see, he understood the phonautograph as a machine for literally transforming sound into writing. In this respect, Scott's phonautograph was one in a long line of nineteenth-century attempts to write sound. But it was set apart from its predecessors by being a writing device explicitly modeled on the middle ear. Bell and Blake understood this: their 1874 ear phonautograph took Scott's metaphor literally. They thought that using the human ear instead of a synthetic diaphragm would advance their quest to get ever closer to the processes of the human ear itself. Hence the name of their peculiar machine—the *ear* phonautograph. As innovators, all Bell and Blake really did was change the recording surface (to smoked glass) and replace the diaphragm with the human ear on which it was modeled.

Bell's interest in the phonautograph is distinguished from others' in that he sought to divert a line of acoustic research toward a wholly different enterprise: the education of the deaf. Scott's phonautograph presented a possible new solution to a pedagogical problem for Bell: teaching the deaf and mute to speak as if they could hear. Alexander Graham Bell had been a major advocate in the Americas for visible speech, a method of elocution

Figure 2. Leon Scott's phonautograph (courtesy Division of Mechanisms, National Museum of American History)

designed by his father, Melville Bell. Visible speech was an attempt at a purely phonetic alphabet: "invariable marks for every appreciable variety of vocal and articulate sound . . . with a natural analogy and consistency that would explain to the eye their organic relations."[5] In other words, visible speech was a set of signs for sounds. The idea was that, if speakers followed the written instructions perfectly, they would be able to reproduce the sounds so notated perfectly. Following his father's lead, Alexander Graham Bell had hoped to demonstrate the utility of visible speech for training the deaf and mute to speak. Bell had met with some limited success with this method, but visible speech did nothing to teach the deaf to modulate their voices like hearing people. Visible speech depended on the faculties of speech and hearing for it to work as an elocutionist's script.

As we will see, this orientation to the reproduction of sound was fundamentally concerned with the reproduction of the mechanism producing sound. Visible speech aimed to train speakers to become machinelike in their abilities to reproduce sound. John Peters calls the Bell family's work in visible speech "the primal scene of the supercession of presence by programming" because it was an attempt to enact communication without interiority, "a code that can pass as an adequate substitute for the original."[6] It is certainly true that visible speech did not require a speaking subject, only a person following instructions to make sounds with his or her voice. It did, however, require a subject who could *hear* and make sense of the available sounds.

George Bernard Shaw's famous *Pygmalion* builds on this premise, where visible speech allows for the possibility of purifying the speech of impure-English speakers. Inspired by Melville Bell and other "phoenetic experts" whom he encountered in the 1870s, Shaw wrote a play about social mobility through the transformation of dialect. His Professor Higgins works to correct the working-class dialect of the flower girl Eliza Doolittle. The phonograph and a laryngoscope allow Higgins and Doolittle to treat her speech as an effect to be modified. Her speech is a matter of technique, her voice an instrument to be worked on. Shaw's vision of social mobility through the transformation of speech thus does for the hearing what Bell had hoped to do for the deaf. A machine hears for the speaker, who can then modulate his or her speech until it is perfect. Like Bell, the aspiration behind Shaw's tale is the eradication of cultural difference through the perfection of technique. In *Pygmalion,* linguistic difference is a kind of disability to be cured through externalization: "For the encouragement of people troubled with accents that cut them off from all high employment,

I may add that the change wrought by Professor Higgins in the flower-girl is neither impossible nor uncommon.... [But] ambitious flower-girls who read this play must not imagine that they can pass themselves off as fine ladies by untutored imitation. They must learn their alphabet over again, and different, from a phoenetic expert."[7] Machines and experts "drill people in general, and flower girls in particular, to adopt a pronunciation purified by written language."[8] In visible speech and *Pygmalion,* proper speech can be mastered with some practice and technical coaching, but it is still the province of the hearing.

Since Bell's ultimate goal was training the deaf to speak, he began to seek alternatives to the methods of visible speech. Scott's phonautograph presented itself as one such alternative because it rendered speech visible through a representation of the waveforms produced by speech, rather than through a representation of positions of the mouth.[9] In other words, it treated sound reproduction as a problem of reproducing effects, rather than reconstructing causes. The phonautograph sought to imitate the activity of the middle ear, not the positions of the mouth. It reacted to changes in air pressure in a manner analogous to the actions of the tympanum and small bones in order to render an indexical, visible record of the sound waves: "My original skepticism concerning possible speech reading had one good result; it led me to devise an apparatus that might help children ... a machine to hear for them, a machine that would render visible to the eyes of the deaf the vibrations of the air that affect our ears as sound."[10]

The device would allow deaf people to *see* the sounds that they were making with their voices, thereby allowing them to modulate the sounds they made until they matched the tracings of vowels or consonants spoken by a hearing person. Bell's descriptive locution suggests the significance of the machine in my own narrative—as a supplement to and stand-in for the human auditory faculty. Although sound-reproduction technologies would be thought of as talking machines or machines for writing sound, they were, ultimately, *hearing machines.* "A machine to hear *for* them" goes beyond the amplification of hearing, the extension of the sense. Extending hearing had been possible for aeons with the aid of ear trumpets and other hearing aids. Bell's use of the phonautograph suggested instead the *delegation* of hearing to a machine and the isolation of the tympanic principle as the basic mechanical function of the ear. Although Bell's planned practical application of the phonautograph would never come to fruition, it also implied a program for the use of the phonautograph's mechanical descendants by people who were not deaf. The phonautograph's mechanical successors would indeed become auditory surrogates. The telephone, phonograph, radio, and other tympanic sound-reproduction technologies could all be described, at their base, as "machines to hear for them."

For his pedagogical purposes, Bell also experimented with the manometric flame, a tympanic device developed by Rudolph König after he experimented with Scott's phonautograph. The manometric flame consisted of a speaking trumpet with a tube that led to small box called a *manometric capsule*. The box was divided in two parts by means of a rubber diaphragm. Lighted gas would flow through one part of the chamber. As the sound waves went through the speaking tube and vibrated the diaphragm, they would produce movements in the gas flame corresponding to the vibrations of the diaphragm. The box was then placed in a cube lined with mirrors on four adjacent sides—the cube could be rotated with a handle, and the effects of the vibrations on the flame could easily be seen and even photographed.[11] The phonautograph, the manometric flame, Bell's whole conceptual schema for his pedagogical approach, all relied on a basic abstract principle: they treated sound as an effect of the vibration of a diaphragm. Because the manometric flame was a tympanic technology, it could transduce audible vibrations into visible phenomena. While the manometric flame turned auditory data into a visible analogue over time, the phonautograph offered a physical record across the space of its tracings. The former offered synesthetic simultaneity, the latter synesthetic durability—a durable "record."[12] Because of the phonautograph's durable tracings, Bell spent more time experimenting with that machine. For our purposes, however, the product is incidental to the process: the phonautograph and the manometric flame were both tympanic technologies modeling human hearing to transform and manipulate sound.

Bell's approach to deafness was really about the eradication of linguistic differences. In fact, Bell's plans for the phonautograph have to be understood in the larger context of his opposition to deaf culture as such. While Bell married a deaf woman and considered himself a friend of the deaf and a committed teacher, historians of the deaf paint him with a different brush. Bell developed an enduring interest in eugenics, which led him to advocate the full integration of deaf people into mainstream American culture; he was opposed to the "formation of a deaf variety of the human race."[13] Concurrent with those beliefs was his stand against deaf people marrying one another and having children of their own. Bell understood deafness, fundamentally, as a human disability to be overcome, not as a condition of life.[14] Edwin Miner Gallaudet, on the other hand, was an advocate of deaf-specific institutions and culture, such as the teaching of sign language. To this day, the Bell-Gallaudet division exists in approaches to deaf culture and deaf pedagogy. As a result, Bell most often appears as a villain in cultural histories of the deaf since he is (correctly) seen as seeking to eradicate deaf culture altogether.

Bell's seemingly "practical" goal of teaching the deaf to speak thus loses some of its apparent simplicity. Oralists, Alexander Graham Bell being one of their most famous figures, sought to eradicate any cultural trace of deafness by teaching deaf children to read lips and speak so that they would be indistinguishable from hearing children. As Douglas Baynton argues, oralist positions like Bell's were at least partly rooted in scientific racism. Manualists like Thomas Gallaudet advocated teaching deaf children sign language so that they could effectively communicate with one another.[15]

Behind Bell's practical task lay a very particular notion of language, speech, and what it means to be human. The idea that speech is one of the essential characteristics of humanity—what separates humans from animals—has a long history dating back at least to ancient Greece, but it attained a new currency in the late nineteenth century. The oralists used this philosophical privilege of speech to attack manualists as encouraging primitivism in deaf communities by teaching their children to sign rather than to speak. According to Baynton, "The value of speech was, for oralists, akin to the value of being human. To be human was to speak. And in that formulation, an unfortunate byproduct of evolutionary theory, lies much of the reason for the decline of manualism and the rise of oralism in the United States." Baynton quotes one oralist as writing, "Savage races have a code of signs by which they can communicate with each other. Surely we have reached a stage in the world's history when we can lay aside the [tools] of savagery." Oralists treated speech as the mark of civilization.[16] But, as Lennard Jeffries argues, to treat deafness as a linguistic disability is somewhat inaccurate. Sign language is a perfectly adequate form of verbal communication. It simply does not make use of sound.[17]

Sound reproduction thus arose, in part, from an attempt among hearing people to "solve" or at least contain the cultural problem of deafness by training the deaf to pass as hearing people through their speech. The ear phonautograph was more than a supplement to their hearing; it was a delegate to hear for them. Bell hoped that it would stand in for hearing to produce deaf people who would speak as if they heard. This tactic ultimately failed, and deaf culture thrives today. But so do machines to hear for people.

If after much effort Bell failed with the deaf, he succeeded with the hearing with almost no effort. Sound-reproduction technologies depend on us delegating our hearing to machines that hear for us. Instead of eradicating the cultural status associated with deafness, Bell's pedagogy actually fetishized it. To paraphrase Friedrich Kittler, deafness was at the very beginning of sound reproduction. It directed Bell's work leading up to the telephone and haunted phonography as well: the Frenchman Charles Cros, who composed plans for a phonograph shortly before Edison's invention, worked at a school for the deaf and mute. Edison himself was hard of hearing. The bite marks on some of his experimental phonographs demonstrate a mode of hearing twice in need of supplementation—once from the machine and once from the bone conduction of his jaw.[18]

Although the ear phonautograph wound up being a dead end in Bell's pedagogy of the deaf, it contributed to the acoustic research that eventually led to the telephone, which was in turn derived from a longer trajectory of acoustic studies. Bell and Blake's innovation of Scott's phonautograph comes late in a long history of nineteenth-century acoustics. Bell had been following various Europeans' experiments with sound and sound reproduction—especially the work of Hermann Helmholtz (who had also influenced Blake). Leon Scott's phonautograph was a significant part of this field. The ear phonautograph, among other things, would teach Bell that a combination of complex sound vibrations could be transmitted through a single point and represented visually. The ear in the ear phonautograph accomplished this with a very thin membrane acting on heavy bones, thus inspiring Bell to simplify his model of the telephone, allowing a simple membrane to vibrate a relatively heavy piece of iron. This is the significance usually accorded the ear phonautograph when it is even mentioned in histories of the telephone.[19]

Scott, however, built his phonautograph for neither deaf pedagogy nor insight into the nature of sound itself. He was interested in using the phonautograph generally to make sounds visible to the eye and specifically to create a form of automatic sound writing. This quest is something of an obsession in nineteenth-century science, and the phonautograph appeared in the middle of a much longer history. Certainly, one could argue that writing and musical notation are attempts to visualize sound that stretch back centuries. But these writing systems bear a largely arbitrary relation to the sounds that correspond to them. The same could be said for pictorial representations of sound. A 1672 book announcing the "Tuba Stentoro-

Phonica," basically a megaphone, included detailed illustrations of the author's theory of sound waves as they made their way through his invention, along with the following description:

> In like manner, as to the Nature of Sounds and Voices; I must confess, that the circular Undulations of a Vessel of Water, by the percussion of any part of its Superficies, and the reverberations of those Undulations when they meet with opposition by the sides of such vessels, make it seem more than probable, that the percussion of the Air by any Sound, spreads and dilates it self by a spherical Undulation (greater, or less, according to the strength and virtue of that percussion) till it meet with some opposition, and so echoes back again. And there is great reason to believe that Voices being first modulated and articulated by the Glottis of the Larinx, and the several parts of the Mouth, make spherical Undulations in the Air, till they meet with the Acoustick Organ.[20]

The water analogy is apt here—the author clearly understood that sound functioned as a wave and therefore was able to represent sound graphically as a wave (figure 3). This was as much a depiction of sound's action as a written description—the images in *Tuba Stentoro-Phonica* are clearly imaginative renderings.

Over the course of the nineteenth century there emerged another kind of visual representation of sound. To use the language of C. S. Peirce, these were "indexical" images of sound, where the sound bears some kind of causal relation to the image itself (and, therefore, the image does not have a wholly arbitrary relation to the sound that conditioned it). These images were artifacts of devices that could be affected by sound and thereby create images ordered in part by sonic phenomena. The use of these devices reflected an emergent interest in the scientific use of graphic demonstration and automatic inscription instruments, a practice that developed slowly in the last quarter of the eighteenth century and did not become prevalent until the nineteenth century. Graphs, and later automatic recording devices, represented to their users a new kind of scientific "natural language," where images would reveal relations hitherto unavailable to the senses. Attempts to represent sound visually were themselves artifacts of a larger process through which sound was isolated as a phenomenon and by means of which it would become an object of theoretical and practical knowledge in its own right. In fact, modern acoustics was very much shaped by this reliance on automatic imaging devices and the assumptions that this reliance embodied.[21]

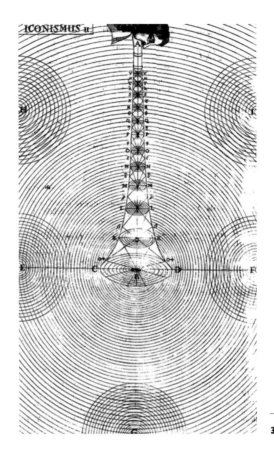

3

Attempts to visualize sound thus coincided with the construction of sound as an object of knowledge in its right: where speech, music, and other human sounds were reduced to special categories of noises that could be studied by the sciences of sound. In acoustics, frequencies and waves took precedence over any particular meaning that they might have in human life: "Frequencies remain[ed] frequencies regardless of their respective carrier medium."[22] Ernst Florens Friedrich Chladni's work at the turn of the nineteenth century is considered to be the founding moment of modern acoustics, and it embodies this connection between objectification,

Figure 3. Drawing of sound refraction from S. Morland, *Tuba Stentoro-Phonica: An Instrument of Excellent Use, as Well at Sea, as at Land; Invented and Variously Experimented in the Year 1670 and Humbly Presented to the King's Most Excellent Majesty Charles II in the Year 1671* (London: Printed by W. Godbid and Sold by M. Pitt, 1672)

visualization, and the reversal of the general and the specific in theories of sound. Trained first as a lawyer and then as a mathematician and physicist, Chladni turned his attention to acoustics when he found the extant music theory lacking in mathematical rigor. Today, he is most well-known for his visually striking "Chladni figures." To create his sand figures, Chladni spread sand over glass plates of various shapes and sizes. When he then ran a violin bow against the plates' edges, the plates would vibrate in such a way as to distribute the sand in regular patterns. By changing the location of the bow, or the shape or size of the plate, the figures would change. Chladni's approach provided insight into the conditions of vibrating solids and the physics of sound waves, and his work adapted research in other areas of physics to the problem of sound: his figures were the acoustic analogue of Georg Lichtenberg's electrostatic figures, which were produced by collecting dust particles on a charged cake of resin.[23] In other words, Chladni correctly constructed an analogy between sound and magnetism as waves as a prior condition to undertaking his experiments. Later researchers would attempt to replicate Chladni's methods for other purposes. For instance, through applying Chladni's method to the tympanum, the French physicist Savart discovered that the perception of low and high tones is significantly affected by the size and thickness of and the elasticity and degree of tension in a membrane. His discovery could be read as a forerunner of Alexander Graham Bell's insight that a tiny membrane can convey sound through a relatively large and heavy surface; Savart's research would pave the way for physiologists later in the century.[24]

Acoustics developed through a host of other physical investigations in the late eighteenth and early nineteenth centuries, each treating sound itself as an object to be studied. For instance, Chladni's contemporary M. Perrole also conducted important investigations into the manners in which solids conduct vibrations. What set apart Chladni's work, along with that of other acousticians like Thomas Young, was that they used sounds to create images that they could then study. Young was the first to use a stylus for tracing the vibration of sounds: "If we fix a small pencil in a vibrating rod, and draw a sheet of paper along, against the point of the pencil, an undulated line will be marked on the paper, and will correctly represent the progress of the vibration."[25] Young, and later physicists like Charles Wheatstone, also produced devices that made use of the persistence of vision to create afterimages of vibrating bodies. Visualizing sound as a species of vibration was a central task of the new science of acoustics. Visual sound has a symbiotic relation with quantification. Sound had, according to the accepted techniques of science, to be seen in order to be quantified, measured, and recorded; at the same time, some quantified and abstracted notion of sound had to be already in place for its visibility to have any scientific meaning.[26] Again, the product is an artifact of the process: visual sound required the simultaneous construction of sound as a discrete object of knowledge.

Scott's phonautograph built on this longer line of experimentation in acoustics; even his locution for the phonautograph built on the prior fifty years' worth of automatic imaging technologies in acoustics. He called the phonautograph an "apparatus for the self-registering vibrations of sound." Like his predecessors, he hoped that the phonautograph would yield insight into a true, "natural" language of sound through its script: "to force nature to constitute herself a general language of all sounds." Scott described the function of the phonautograph almost wholly in terms of writing:

Is it possible to achieve for sound a result analogous to that attained presently for light by photography? Can one hope that the day is near when the musical phrase escaping from the lips of the singer will come to write itself . . . on an obedient page and leave an imperishable trace of those fugitive melodies that the memory no longer recalls by the time it searches for them? Between two men joined in a quiet room, could one place an automatic stenographer that preserves the conversation in its minute details. . . . Could one conserve for future generations some trains of diction of our eminent actors, who now die without leaving after them the feeblest trace of their genius? This improvisation of the writer, when she rises in the middle of the night, could she recall the day after with all her freedom, that complete independence of the pen so slow to translate an ever-fading thought in her struggle with written expression?[27]

This long quotation manifests a variety of aspirations for the phonautograph that would come to be attached to phonography later in the century. The significant difference is that Scott maintained a monomaniacal emphasis on *writing* as the aid to preservation and recall. It was because the phonautograph *wrote* that it would be able to preserve instantaneously and thus aid in recall. Scott sought to produce a "natural stenography" that would smash the distinction between orality and literacy because sound could literally write itself—hearing and speaking would become equivalent to reading and writing. Writing was the ultimate goal for Scott. Twenty years after his work on the phonautograph, Scott would evaluate Edison's phonograph as a failure because it "merely reproduced sound—it was not a *soundwriter.*"[28] Writing was for Scott of greater significance for civilization.

Of course, Scott's plan does not hold up well to logical scrutiny. In essence, he was simply suggesting a different *kind* of writing rather than the abolition of writing itself. Sound writing would bear an indexical relation to speech, rather than the abstract and arbitrary relation to speech that typography was said to have. But it was not a direct representation of speech. As Derrida and others have noted, to treat writing as simply a representation of speech is to efface its own social character.[29] In this way, Scott's plan was simply to have the phonautograph replace one form of phonetic writing (stenography) with another. His inability to see even the Edison phonograph as a major improvement on his own device was an artifact of a monomaniacal focus on writing, on the *product* of the machine, overlooking its more significant processural dimensions. The phonautograph submitted sound to a tympanic process in order to transform it. This was Scott's major contribution to the practices of reproducing sound.

Sound reproduction is thus artifactual of a transformation in *process* where sound and the tympanic mechanism are isolated as phenomena that can be studied, translated, and operationalized. Yet, long into this century, theories of sound reproduction have emphasized its affinity to writing as a practice. Indeed, many believed that the scripts produced by phonautographs contained secrets to a more fundamental natural language. As a result, the promise of sound-writing remained seductive to nineteenth-century thinkers after Scott. Bell's more modest—yet equally unsettling—program for the phonautograph still led him to comment on its tracings in his 1877 speech to the Society of Telegraph Engineers, and the published version of that speech provides diagrams of different sounds as recorded by the phonautograph.

While for Bell the production of phonautograms was tied to the immediate goal of training deaf people to speak, and for Shaw the goal was to train the hearing to speak "better," there were a variety of other ideas as well. When Emile Berliner provided an illustration of sound-writing in the 1888 address introducing his gramophone, he did so with only a more general gesture to "scientific research."[30] The mechanism of Berliner's gramophone differed significantly from that of earlier phonographs and graphophones: it looked and worked much more like a twentieth-century phonograph (figures 4–5). Instead of using a rotating cylinder on a verti-

4

5

cal spindle, Berliner's chosen recording surface was a flat disk that rotated on a horizontal plate. The recording consisted of grooves etched into the plate—on close inspection, the disk looked like a spiraling script. Since the disk was flat, it could be viewed like a page out of a book. Berliner's uncertainty as to the significance of sound-writing was transformed back into a suggestion of pedagogical possibilities and an appeal to aesthetics in an

---

Figure 4. Edison's tinfoil phonograph (courtesy Division of Mechanisms, National Museum of American History)

Figure 5. Berliner's gramophone (courtesy Division of Mechanisms, National Museum of American History)

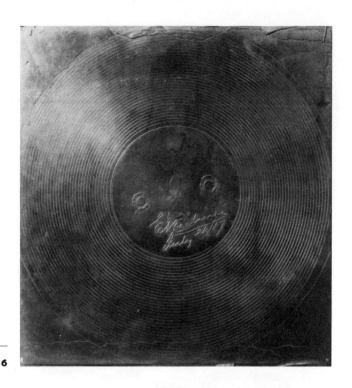

1895 pamphlet accompanying gramophones for sale by a Philadelphia firm. The pamphlet declared that the "voice may be analyzed by studying the beautiful record curves which they show in phonautograms printed from original record plates" (figure 6).[31] Scott had already anticipated Berliner's aesthetic move; he suggested that its tracings would be worthy of aesthetic contemplation.[32] The visual representation of sound was a recurring theme in radio as well—ranging from Marconi's use of a Morse telegraph's register to record dots and dashes on a strip of paper to corporate and military efforts to record radio signals visually.

A particularly sustained attempt to study the curves of sound recordings visually was conducted by Edward Wheeler Scripture (perhaps the most aptly named figure in the history of sound technology). Scripture's hubris and ultimate failure are instructive because his thought typified the emphasis on the product—writing—over the process of reproducibility itself. In prose soaking with the metaphysics of presence, Scripture wrote that his investigations "had their origin in an attempt to use the methods

Figure 6. Early gramophone record (courtesy Division of Mechanisms, National Museum of American History)

of natural science in studying the nature of verse. The only true verse is that which flows from the mouth of the poet and which reaches the ears of the public; printed verse is only a makeshift for the verbal communication. It is evident that the only way to undertake a scientific study of verse is to get it directly as it is spoken and then to use the methods of analysis and measurement."[33] Scripture thus sought to replace arbitrary writing with a true, natural sound-writing that would capture the voice in its verity and fullness. If print was a poor supplement to verbal communication, Scripture hoped that his version of natural stenography would provide the essence of poetry. Here we find, encapsulated, that distinctly conflicted modern attitude toward writing: insofar as it is arbitrary, it is a fallen sign, an empty container where speech once lived. Yet Scripture believed that automatic or indexical writing contained the possibility of a truer, hidden code—the very secret of existence. In this respect, he followed a much longer tradition of searching for a "true" plane of writing. His work extended Scott's project of "natural" stenography and the search for a universal code for speech.

Scripture fashioned a device to trace out and magnify the engravings in a gramophone disk. He hoped thereby to analyze the curves and deduce the "laws of verse," although this quickly proved to be impossible. Instead, he hoped that he would be able to establish the "laws of combination of sound" in speech and thereby build a scientific arm of phonetics and philology; "this is the way the natural sciences have traveled," he wrote.[34] Scripture's book is full of methods by which to acquire and analyze the tracings. His work thus lies in a curious in-between space: on one side lay Scott's hope for a truly natural sound-writing or stenography; on the other would lie modern speech science, which does make extensive use of imaging tools. Scripture's desire to read visual sound as writing demonstrates the persistence of the ideology of natural writing. Although the technology was different, the hope was the same as it had been for centuries. "Natural writing is immediately united to the voice and to breath. Its nature is not grammatological but pneumatological. It is hieratic, very close to the interior holy voice of the *Profession of Faith,* to the voice one hears upon retreating into oneself."[35] Modern speech science images the voice; Scripture hoped that sound reproduction would prove a more scriptural technology.

Almost thirty years after Scripture's failed efforts to decipher soundscripts, Theodor Adorno would still be speculating on the potential of physically reading a gramophone record. Adorno believed not only that through the gramophone recorded music approached its true character as writing, but that eventually people could be trained to read acoustic grooves in a record as a musician could read a score.[36] This history continues down to the present, where iconic visual representations of sound play an important part in multitracking, sound mixing, and other forms of sound manipulation. Put simply, now forgotten audiovisual technologies like Chlandi's glass plates and Scott's phonautograph subject visual phenomena to the orderings of sound.[37] Through modern physics and acoustics, and through the new relation between science and instrumentation, auditory and visual phenomena could be first isolated and then mixed or made to stand in for one another. Scott's discourse on the phonautograph and its successors suggests that this kind of synesthesia—of mixing codes and perceptible material—is a constitutive feature of technological reproduction of sound and image.

This synesthesia also directs us toward another tributary current in this history of sound. The names for these machines were all hybrids of one sort or another: *phonograph, graphophone,* and *gramophone* suggest a mixture between speech and writing; *telephone* suggests the throwing of speech; *radiotelegraphy* and *radiotelephony* suggest the radiation of waves out from a single point.[38] At the core of all these transformations (alongside many others) is the isolation, separation, and transformation of the senses themselves. This history of the senses is simultaneously a history of a body—a body made of functions like the tympanic that could be isolated, transposed, replicated, and *put to use.* Especially in the wake of Derrida's work, theorists of sound have sometimes been tempted to use a deep, processual description of writing as the play of difference in language to explain the power and significance of sound reproduction as a whole. While this deep notion of writing holds some explanatory power, it remains incomplete. In addition to fetishizing sound recording over other forms of sound reproduction, this is to mistake product for process. Privileging *recording* as the defining characteristic of sound's reproducibility privileges a certain kind of temporality over an analysis of the transformation of acoustic space and its inhabitants. Theorizing sound reproduction as a historical subspecies of writing also suggests a certain disembodiment of sound in the process of sound reproduction. In contrast, my history suggests that the tympanic mechanism—the mechanical function that lies at the heart of all sound-reproduction devices—points to the resolutely embodied character of sound's reproducibility. But this is not a transhistorical, transparent, and experientially undifferentiated body. The history of sound reproduction is the history of the transformation of the human body as object of knowledge and practice. Alongside the problematization of sound, the abstraction of auditory perception and its condensation into a tympanic function defines sound-reproduction technologies as we know them today. Recall that Scott's machine was distinguished from its predecessors because he based it on his understanding of hearing as a mechanism. For that reason, we now turn our attention to the history of hearing itself.

Otology, Physiology, and Social Ontology

The model of the ear on which Scott based his phonautograph emerged over the first half of the nineteenth century. In that time, hearing became a distinct object of knowledge. It became a scientific problem in its own right. With this problematization of the ear came a new branch of medical science: otology (or ear medicine). The human ear affixed to the ear phonautograph's chassis thus offers a route into another tributary current in the history of sound reproduction. The ear phonautograph would have been much more difficult to build even twenty years earlier. It is an artifact of otology's institutionalization and, with it, a new orientation toward hearing and the ear itself.

The late eighteenth and early nineteenth centuries were very fertile periods for the sciences, and alongside acoustics developed a whole set of sciences of hearing. The new science of otology or ear medicine constructed diseases of the ear as a problem separate from the eye, advances in anatomy yielded new knowledge about the morphology of the ear, and physiology advanced theories of the function of hearing and the distinctness of the senses. All these sciences depended on a new institutionalization of medicine, the use of scientific and medical instruments, and the ready availability of human bodies for study and dissection. These fields were also, of necessity, intermingled—advances in one provided the bases for advances in another. Physiology depended on anatomic research into the form of the ear and physical research that applied instruments to study and model the senses. In turn, physiological research into the functional aspects of hearing fed back into otology and ear science.

As these fields grew, they sparked three key developments in the history of hearing. When the ear became a discrete object of study, it became measurable. Specific quantities and qualities of the ear's shape and function could be isolated and measured. One of those regions of shape and function was the middle ear and, with it, the tympanic function embodied in the tympanic membrane. Once that function had been isolated, it could be abstracted from the body and defined by physiologists and others in almost purely mechanical terms. This abstraction was both figural and literal, as in the case of cutting ears out of the heads of cadavers. When Blake attached an excised ear to the chassis of the experimental phonautograph he shared with Bell, he was operating at all these levels of abstraction simultaneously. The ear could be abstracted from the body, the tympanic function could be abstracted from the ear, and the tympanic function itself could be actualized as a purely mechanical operation.

Let us then consider the conditions under which a human ear came to be fixed to Bell and Blake's improvement of the phonautograph. In his early experiments with the phonautograph, Bell was struck by its structural similarity to the human ear and sought to better imitate that ear function. When he reported this idea to his friend Clarence Blake, a Boston otologist, Blake suggested using an actual human ear for the machine. Blake had studied hearing and perception and taught Bell the workings of the human ear.[39] Taking advantage of his connections with the Harvard medical school, Blake procured two ears, one for Bell and one for himself. Each ear was then affixed to a machine, and the ear phonautograph was born. Bell and Blake spent the spring of 1874 experimenting together with the ears of two medical school cadavers. Years later, Bell would reflect on those experiments as "one of the most joyous scientific experiences of a lifetime."[40] Having already followed Bell into the history of deaf pedagogy and physical acoustics, we now follow Bell and Blake into the history of medicine and the sciences of the body.

Blake's texts concerning the ear phonautograph explain its construction in painstaking detail. The conventions of nineteenth-century technological discourse (as manifested in semipopular journals such as *Scientific American* and *Electrical World* as well as the medical journals for which Blake wrote) required detailed explanations of the construction and function of any technical apparatus so that readers could both gain a practical understanding of the device under consideration and have the necessary knowledge to build it themselves. In this respect, the ear phonautograph is remarkable in its typicality: Blake wrote as if there was nothing unusual about procuring a human ear and nailing it to some wood:

In preparing the ear for use as a phonautograph, the roof of the cavity of the middle ear is first cut away; through this opening a narrow-bladed knife may be introduced to divide the tendon on the tensor tympani muscle and the articulation of the incus with the stapes. By means of a hair-saw a section of the middle ear is then made from [the front] backward through the divided articulation. The section removes the inner wall of the middle ear cavity with the portion of the bone containing the internal ear and exposes the inner surface of the drum membrane, with the malleus and incus attached. . . .

In using a preparation of the ear as a phonautograph, a stylus made of a single fibre of wheat-straw is glued to the descending part of the small bones, parallel to the long axis of the bone. With this, tracings may be made upon a plate of smoked glass, sliding upon a glass bed at a right angle to the line of excision of the drum membrane, and moved by clock work or a falling weight, as in the apparatus mentioned by Professor Bell.[41]

Certainly, modern medicine has depended on the acquisition of bodies for medical examination, experimentation, and pedagogy. But the strangeness of carrying around a machine consisting in part of a human ear was not lost on its inventors. As was his practice, Bell spent the summer of 1874 with his parents in Brantford, Scotland. He brought his machine with him, and word quickly got around town that he had a machine with a dog's ear or a pig's ear affixed to it. One biographer speculates that Bell himself propagated these rumors in order to prevent gossip about a human ear.[42]

Blake's more casual attitude toward the ears in the phonautograph likely came from his professional milieu; in this way, Blake's work marks another set of changes in nineteenth-century understandings of hearing: the construction of the ear as a discrete object of medical knowledge and the growth of otology as a field of medical science. When Blake set off to Vienna for graduate study in otology in 1865, there were perhaps four people in the United States who had more than a passing acquaintance with diseases of the ear. While there were works on ear medicine available in English in the United States, there was no specialized training available in otology, even after the rush to medical specialization following the Civil War. Often considered an adjunct to eye medicine, the state of ear medicine was generally regarded as inferior to almost every other form of medicine. In his 1672 discourse on the Tuba Stentoro-Phonica, S. Morland wondered about the actual physiology of hearing and gave up: "The more we torment our thoughts about it, the less we understand it, and are forced to confess our ignorance."[43] A 1713 treatise on the structure and diseases of the eye apologized for its appendix on the ear: "The reason I here take notice of diseases of the *Ear* is, because of the mutual Communications of some Diseases of the said Organs of Sight and Hearing."[44] The author of a British manual on aural surgery (ca. 1843) also felt compelled to defend his choice of object:

> We daily hear and read, and it has been reiterated from mouth to mouth, and copied from work to work, that the treatment of such affections [of the ear] is an opprobrium to the healing art. . . . Now notwithstanding the injudicious treatment by quacks and nostrummongers, the neglect of patients, and—as in many instances we know it is—the total abandonment of all treatment by the general practitioner, still, were the statistics of all diseases carefully collected, it would be found that there were among them as many curable cases of affection of the ear as there are among the severer maladies of the eye, or among diseases of the chest, the brain, the liver, or any other organ. Up to a very recent period, from well-educated men in this country either considering it beneath their station or acquirements to treat so insignificant an organ specially, or not finding in the direct cultivation of aural surgery a sufficient remuneration for their time and talents, this branch of the healing art remained in the state in which ophthalmic surgery was half a century ago.[45]

As in philosophy, so it was in surgery: the eye enjoyed greater status and prestige than the ear.[46] This was no doubt in part a practical matter. The ear is a tiny structure full of still tinier structures, surrounded by bone. As of 1843 it was difficult to study visually even in death, let alone in living patients. But there was also a matter of professional attitude, of prestige. Physicians of the eighteenth century did not often wish to be thought of as "aurists," and only gradually did this attitude shift in the nineteenth century.

Anton von Tröltsch, a pioneer in German otology, made a similar assessment in the course of explaining why he was writing a textbook encompassing the whole field of ear medicine based mostly on his own work: "I scarcely need to apologize to my professional brethren for the attempt here made to present a text-book which should embrace the whole field of aural medicine and surgery, and be chiefly founded upon my own observations and investigations. If I required any justification for this endeavor, it may be found in the dissevered position which Otology still holds, both in science and practice, as well as in the rarity of strictly scientific and independent labors in this field." Tröltsch added that many of his predecessors erred in their exclusive attention to the anatomy of ears in the deceased, thus giving rise to many misconceptions about the membrana tympani and the mechanism of hearing.[47] As a Harvard medical student, Clarence Blake confronted the same kinds of hostility to ear medicine. In 1864, when he was "house officer" at the Boston City Hospital and still a student at Harvard, Blake told an unnamed visiting surgeon that he intended to go to Europe to study diseases of the ear. The visiting physician offered this dismissal: "All that you can do for the ear, you can do with a syringe."[48]

Otology was the runt of the anatomic litter. While other fields of medicine had grown throughout the nineteenth century, otology remained small and weak into the 1850s. Otologists' professional concerns mirrored a larger intellectual concern: how to make the ear perceptible and knowable. Otologists understood their field's lag behind other medical subfields in terms of the difficulty of the most basic empirical research, especially through dissection. Still, there was a sense that this difficulty had perhaps been overstated. Joseph Toynbee—Tröltsch's British counterpart—wrote that the ear was no *more* concealed or mysterious than any other internal organ:

It is a question, however, whether the inherent difficulties of Aural Surgery are of a nature to prevent its being as thoroughly understood as other branches of Surgery. This question has been answered in the affirmative by some, on the ground of the deep and hidden situation of the larger part of the organ, and the extreme intricacy of its structure. But surely the organ of hearing is not so much concealed from view as several of others (the heart, for instance), of whose diseases we have a very clear knowledge; nor is its structure more complicated than that of the eye. The result of my own experience, and I think also of those who have carefully attended to my practice at St. Mary's hospital is, that the diseases of the ear are not more difficult to diagnose, nor are they on the whole less amenable to treatment, than those of the eye, the joints, or almost any other organ that can be named.

Toynbee's optimism for the future of ear medicine was based on his belief that the ear could be dissected, made visible, and analyzed. The introduction to his textbook on diseases of the ear concludes with a detailed description of options for removing the middle ear from the head of a corpse and for its careful study.[49] Later writers would agree with Toynbee that advances in knowledge of the ear in general and otology in particular were related directly to advances in methods of dissection. Although Tröltsch had rightly cautioned against relying exclusively on the ears of the dead, dissection remained an important resource for anatomic and pathological knowledge.

This is the milieu in which Clarence Blake was educated. His sensibility was shaped by an emphasis on dissection as the route to medical knowledge of hearing and the ear. So it is not surprising that he had a strong investment in dissection. He translated an atlas of the osseous anatomy of the human ear into English in 1874. The manual is significant both for its application of photography to the study of hearing and for its aestheticization of dissection. Blake's translation is an early example of the application of photography—which had been gaining importance in medical and anatomic pedagogy more generally—to the ear for pedagogical purposes. But Blake's aestheticization of dissection served very important professional and intellectual purposes: it is a lot easier to think of hearing abstractly if you can physically abstract the ear from the rest of the body. Dissection was the physical ground for this philosophical move. Blake's notes suggest, therefore, an appreciation of dissection both as a skill central to the procurement of knowledge and as an aesthetic practice:

The specimen represented in this Plate was prepared in the same manner as that of Plate V, and a portion of the walls of the semicircular canals of the vestibule and cochlea then removed by careful use of the file; a work requiring considerable caution, as the labyrinth walls, in all parts fragile, are especially liable to break when even a small portion has been removed. The specimens represented in Plates V, VI, and VII, bear especial evidence to the patience and mechanical skill of the author. The bony ridges on the under surface of the osseous spiral lamina are particularly well shown in this Plate, forming as it were a series of braces for the support of the spiral lamina.[50]

Blake could conceive of dissection as an art because it signaled a kind of virtuosity; it promoted the central professional virtue of otology: the abstraction of the ear as a discrete set of forms, functions, and mechanisms. His text simultaneously aestheticized the extracted ear and the act of cutting it out of a corpse's head. The fascination with technique rested alongside the fascination with the ear as a technology, as a mechanism. In this way, the use of a human ear in the phonautograph was symptomatic of a more standard professional disposition. Once again, audition and visibility were interconnected in the construction of knowledge about ears.

In Vienna, Blake studied with Adam Politzer, the first professor of otology at the University of Vienna. Blake worked as his assistant in the clinic as well as the laboratory. Politzer was also the first to use the human ear in obtaining tracings of the membrana tympani's vibrations; when Blake proposed the use of an actual human ear in the phonautograph, he knew that it could be done.[51] Politzer in turn developed his appreciation for dissection during a visit to Joseph Toynbee in London. Subsequently, dissection became one of Politzer's distinguishing skills.

Most famous, perhaps, was Politzer's contribution to the Austrian exhibit at the 1876 Philadelphia International Exposition (the same event where Alexander Graham Bell would first demonstrate the telephone in public). Politzer's collection—which he had amassed from working with thousands of patients during his time in Vienna—included forty-four temporal bone dissections, fifteen enlarged plaster models of membrana tympani, and an atlas of twenty-four watercolor sketches of "various conditions of the membrana tympani." These were exhibited alongside glass, pottery, furniture, clothing, musical instruments, and other distinctly Austrian items under the slogan "the best we can do." Politzer's display attracted local and national press attention and won several awards, later winding up in the Mutter Museum of the College of Physicians in Philadelphia. Politzer was also interested in the use of photography in otology. Together with his student Alexander Randall, Politzer began to build up a collection of photographs of excised human ears, mounting his favorites on the walls of his clinic.[52]

Knowledge of the ear was intimately connected with the physical and analytic abstraction of the human ear from the body in this period. The use of human ears in experiments was, thus, intimately tied to a mechanical understanding of the ear and hearing. The ear could get attached to machines in part because ears were already being treated as mechanisms. Inventions such as the auriscope were also part of this transformation in knowledge of the ear. Invented in the early 1860s, the auriscope (or otoscope) used a speculum to open up the ear cavity and then focused a light from a candle or lamp through a funnel and into the ear. This light was then reflected on a mirror so that the physician could get a clear view of the middle ear. Later improvements included a magnifying lens. The auriscope allowed doctors a much clearer view of patients' ears and aided in diagnosis.[53] As we have already seen, instrumentation was a key to many forms of nineteenth-century knowledge of sound and hearing. The connection between instrumentation and mechanical models of human hearing is particularly acute, however, in the development of auditory physiology and physiological understandings of hearing.

Physiology connected with acoustics and otology in its development of mechanical models of hearing and tympanic machines. Georg Békésy and Walter Rosenblinth's seminal work categorized the history of hearing research into five periods on the basis of techniques of and instruments for empirical observation:

—a first period of pure speculation in the absence of observation (recall the author of *Tuba Stentoro-Phonica* giving up in dispair);
—a second period in which observation of the ear was based on the shattering of the temporal bone;
—a third period in which a forceps and a file were used in anatomic investigations (this is the beginning of modern dissection);
—a fourth period in which auditory physiology was linked most directly with microscopic observation; and
—a fifth period (contemporary at the time Békésy and Rosenblinth's article appeared) characterized by the use of a dental burr, experiments with living animals, and the recording of electric effects.[54]

While this characterization of historical change is quite technologically deterministic, it does underscore the importance of instrumentation, technique, and observation in both medical knowledge of the ear and histories of that knowledge: the understanding of the ear and its function was closely tied to the instruments allowing access to ears.

Instrumentation therefore plays a double role in this history. Although it may seem to be almost circular reasoning at first, the relation between instrumentation and knowledge of all things auditory follows a clear logic. As Soraya de Chadarevian (and later Thomas Hankins and Robert Silverman) argues, modern acoustics and auditory physiology were very much products of changes in scientific attitudes about instrumentation. As we have already seen, changes in the use of instruments were an enabling condition for modern acoustics starting with Chladni. The same can be said for physiology, where physiologists essentially transformed the field into an "exact science" through their use of self-recording instruments.[55] Researchers' use of instruments allowed for new phenomena to be observed, which in turn led to conceptualizations of the human senses *as* and *through* instruments. Through instrument-based physiological research, the human senses came to be understood as mechanisms themselves.

The physiological thought of the nineteenth century was distinguished from that of earlier centuries because it became a field intellectually separate from anatomy. Today, anatomy is understood to be concerned with the form of living matter and physiology with its function. This was not a salient distinction until the nineteenth century. Earlier anatomic writing often freely moved between form and function. Typical eighteenth-century writings on hearing did not separate functional questions from physical description of the ear as an organ. For instance, Alexander Munro, one of the pioneering figures in otology, devoted almost the entirety of his textbook to the anatomy of the ear and, in particular, the cochlea, which fascinated him. For him, function clearly followed form. After his work with the wild child Victor (discussed in the introduction), Jean-Marc Gaspard Itard became one of the pioneers of French otology. His work combined anatomic and physiological knowledge, but it was fundamentally concerned with treating maladies of the ear.[56]

Physiological questions began to emerge through the use of instruments for investigating hearing. A 1788 treatise on hearing devoted an entire chapter to physiology. The author, one Peter Degravers, spent most of his time debunking others' work, but he did offer a functional theory of hearing, arguing that sounds were changes in air pressure that would affect the membrana tympani, which in turn transmits the vibrations through the middle ear to the cochlea, "where it shakes the delicate nervous filaments spread very thick in the membranes of the cochlea, and produces a sensation, carried or conveyed along the portio mollis to the *grand focus of sense.*"[57] Much is missing from this description of the physiology of hearing, but much is also there—the notion of sounds as vibrations transmitted through the air, the ear vibrating in sympathy with those airborne vibrations, and the transmission of this vibration to the cochlea. Degravers attributes the knowledge that he does have to experiments with sound-creating instruments, especially a violin.

For our purposes, the significance of modern physiology is twofold. The modern physiologists advanced a doctrine of the separation of the senses, according to which the same stimulus could excite different effects in different senses. At the same time, they developed the peculiar mechanical theory of hearing that would be embodied in tympanic sound-reproduction devices. While physiological studies of the eighteenth century such as Degravers's remained isolated, they gave way to a more systematic and experimental approach in nineteenth-century Europe.

Charles Bell, a Scottish surgeon and physiologist, is generally credited with first distinguishing between motor and sensory functions in nerves, arguing that the anterior roots of the spinal nerves are motor and the posterior roots sensory. Bell was also the first to connect specific nerves with specific senses, essentially arguing that the same stimulus (such as electricity) would result in different sensations in different nerves: "The key to the system will be found in the simple proposition, that each filament or track of nervous matter has its peculiar endowment, independently of the others which are bound up along with it; and that it continues to have the same endowment throughout its whole length."[58] In other words, to borrow a phrase from Jonathan Crary, Bell was the first to put forth the hypothesis of the "separation of the senses."[59] The German physiologist Johannes Müller would expand on this thesis.

Müller is often regarded as the founder of modern physiology. Müller's physiology of hearing developed insights into acoustics and otology through experimentation, and he offered functional explanations for all parts of the external, middle, and inner ears across different species. His work is important for our purposes because he proposed that each sense is functionally distinct from the others, can be stimulated by a variety of internal or external stimuli, and therefore can be conceptualized functionally. Müller's discussion of hearing appears in several places in *Elements of Physiology,* his most systematic elaboration of human physiology. At each juncture where he discusses sensation, he is careful to discuss all the senses in turn; my emphasis on hearing in this discussion should be read in purely heuristic terms. But the reason that he attends to all the senses is in fact the key to his argument: everything on sensation in the *Physiology* follows from the basic premise that each sense is functionally and mechanically distinct from the others. In contrast to his predecessors, who (he claims) attributed to each nerve a "special sensitivity" to different phenomena, Müller argued that "each peculiar nerve of sense has special powers or qualities which the exciting causes merely render manifest. *Sensation, therefore, consists in the communication to the sensorium, not the quality or state of the external body, but of the condition of the nerves themselves, excited by the external cause.* . . . Sound has no existence but in the excitement of a quality of the auditory nerve."[60] Like Bell, Müller posited that each sense is separate because its data travel down separate nervous highways.

Müller followed up with the argument that sensation is actually sensation of the states of nerves and not necessarily external phenomena. As it was in acoustics, so it was in physiology: sound was conceptualized as an effect, a particular state of things. The external cause or stimulus for a sensation is of purely instrumental interest to Müller—it is simply a means to sensation, not the sensation in itself. Like Bell, he used the electricity example to argue that it can be seen as light, felt as heat, or heard as buzzing: "Volta states that, while his ears were included between the poles of a battery of forty pairs of plates, he heard a hissing and pulsatory sound, which continued as long as the circle was closed." For Müller, the differences among the senses are almost entirely chemical and mechanical. The senses simply perceive and convey differently: "The sensation of sound, therefore, is the peculiar 'energy' or 'quality' of the auditory nerve."[61] Sound is the *effect* of a set of nerves with determinate, instrumental functions.

Not only are the senses separate and mechanical, but they are also almost purely indexical. That is to say, *any* stimulus of the nerves of sensation can register as a sense datum. Müller argues that there is no fundamental difference between interior and exterior sensation and that the nerves of hearing can be excited by several causes:

1. The mechanical influences, namely, by the vibrations of sonorous bodies imparted to the organ of hearing through the intervention of media capable of propagating them.
2. By electricity.
3. By chemical influences taken into circulation; such as the narcotics, or alterania nervina.
4. By the stimulus of blood.[62]

As Crary writes of Müller's theory of sight, so it was for Müller's audition: "Müller's theory eradicated distinctions between internal and external sensation," resulting in a mechanical, rather than a spiritual, ground for sensation.[63] Whatsoever stimulated the nerve could cause the sensation. Müller's conception of audition is, therefore, as antithetical to romantic notions of inner perception or even orality as possible. While the latter approaches imagine a willful subject immersed in a world of sensuous experience, Müller's sensing subject is more like an amalgamation of perceptual events connected to both internal and external stimuli.

The importance of Müller's hypotheses for sensation can hardly be overstated. Looking backward, his constructs of the senses can be thought of as *media* in at least two senses of the word. They mediate between the stimulus and the mind (or "sensorium"), and they transmit only certain sensations. It is, therefore, possible to read Müller's theory of hearing anachronistically as a "telephonic" theory of hearing, where only certain vibrations become perceived as sound and vibrations are transmitted down the line as impulses, to be decoded in the brain as sound. Moreover, the auditor will not necessarily be able to distinguish between noise on the line and noises on the other end. Hearing, in other words, is already an instrument. More important, it is for Müller a specific kind of instrument, a transducer. Transducers, like microphones and speakers, change audible vibrations into electric impulses and back again.[64]

Müller's most detailed analysis of hearing bears out this interpretation. It also demonstrates the connections between physics, physiology, and otology. His full analysis of hearing begins with a theory of vibrations derived from the physics of Chladni and his followers. Having earlier made the point that, without hearing, there would simply be vibration and not sound (and having reminded his readers that vibration can also be perceived by sight and touch), Müller moves forward to discuss the specific characteristics of vibration as it affects the sense of hearing. From there, he moves to a detailed anatomic description of the "auditory apparatus"—an especially good name for his mechanical conception of the ear—highlighting the different forms of ears in lower and higher animals. Finally, the section concludes with a lengthy discussion of the relation between the form of each part of the ear and its function. For instance, he argues that our hearing is conditioned by the relative laxness of the tympanic membrane, which allows it to convey vibrations more effectively than a membrane with greater tension. He also claims that the labyrinth has particular acoustic properties that help shape our hearing. In other words, form is still related to function, but it is now *function* that is privileged in the theory of hearing.[65] Müller thus managed to develop an entirely functional and mechanical theory of hearing, one that separated it from the other senses and defined it as a complex mechanism.

Bell and Müller's contributions seem simple enough, but they mark a turning point in the history of ideas about hearing. The separation of the senses posits each sense—hearing, sight, touch, smell, taste—as a functionally distinct system, as a unique and closed experiential domain. Each sense could be abstracted from the others; its peculiar and presumably unique functions could be mapped, described, and subsequently modeled. Physiology moved questions of hearing from morphology to function and technics. Audition became a mechanism that could be anatomically, processurally, and experientially abstracted from the human body and the rest of the senses.

Despite my emphasis on Müller thus far, the work of Hermann Helmholtz probably represents the most influential account of auditory perception in the nineteenth century. While his anatomist predecessors understood the ear as a unique sound appliance and his physicist predecessors understood sound to be a set of organized vibrations, Helmholtz synthesized these two premises with the physiologists' attention to the separation of the senses. Hearing was an amalgamation of the acoustic properties of sound, the shape and mechanics of the ear, *and* the determinate function of the nerves. The work of Bell and Müller provided the foundation for Helmholtz's theory of hearing, but his synthesis of physiology with these other fields distinguishes his work. In fact, the first chapter of his *On the Sensations of Tone as a Physiological Basis for the Theory of Music* begins with a restatement of the separation of the senses:

> Sensations result from the action of an external stimulus on the sensitive apparatus of our nerves. Sensations differ in kind, partly with the organ of sense excited, and partly with the nature of the stimulus employed. Each organ of sense produces peculiar sensations, which cannot be excited by means of any other; the eye gives sensations of light, the ear sensations of sound, the skin sensations of touch. . . . The sensation of sound is therefore a species of reaction against external stimulus, peculiar to the ear, and excitable in no other organ of the body, and is completely distinct from the sensation of any other sense.[66]

Helmholtz's theory of auditory perception begins with the separation of the senses as a first premise. In fact, he can even parse out the meaning of the sense of hearing further than his predecessors. In bringing together several varieties of acoustics and aesthetics, Helmholtz sought to distinguish his inquiry from those that had come before him: "Hitherto it is the *physical* part of the *theory of sound* that has been almost exclusively treated at length, that is, the investigations refer exclusively to the motions produced by solid, liquid, or gaseous bodies when they occasion the sounds which the ear appreciates." Essentially, insights in physiological acoustics had to that point often been side effects of more general investigations into vibrating bodies. The ear was merely a convenient location for the study of vibration. But Helmholtz sought to study the ear as itself a phenomenon; the aim of physiological acoustics was to "investigate the processes that take place within the ear itself." This was, for Helmholtz, the key to connecting the science of hearing with the aesthetics of music. In particular, he would argue that "it is precisely the physiological part in especial— the theory of the sensations of hearing—to which the theory of music has to look for the foundation of its structure." In other words, while physical acoustics explained the movement of vibrations from their source to the ear, physiology would explain the means by which sensation itself was caused. Through investigating this physiological domain, "within the ear itself," Helmholtz would elaborate Müller's theory of hearing. While

Müller had essentially offered a dualistic theory of sense—with the sense and the stimulus—Helmholtz offered a tripartite schema where the stimulus, the sense, and the sensory perception were three different elements.[67]

Helmholtz's conception of "the ear itself," however, was in part a product of advances in otology and the anatomy of the ear. In particular, chapter 6 of *On the Sensations of Tone* contains lengthy discussions and detailed illustrations of the various components of the ear. This physical abstraction of the ear from the body both accompanies and conditions the physiological abstraction of hearing from the other senses. As we will see shortly, function still loosely follows form in Helmholtz: "Now, as a matter of fact, later microscopic discoveries respecting the internal construction of the ear, lead to the hypothesis, that arrangements exist in the ear similar to those we have imagined. The end of every fibre of the auditory nerve is connected with small elastic parts, which we cannot but assume to be set in sympathetic vibration by the waves of sound." Helmholtz concludes that "the essential result of our description of the ear may consequently be said to consist in having found that the termination of the auditory nerves everywhere connected with a peculiar auxiliary apparatus, partly elastic, partly firm, which may be put in sympathetic vibration under the influence of external vibration, and will then probably agitate and excite the mass of nerves." The ear is a mechanism of sympathetic vibration, and it is the ways in which the ear conducts and organizes this vibration that make possible the sensation of hearing. It is, therefore, no surprise that Helmholtz discusses Scott's phonautograph and Politzer's experiments with the auditory bone of a duck, where elements of the middle ear—the tympanic membrane and the small bones—are essentially conductors of vibration.[68]

One of Helmholtz's most lasting contributions was his theory of upper partials or overtones—a principle still widely applicable every time someone listens to a telephone. Any given sound is made up of a wide range of frequencies of vibration, potentially from the lowest to the highest ranges of human hearing. It contains a lower partial (now called a *fundamental*) and a series of harmonic overtones that determine its sonic and timbral character. Through his research, Helmholtz learned that sounds could be best distinguished from one another by their upper partials, that is, through their higher frequencies. Thus, while telephone receivers do not produce the entire range of audible sound, we can recognize the voice at the other end because we can hear the upper partials. Our brains then perform a little psychoacoustic magic, and we hear the rest of the sound, including the very low tones. In addition to telephony, this principle accounts for a major dimension of twentieth-century music. Helmholtz's emphasis on timbre in his theory of musical perception foreshadows distorted styles of guitar playing (heavy metal, hard rock, grunge, etc.) by about a century. As Robert Walser argues, much of the musical force from "power chords" on guitar comes from a lower note that is essentially synthesized when two higher notes a fourth or fifth apart are played. Essentially, upper partials create a lower tone.[69]

The theory of upper partials is important because it treats sound fundamentally as an effect that can be reproduced, rather than something that is tethered to a specific and local cause. Because sounds are made up of a range of frequencies, Helmholtz reasoned that it would be possible to synthesize almost any sound through the production of the right harmonic overtones. As John Peters writes, "Helmholtz levels all modalities and is indifferent to bodily origins: sound is sound is sound. What matters is the wave form and not the source (though, in practice, some sources are extremely hard to mimic, the voice above all)."[70] Frequencies are frequencies. For Helmholtz, sounds are *effects* because (1) sounds can be synthesized and (2) sound is a process that takes place "within the ear itself." If you can get the same reaction in the nerve, you create the same sensation. The cause is irrelevant.

This is a very important condition for sound reproduction as we know it. Since sound is an effect indifferent to its cause, the various processes of hearing can be simulated (and, later, reproduced) through mechanical means. Instrumentation was, in fact, central to Helmholtz's hearing research. His resonators were machines built to embody and test his resonance theory of hearing: these were glass bottles with openings at both ends, covered with pigskin membranes, shaped so that each would resonate at a different pitch. Once trained to hear the various upper partials, the listener could conceivably pick them out from a potentially infinite number of sounds. But it becomes difficult to parse out what is a model and what is a copy in Helmholtz's engineering. The nervous system itself becomes "an analogic extension of media," just as the instruments in Helmholtz's studies become analogical extensions of the middle ear.[71] At one point or another, pigskins, pianos, and telegraphs all become for Helmholtz analogues of aspects of human hearing. Conversely, he is at crucial moments also able to substitute the human ear for its simulation, for instance, by adapting one of the holes in the resonator "for insertion into the ear" and thereby substituting his tympanic membrane for the pigskin.[72]

Helmholtz's piano theory of hearing, which held sway into the twentieth century, is a curious combination of this instrumental (in both senses of the word) understanding of hearing and an extension of the separation-of-the-senses hypothesis. Essentially, Helmholtz argued that the tiny hairs inside the cochlea were like the strings of a piano, each tuned to perceive a particular frequency. As combinations of tones, sounds excited particular hairs in the cochlea and, in turn, produced unique and determinate sensations: "This is a step similar to that taken in a wider field by Johannes Müller in his theory of the specific energies of sense. He has shown that the difference in the sensation due to various senses, does not depend upon the actions the excite them, but upon the various nervous arrangements which receive them. . . . The qualitative difference of pitch and quality of tone is reduced to a difference in the fibres of the nerve receiving the sensation, and for each individual fibre of the nerve there remains only the quantitative differences in the amount of excitement."[73] So, for Helmholtz, it is a separation of the senses all the way down to the partial tones that makes up a single sound. In fact, this approach would lead several later researchers to believe that it would be impossible to reproduce the human voice since doing so would require an instrument with as many fine gradations of pitch as the hairs in the ear itself. Alexander Graham Bell would attempt to build "'a sort of piano-sized musical box-comb with between 3000 and 5000 tines to replicate the hair-like organs of Corti within the human ear.' . . . With Bell we have the effort not just to envision the ear as a piano but to build a piano *as* an ear."[74] Later, Emile Berliner would write that Helmholtz's piano theory of hearing nearly derailed a line of research leading up to the telephone and phonograph because it posed such a significant obstacle to synthesizing the human voice.[75]

Contra Berliner, Helmholtz's research fits nicely within the longer history of the tympanic function that I am describing here. Helmholtz took the earlier physiological hypothesis of hearing's functional uniqueness and developed it into a processural theory of sensation. He treated sound as a determined *effect* that could be created irrespective of its cause, and he offered a theory of hearing as sympathetic vibration that would be borne out in later sound-reproduction technologies. In fact, Helmholtz understood that the tympanic membrane worked to focus and direct sound into and through the middle and inner ear.

Tympanic machines would rely on this same principle. Sound is first focused and directed into the machine through a microphone or recording diaphragm and stylus and then forced out of the machine, thereby vibrating the diaphragm in the speaker, which sets our own eardrums in sympathetic vibration. Hearing is thereby tripled—once by the machine hearing "for us," a second time by the machine vibrating a diaphragm in reproducing the sound, and a third time in vibrating our own tympanic membranes so that the sound may be conveyed into the inner ear. Helmholtz physically abstracted the ear from the body (as is illustrated by his extensive use and discussion of anatomic drawings in his work). He conceived of it as physiologically abstracted and separated from the other senses; he treated hearing as a physiological effect rather than as the result of a particular external cause. In this way, Helmholtz's work marks a crucial conjuncture in the history of hearing. His interest in hearing as a pure function abstracted from the practical research of acousticians, otologists, and anatomists.

Politzer and his students would reconnect Helmholtz's physiological insights with the more practical orientation of otology, cutting ears out of corpses as they went along. Blake's work built on that of his teacher Politzer, who built on that of his teacher Helmholtz, who built on the work of physiologists, physicists, and anatomists. Blake rendered the ear as a functional mechanism within the body, but one that could be extracted, examined, and made operational independently of the rest of that body. This is the intellectual history of the ear phonautograph. But the history of the ear phonautograph, and the entry that it offers us into the history of sound reproduction, is not purely a history of ideas.

Above, I argued that the theoretical abstraction of the ear required its physical abstraction from the human body. Dissection was a key to medical knowledge, and dissection was a hotly politicized practice. The aesthetic, professional, or scientific tones of anatomic and physiological texts theorizing hearing performed a usefully euphemistic function. Despite the resolutely sober tones of the scientific and medical texts that we have been examining, science and medicine were eminently social and political practices. This is to say that the theoretical, practical, and physical abstraction and extraction of the ear from the rest of the human body has a distinctly political valence—a valence rendered most clearly in the history of dissection. As Paul Starr has argued, the creation of professional organizations, the growth in size and prestige of medical schools and hospitals, the unification of the industry through the reorganization of the American Medical Association, and the standardization of licensing all played a part in the institutional growth of medicine. Medicine became more politically organized, more respectable, and more prestigious. The boat of otology rose with this tide.[76]

Clarence Blake's career illustrates the state of medicine in the 1860s and 1870s. Blake's European education would allow him to return to Boston and take part in this larger process. He eventually became Harvard's first professor of otology and would play a part in the promotion and advancement of the field as a whole. When he returned to the United States from Vienna in 1869, he also worked at the Massachusetts Eye and Ear Infirmary. While the establishment's name suggests work on otology, it was really a clinic of ophthalmology that had only reluctantly branched out into otology, largely because patients with afflictions of the ear were in the habit of going to ophthalmology clinics to seek help. Over the next few years, Blake turned the Infirmary's Aural Clinic into a center for research as well as treatment.[77] Understandings of the ear were thus closely tied to the institutions (as well as the technologies) that allowed access to the human ear. Access to the human ear meant access to both living ears in living patients and a steady supply of corpses for medical research. Dissection played an important part in medical education, and that meant that the profession needed access to corpses. The sources of the ears for the two ear phonautographs are worth considering for a moment.

Dissection and anatomy have been central parts of medical education since the late eighteenth century and date back to the thirteenth century. As in England (where medicine was more developed throughout most of the nineteenth century), early American medicine required many more bodies than it could get through legal means. Executed criminals were a common legal source of bodies for dissection, but, through the better part of the nineteenth century, grave robbing was the most common means of acquiring bodies for medical students and researchers. In some cases, the students themselves were the grave robbers. Needless to say, this did little to enhance medicine's public reputation. The historians Ruth Richardson and Suzanne Shultz have both documented numerous instances of crowds descending on medical schools in response to the discovery of an empty grave.[78]

Over the course of the nineteenth century, anatomy acts became the solution for medical schools in need of bodies. By providing a plentiful and legal source of corpses for dissection, they were designed to curb grave robbing, enhance the public image of medicine, and, in almost every case, assure middle-class and upper-class citizens that they would no longer have to worry about being disinterred. Prior to the acts, people from all classes could fear grave robbers for several days after a burial: it was a textbook case of Ulrich Beck's argument that risk does not necessarily correspond with social class.[79] The anatomy acts compensated for this by connecting medicine with the state-based enterprise of burying the poor. Although no act could guarantee a sufficient supply of bodies (and, therefore, the acts did not entirely stamp out grave robbing), they did provide a steady supply.[80]

Most American anatomy acts were modeled on the British Anatomy Act of 1832, which offered to medicine any corpse that would otherwise have to be buried by the British state: people who died in workhouses or who would otherwise receive a parish funeral. In the United States, since workhouses were not as widely institutionalized, this simply meant that unclaimed corpses or the bodies of people whose families could not otherwise afford a funeral were now offered up to medical science. Ruth Richardson understands the act as a form of class warfare on the poor: "It paved the way for the systematic dismantling of older and more humanitarian methods of perceiving and dealing with poverty."[81] Both the British and the American acts made the bodies of the poor the raw material for medical knowledge.[82]

Since Blake acquired his bodies for study from the Harvard medical school, he was likely a beneficiary of the Massachusetts Anatomical Act, which in 1831 was the first such act in the United States. After discussing with Bell the virtues of using a real human ear in the phonautograph, Blake "went to the Harvard Medical School to get it." In fact, he got two—one for Bell and one for himself.[83] Thus, the construction of the ear phonautograph—as an event—is made possible by a distinct set of social relations. The expropriation of anonymous corpses as fixed capital for the production of knowledge is illustrated nowhere better than in the history of an ear attached to a machine. The medical historian Charles Snyder casts the "donors" of the ears in Bell and Blake's experiments as the "true heroes" of the research. The part played by these people was almost certainly involuntary, and their lesson is less about heroism and scientific progress than about the social relations on which science and technology depended for their existence. This was a human sacrifice of the second order: although death was the result of natural causes, the bodies of the dead were to be sacrificed at the altar of experimentation and medical education. All achievement in history is piled on top of anonymous bodies;[84] the ear phonautograph is rare in that it gives us a glimpse of what lies beneath it. A certain distanced brutality underlies the fundamental mechanism in sound-reproduction technologies: "Wherever phones are ringing, a ghost resides in the receiver."[85]

The ear on the phonautograph did not simply emerge from an abyss of ignorance to become an object and an instrument of knowledge; it had to be put there. The presence of the ear on the phonautograph depended, not only on practical understandings of the human ear as a mechanism, but also on the pedagogies and institutions of human dissection, which themselves relied on the class structure of nineteenth-century American society. In speaking of the abstraction of hearing, then, we are really speaking of a set of related developments in physics, physiology, philosophy, and medicine. The ear on Bell and Blake's phonautograph thus directs us back to a whole range of tributary historical currents. Otology and its attendant knowledges, pedagogies, and procedures, institutions and professional networks; advances in physical, physiological, and anatomic research that allowed for greater attention to the physical ear; and a secondary form of human sacrifice—together these isolated the human ear as a problem, a mechanism, and an object of knowledge.

Diaphragms, Vocal Organs, and Sound Machines

If only for a moment, the ear phonautograph crystallized the wide and sweeping movements of nineteenth-century sciences and medicines of sound. It also directs our attention to a divide between two different understandings of sound reproduction. We have already seen how mechanical models of the ear and understandings of sound as an *effect* are necessary preconditions for the technology of sound reproduction. Understandings of sound reproduction as attempts to reproduce an effect were only one current of sound history. A whole other history of attempts to reproduce *sources* of sounds ran prior to and in some cases parallel to the history of the tympanic mechanism that I have been describing. On mental maps of the body, this could be considered the difference between privileging the ear in the case of effects and privileging the mouth in the case of sources.[86] In fact, this divide is itself part of the story that I am recounting here.

Early accounts of telephony and phonography are full of historical narratives that attempt to connect then-contemporary inventions with earlier attempts to preserve or imitate sound (even the simplest instruction books for early telephones and phonographs would often have a historical narrative attached as a preface). According to these tales, which demonstrate little formal variation (although credit is meted out differently depending on the author's favorite inventor), earlier inventors sought to freeze or contain sound itself or to construct "automata" to imitate speech that imitated the processes by which sound is produced. Later inventors, such as Bell, Edison, and Berliner, were credited with the innovation of switching from machines modeled on the production of sound through speech or music to machines based on the production of sound at the perceptual end—the middle ear's transduction of vibrations into perceptible sound.

Automata and tympanic machines reproduce sound through two totally different processes. Automata privileged speech and the human voice; they took particular instances of sound production and attempted to re-create them. Tympanic machines treated hearing and sound as general problems and were oriented toward the human ear. Once again, inversions of the general and specific in theories of sound developed by Chladni, Scott, Helmholtz, and others prove crucial. For automata, sounds were the result of sound-production devices such as mouths. For tympanic machines, frequencies were frequencies—to be heard by ears; speech and music became specific instances of sound, which was itself a reproducible effect. The new sound-reproduction technologies were all based on the tympanic principle and the use of diaphragms, and they were hailed as revolutionary on that basis. As one corporate history from 1900 put it:

Faber [who created an elaborate automaton] and his predecessors were on the wrong track in attempting to solve the problem of sound reproduction in this manner, on its physical side. Faber sought a cause; Edison saw an effect, and said, "The Thing is there, it has but to be found." Faber started from the *source* of the sound, and built a mechanism, reproducing the *causes* of the vibrations that made articulate speech. It remained for Edison to start from the vibrations; to obtain the mechanical *effects* of such vibrations; to record them on a pliable material and then to reproduce them.

Faber copied the movements of the vocal organs, Edison studied a vibrating diaphragm, and reproduced the action of the ear drum when acted upon by the vibration *caused* by the vocal organs.[87]

Apart from misattributing a long line of experimentation wholly to Thomas Edison, this account is fairly representative of late-nineteenth-century understandings of what was new in the phonograph and the telephone; it is also more or less correct in its understanding of the nature of the technological innovations later embodied by the telephone and the phonograph. Their common ancestor, the phonautograph, marks a shift in abstract understandings of the nature of mimetic sound among scientists and inventors. It represents a different understanding of the nature and the function of the ear. Even Bell understood this to be the essential lesson of the ear phonautograph—this is why he credits this machine with giving him "the clue to the present form of the telephone" in early speeches on telephony.[88] Thus far, this chapter has tracked the history of the tympanic function and the machines that used diaphragms to reproduce sound that arose from the isolation of this function. It will be useful to contrast these developments with the history of automata.

Automata were not necessarily sound machines; the term refers to a whole class of "automatic" machines. Automata held philosophers' interest for centuries—more or less because they worked. They automatically simulated human or animal behaviors and, in so doing, were supposed to offer some insight into the functioning of nature. Francis Bacon was fascinated with the causes of natural phenomena, and his utopic *New Atlantis* was a fantasy work in which natural phenomena were imitated through artificial devices. Among apparatuses for re-creating foods, textiles, colors, smells, and all other levels of experience, we find "sound houses where we practice and demonstrate all sounds and their generation. . . . We represent and imitate all articulate sounds and letters, and the voices and notes of beasts and birds. We have certain helps, which set to the ear do further the hearing greatly."[89] Here, sound was to be reproduced through the reproduction of the mechanism by which it was caused. Bacon's utopia was one in which every natural phenomenon and experience could be reproduced through the reproduction of its *source*. This causal imitation was meant to suggest a level of understanding of and mastery over nature.

Automata also had an important connection to mechanistic philosophy. Derek J. de Solla Price argues that mechanistic philosophy depended on automata; philosophical principles were derived from machines "whose very existence offered tangible proof, more impressive than any theory, that the natural universe of physics and biology was susceptible to mechanistic explanation."[90] Mechanistic explanations of human thought and perception have a long history—dating back at least to Aristotle. Descartes is usually credited with coming up with a mechanistic theory of sensation where the mind perceived external phenomena through the mediation of the nerves; Cartesianism introduces mechanism into modern philosophy.[91] But Descartes's interest in mechanism went further—he was, in fact, fascinated with automata. Price reports rumors that Descartes planned to build a dancing man, a flying pigeon, and a spaniel that chased a peasant and that he did build a blonde automaton named "Francine" that was discovered in her packing case aboard a ship and summarily thrown overboard by a captain frightened of witchcraft. No description of "her" function accompanies the story, so we are left to imagine what this fictional Descartes wanted of a mechanical woman.[92]

We do know that Descartes's interest in automata is very much linked with the "Cartesian dualism" so widely cited in Western philosophy. The whole of Descartes's *Treatise of Man* discusses an imaginary species of automaton, a fictional analogue of the human body as a machine: "I assumed their [the automatons'] body to be but a statue, an earthen machine formed intentionally by God to be as much as possible like us."[93] Throughout the book, he refered to his version of the human body as a "machine." Although he leaves unfulfilled his promise to discuss the relation of the body and the soul, it is clear that Descartes believed that it was the soul that separated human beings from plants and animals; he usually portrayed plants and animals as soulless automata. The mechanism was there all the same. His depiction of the human body in the *Discourse on Method* is much the same: but for the soul, the body is nothing but an automaton.[94]

Projects that sought to mimic the circulation of blood were often connected with attempts to imitate the voice. Jacques de Vaucanson, a famous eighteenth-century inventor, and Louis-Bertrand Castel were both asked on a number of occasions to resolve debates among anatomists and physiologists by building automata that could simulate the human voice. In the 1730s, Vaucanson built a flute player that controlled a real flute with automatic lips and fingers, a tabor and tambourine player built on a similar principle, and an artificial duck that was capable of "eating, drinking, macerating the Food, and voiding Excrements, pluming her Wings, picking her Feathers, and performing several Operations in Imitation of a living Duck."[95] Vaucanson's work was derived from scientific interests in physiology. He proposed to build a group of automata that would imitate the "natural functions of several animals" and also hoped to build a machine that would mimic the circulation of blood. Claude-Nicolas Le Cat, Vaucanson's acquaintance and sometime promoter, also hoped to build a working model of the circulatory system in order to resolve a debate about the utility of therapeutic bleeding.[96]

Between 1770 and 1790, four persons in Europe built working speaking machines, apparently without knowledge of one another. All these inventors modeled their speaking automata on the human organs of speech.

The abbé Mical lived in Paris and created first a ceramic head that could utter a few phrases and then a pair of heads that exchanged sentences praising the king. Mical hoped that the heads would have some scientific use, but he was ultimately interested in financial gain. Christian Gottlieb Kratzenstein, a member of the Imperial Academy at St. Petersburg, constructed a machine that could accurately simulate all the vowel sounds. He based his own work on that of anatomists who had been concerned with the voice, constructing a table of positions of the larynx, tongue, teeth, palate, and lips for each vowel. Wolfgang von Kempelen, a member of the Viennese aristocracy, was most famous for an automaton chess player that he built in 1769. Later widely exhibited by Mälzel, it became known that the chess player was a hoax—it included a compartment where a small person could fit. Kempelen's speaking machine could say a few short words and phrases like *papa, mama, Marianna, astronomy, Romanum Imperator semper Augustus,* and *Maman aimez-moi.* Contemporary observers expressed surprise that Kempelen's automaton did not have human form—it was made from bellows and boxes with hinged shutters. Erasmus Darwin, meanwhile, became interested in building a speaking machine through his research into the origins of language. After inserting rolled balls of tinfoil into his own mouth to ascertain the positions of the mouth for each vowel and consonant, Darwin built a machine with a wooden mouth, leather lips, valves for nostrils, and a ribbon for a tongue. When air was blown with a bellows, the mouth could sound out the letters *b, p, m,* and *a.*[97]

Alexander Graham Bell and other inventors had an acute interest in this history. The quotation reproduced below, from John Bulwer's *Philosophicus* (1648), appears in Bell's files. It was likely of interest because it hints toward the long duration of a history of reproduced sound while at the same time marking the difference between earlier attempts and the work of Bell and his contemporaries:

Frier *Bacons* brazen Head, and that Statue framed by *Albertus Magnus* which spake to *Thomas Aquine,* and which he mistaking for a Magic Device brake, was certainly nothing else but Mathematical Inventions framed in *imitation* of the *motions of speech* performed by the Instruments in and about the Mouth. As for that leaden Pipe which *Baptista* Porta in his *magia naturali* speaks of as effectual to this purpose; or that of *Walchius* who thinks it possible intirely to preserve the voyce or any words spoken in a hollow Trunke or Pipe, and that this Pipe being rightly opened, the Words will come out of it in the same order wherein they were spoken, they have not as substantial a way for such a Discovery.[98]

Although nineteenth-century inventors sought to establish themselves as within a centuries-old historical stream, for them the phonautograph was significant because it embodied a switch from the mouth to the ear in efforts to understand, control, and reproduce sound.

By the nineteenth century, these scientific instruments had become, essentially, amusements or demonstrations for children. So, while it took a Viennese baron to build a speaking head in 1783, the male children of a well-known elocutionist could do it by the middle of the next century. Bell tells a story of childhood experiments in which he and his brother set out to construct a speaking automaton. This project itself was inspired by a visit to Sir Charles Wheatstone, where Bell and his brother first saw a reproduction of Kempelen's speaking automaton. Bell recounts his fascination with hearing it speak a few words. Wheatstone lent the elder Bell the instructions, which were the basis for the machine described here: "Stimulated by my father, my brother Melville and I attempted to construct an Automaton Speaking Machine of our own. We divided up the work between us, his special part consisting of the larynx and vocal chords to be operated by the wind chest of a parlor organ; while I undertook the mouth and tongue." For his part, Bell attempted to copy "Nature herself," using a cast made from an actual human skull as his point of departure. The goal was an "exact copy of the vocal organs." Although Bell describes the physical makeup of the machine at great length, his treatment of its function is most revealing:

We could not wait for the completion of the tongue: we could not wait for the arrival of the organ bellows. My brother simply fastened his tin larynx to my gutta percha mouth, and blew through the windpipe provided.

At once the character of the sound was changed. It no longer resembled a reed musical instrument, but a human voice. Vowel quality too could be detected, and it really seemed as though someone were singing the vowel "ah."

I then closed and opened the rubber lips a number of times in succession while my brother blew through the windpipe. The machine at once responded by uttering the syllables "Ma-ma-ma-ma" &c, quite clearly and distinctly. By using only two syllables and prolonging the second we obtained a quite startling reproduction of the word "Mamma," pronounced in the British fashion with the accent on the second syllable.

Well of course boys will be boys and we determined to try the effect on our neighbors.

My father's house in Edinburgh was one of a number of houses and flats that opened upon a common stair. We took the apparatus and made it *yell!* My brother put the windpipe into his mouth and blew for all he was worth, while I manipulated the lips. Soon the stairway resounded with the most agonizing cries of "Mamma—mamma—mamma." It really sounded like a little child in great distress calling for his mother.

Presently a door opened upstairs and we heard a lady exclaim, "my goodness, what's the matter with that baby?!"

This was all that was necessary to complete our happiness: delighted with our success we stole quietly back into my father's house and gently shut the door, leaving the poor lady to make a fruitless search for the now silent child.

I do not think the speaking machine progressed very far beyond this point; but it had undoubtedly been successful in realizing my father's great desire that through its means his boys would become thoroughly familiar with the actual instrument of speech, and the functions of the various vocal organs.[99]

Bell's description of his own adventures provides a rich text: here the "human voice" becomes a purely reproducible mechanical function, the copying of nature by science and, more important, technology and technique; scientific ingenuity mystifies a woman who searches for a crying baby; and, most important for our purposes, the reproduction of the human voice is accomplished through the mechanical reconstruction of the human mouth. But, while this was intended by the elder Bell as a kind of physiological instruction and amusement for his sons, it was no longer at the center of sound research.

By the time the young Bell brothers built their mouthy toy, scientists retained some interest in replicating the human voice through imitations of the mouth and vocal organs, but they had largely moved on to other things. Johannes Müller had experimented with artificial glottises, and Wheatstone had hoped to improve on eighteenth-century speaking automata to the point where they might be able to replicate human speech fully. In fact, Wheatstone may have imagined some kind of speaking telegraph, where speaking automata were connected with a device to transmit instructions for sound at some distance. But none of these projects came to fruition.

In the nineteenth century, automata were more common as entertainments. Mälzel took de Kempelen's speaking automaton on tour, and others attempted to build new and improved automata. Johannes Faber, an astronomer with failing eyesight, built a talking automaton called the *euphon* in the early 1840s. It consisted of a simulated torso and head (dressed like a Turk), a bellows, an ivory reed, and a keyboard the keys of which corresponded to various states of the mouth. Yet these were not terribly successful—apparently nineteenth-century audiences preferred spectacular deceptions like ventriloquism to the bland but real synthesis of speech. Henri Maillardet's Musical Lady, built early in the nineteenth century, worked on a similar principle, using levers to control its piano-playing fingers, reeds to draw air into its chest to simulate breathing, and a clock in its head to switch it on and off automatically.[100]

The main advancements in sound research and in sound reproduction came from abandoning the mouth altogether. Helmholtz's celebrated synthesizer, which used tuning forks and resonators to create vowel sounds, was modeled on the ear. His student, Rudolf König, built a "wave siren" that was composed of several metal rings with jagged edges. The edges corresponded to the waveforms of different vowels. When a jet of air was directed at the edge of the rotating ring, it would emit a vowel sound.[101] Both these machines treated sound as an effect; they treated sound in general, rather than attempting to replicate a specific cause.

One of the clearest elaborations of this thesis of a shift from mouth to ear appears in the writings and speeches of Emile Berliner, who is credited with inventing the gramophone in 1888. Since the gramophone recorded onto disks instead of cylinders, and because he was seeking patent rights and a market, Berliner was careful to construct a distinct genealogy for his invention. Berliner's genealogy was based on this notion of the imitation of the ear and specifically on machines that used vibrating diaphragms. This was both for scientific and for economic reasons. From a scientific point of view, Berliner's account is a reasonable and somewhat representative interpretation of the history of sound-reproduction apparatus from the mid-nineteenth century on. But this account also served an entrepreneurial function for Berliner. By tracing his lineage back through Charles Cros and others, he would be able to argue that his patents were not infringing on Edison or Bell's patents on sound-recording apparatus. Essentially, this genealogy sought to prove that Berliner had, in fact, invented a completely new machine. Berliner's account was likely shaped by purely mercenary concerns; he sought to distinguish and locate his work in the web of nineteenth-century innovation. But his central thread—the diaphragm and the imitation of human hearing processes—demonstrates something else entirely. The characteristics of the modern sound-reproduction technologies that later authors would characterize as revolutionary were themselves embedded in the flow of nineteenth-century ideas and practices.

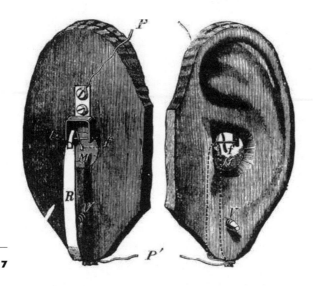

Berliner's genealogy begins with Charles Bourseil's 1854 proposal that two diaphragms vibrating in sympathy could reproduce speech over telegraphic distances. By 1859, a Frankfurt teacher named Philip Reis had constructed an apparatus based on Bourseil's suggestions. Now known as the *Reis telephone*, this machine did effectively transmit some variability of sound (such as the cadences and rhythms of speech)—as if to *mime* the reproduction of speech—but it did not effectively reproduce articulate (which is to say, understandable) speech. Interestingly, Reis used the human auricle as a model for a carved wooden receiver—the functional equivalent of the horn of a phonautograph or phonograph—in an early version of his telephone (figure 7). This proved less than fully effective, and the final form of the machine was functionally tympanic but gave up all visible resemblance to the auricle since the function of the Reis telephone came from its imitation of the middle ear, not the outer ear. Helmholtz's resonators also appear in this period as well. Scott began his work on the phonautograph in France two years earlier, in 1857, and his apparatus is generally credited with rendering sound visible. Berliner, along with the general current of opinion at the time, casts it as a direct predecessor of the phonograph. In fact, Berliner draws a direct link between the phonautograph and the Frenchman Charles Cros's ideas for storing and reproducing sound. Yet Berliner takes the imitation of the human ear quite literally in

Figure 7. Early Reis telephone

his account. Commenting on Bourseil's plan, Berliner critiques him and, later, Bell for *insufficiently* imitating the human ear: "He evidently desired extreme flexibility [in the diaphragm], and diaphragms constructed on that principle proved fatal to the efforts of many subsequent experimenters, even at first to Mr. Bell, who like Bourseil, borrowed the idea from the flexible *tympanum membrani* of the human ear, and who overlooked the important modifications which the vibrations undergo, before reaching the auditory nerve, by the series of muscular hinges in which the various bony accessories of the ear are mounted, and which act as elastic dampers against the *tympanum membrani.*"[102] The failures of earlier attempts to reproduce sound—attempts that were clearly modeled on the human process of audition—appear here as inaccurate reproductions by their own criteria. For Berliner, more ear was needed to reproduce sound. It was not just the membrana tympani, but the way it *focused* sound as part of a whole hearing mechanism, that was important.

Despite his protests to the contrary, the diaphragm is the one common denominator of the technologies that Berliner considers. Helmholtz's and König's acoustic experiments, for instance, appear in Berliner's narrative as detours from the teleology toward sound reproduction.[103] Helmholtz's discussion of combinational tones and the tuning of the hairs of the auditory nerve seemed to render hearing an immensely complex process—almost impossible to simulate—so that "the perusal of their work left a serious doubt in many a student whether there was not something in articulate speech, and its audibility by the human ear, beyond the grasp of the *mechanical* mind of man" (emphasis added).[104] Berliner's complaint rested on Helmholtz's piano theory of hearing—which would have required a machine like Bell's piano-ear to reproduce sound (although, as I have argued above, Helmholtz's more fundamental insights concerning the perception of sound as an *effect* were central to the history of sound reproduction). Likewise, Faber's ca. 1860 automaton appears as an interesting but unnecessary detour in the technological history. This is because his work moves too far from a tympanic understanding of sound reproduction.

The importance of the telephone, for Berliner, was not that it finally transmitted understandable speech over a distance but that it was a relatively simple apparatus (in contradistinction to Faber's automaton) based on the vibration of a diaphragm. Charles Cros's phonograph, which applied some of the telephone's principles to the phonautograph to suggest a method for storing and reproducing sound, is the final stop in Berliner's narrative before he turns to his own invention.[105] Thus, Berliner's history of sound reproduction is a history of tympanic machines—each held a diaphragm that could vibrate and, in so vibrating, invite nearby eardrums to vibrate in sympathy.

It is also worth noting that Bell too understood the relation between this diaphragm principle, telephony, and sound recording. Bell's response to learning of Edison's invention is instructive here. It was not the machine but the deeper principle that was the fundamental innovation: "It is a most astonishing thing to me that I could possibly have let this slip though my fingers when I consider how my thoughts have been directed to this subject for so many years past." Like some of the telephone claimants, he erroneously passed easily from the feeling that he *should* have thought of it to the conviction that, in principle, he had. In his telephone lecture, he had remarked that, if some implement could be made to follow the curves of a phonautograph tracing, it would reproduce the sound that had made the tracing: "And yet in spite of this the thought never occurred to me to indent a substance and from the indentation to reproduce sound." Although Bell's lamentations are also those of a wishful entrepreneur, it is certainly the case that, through the principles of the phonautograph, he and many others (most notably Cros) had grasped the principle behind the phonograph before it was actually invented.[106]

The tympanic principle was everywhere. Even early advertisements emphasized the significance of diaphragms. A Columbia Phonograph Company pamphlet from 1895 described the functioning of the graphophone in four essential parts:

1. The diaphragm, which vibrates in the same way as the human ear drum in response to the air waves made by any sound.
2. The needle, which is attached to the diaphragm, and engraves an impression as the result of the vibration on
3. The cylinder;
4. The arrangement for making the cylinder revolve evenly.[107]

Even in this most basic characterization, the diaphragm and its vibrations are the central functional element of the graphophone, with the other parts of the machine either regulating or channeling that vibration. The figure of the ear is at the center of the machine.

Indeed, the movement from cause to effect might be understood as the singular achievement of the sciences of sound in the nineteenth century. In an 1854 speech, Helmholtz cast Vaucanson's mechanical duck as a "merest trifle" because it "copied animal function as it was observed"; Helmholtz's tuning-fork synthesizer copied what he thought to be its mechanism.[108] Vaucanson's duck and young Alexander Graham Bell's speaking mouth had potential as scientific devices in the eighteenth century. By the nineteenth century, they were more likely to be conceptualized as scientific toys.[109] In the world of Helmholtz, Scott, and, later, Blake, Bell, Edison, and Berliner, sound's reproducibility was based on a mechanistic conception of hearing crystallized in the tympanic function. The goal was to have our ears resonating in sympathy with machines to hear for us. Sound was first and foremost a form of vibration—its particular causes mattered less than the determinate effect that it had on the sense of hearing.

Ear to Machine and Back Again:
The Very Possibility of Sound's Reproduction

The repeated reversals between the ear and its mechanical simulation that began with instruments used in acoustic and physiological research continued with sound-reproduction technologies. There was a certain possibility for interchange between ears and machines to hear for them; the tympanic machine could be mapped onto the ear itself. A working phonograph or telephone, so it was thought, could possibly compensate for or even fix a nonfunctional human ear. The 1890s saw many attempts to use the phonograph and the telephone as cures for or at least solutions to deafness. A *New York Times* article from the same year as the Columbia circular had a Dr. Leech proposing to use the phonograph to "massage" the ossicles in the ears of the deaf: "The principle of treatment employed is the massage, or mechanical stimulation, and consequent reawakening of the sound-conducting apparatus of the ear, by means of vibrating force [of the phonograph]."[110]

Perhaps the most striking example of this phenomenon was J. C. Chester, "the human telephone," a man who wired himself up with a complete telephone assembly (including battery) and marched to Washington, D.C., to patent himself (figure 8): "He has found by many experiments that the dulled nerves of the ear are quickened by these powerful electric appliances and that he does hear." In addition to an earpiece and mouthpiece for an interlocutor, Chester had outfitted his own end with a mouthpiece and an earpiece with a special wire connected directly to his teeth so that the signal could approach his auditory nerve from two directions at once. "A

gentleman meeting this walking telephone upon the road is offered the transmitter and receiver that hang upon the hook. The gentleman places one to the ear and talks through the other, sound being much assisted by the receiver in his ear. When he replies, he speaks through a tin horn connecting with the wires and trusts to the carrying effect of the telephone. In this way he can converse over a space of several feet as easily as any other man, the painful ear-splitting being avoided."[111] The actual effectiveness of this apparatus is questionable. It is true that hearing aids followed the advent of tympanic machines (one history dates the electrically amplified hearing aid back to 1880).[112] While such apparatuses might be of some as-

Figure 8. The human telephone, 1897 (courtesy Division of Medical Sciences, National Museum of American History)

sistance to the hard-of-hearing by focusing sound and channeling it toward a single point (through the use of the receiver in the telephone or an ear tube for the phonograph), they were, however, of no use as cures for deafness. All the same, these early sound technologies were at once supplements to, imitations of, and replacements for the human ear.

The locution *machines to hear for them* suggests a further implication of cultural attitudes about hearing, deafness, and states in between. Put simply, it puts the hearing body in analogical relief against the social body: sound reproduction came to be represented as a solution, not only to the physical fact of deafness or hardness of hearing, but, more important, to the social fact of unaided hearing. As I argue in the next chapter, sound reproduction required a notion of hearing in need of supplementation. In that sense, the treatment of the deaf became a model for the treatment of the hearing.

In Berliner's account, in the narratives presented by Bell, Edison, and those around them, and in everyday representations of the new sound technologies we can say with some certainty that the ears have it. The key element, the defining function, in these early versions of sound-reproduction technologies was the diaphragm—embodying the tympanic mechanism, a principle that connected ear to machine through analogy, imitation, or thumbscrews. This ear function that could be abstracted from the human body, transposed across social contexts, produced, proliferated, and mutated through technique and technology, suggests that the ear (and, specifically, the diaphragm) did not simply come to be a representation of sound reproduction in this period. The ear—in its tympanic character—became a diagram of sound's reproducibility. The ear, as a mechanism, became a way of organizing a whole set of sounds and sonic functions; it was an informal principle by which the mechanics of sound reproduction were arranged.[113]

The tympanic construct, the abstraction of hearing into a mechanism, cut across social relations and social contexts; it became a method for organizing people, forces, matter, and ideas. Yet it is *not* a *deep structure* in Lévi-Strauss's sense of enduring structural relations lying dormant beneath a society that are then carried out through social activity. *The tympanic* is also not an "ideal type" derived through analysis of a normative structure by which to consider a range of multiple and differing practices.[114] At no time was the tympanic a fixed thing that existed outside, above, or prior to the history of sound reproduction. As machines, ideas, and practices changed, the tympanic function and the status of the ear gradually changed over time. While, in its formal characteristics, each sound-reproduction technology exhibited a "family resemblance" to the others through the tympanic function, the function was itself a variable.

Thinking through the history of the tympanic function offers an alternative to triumphalist and ultimately tautological theories of sound reproduction that presume its contemporary form (i.e., sound-reproduction technologies as we now know them) in their definition of the analytic problem; it allows us to move away from presuming and then attempting to adjudicate the widely variable relations among different sounds (original/copy; reality/representation) to a consideration of the social, cultural, and technical mechanisms that open up the questions of those relations in the first place. But this is also precisely the reason for *not* positing the ear and its tympanic function as a stable and timeless deep structure of sound reproduction. To do so would be to suggest that the *prehistory* (to use Adorno's term) of the telephone, phonograph, microphone, and radio entirely determines their subsequent social and cultural significance. While sound-reproduction technologies did not drop out of the sky to create a new sense of the human ear or transform the fact and function of hearing, they certainly would undergo their own transformations as they grew more socially and institutionally established, as they were shaped into media. As the technologies were gradually organized into media systems with their own distinctive industrial and cultural practices—as they became sound recording, telephony, and radio as we know them today—they could in a sense take on a life of their own.[115] The same is true of the tympanic diagram: it begins as an imitation of the human ear, but very quickly the human ear would become but one instance of a more general tympanic function, a function that could begin proliferating through other contexts by virtue of its embeddedness in the institutions of science, culture, and commerce. The tympanic would take on a life of its own.

After the initial construction of the telephone and the phonograph, Bell and others would quickly turn away from literal imitations of the human ear since, for instance, the telephone's diaphragm needed to be heavier than the tympanic membrane because it was used to vibrate iron, not bone. Nevertheless, as late as 1878, Bell, Watson, Blake, and others were still experimenting with human ears, this time with an ear telephone.[116] Blake wrote, "I have been able to carry on conversation without difficulty over a line something more than six hundred feet in length, the ear telephone being used only as the receiving instrument."[117] Yet, already in that same year, the human middle ear was becoming a weak instance of the tympanic function. Thomas Watson wrote to Blake saying that he and Bell had also tried a tympanic membrane among a number of different diaphragms for the telephone: "They all worked, even the real ear telephone, which was, however, the poorest of the lot."[118]

2 Techniques of Listening

A Brandes advertisement from the May 1925 issue of *Wireless Age* bears an imperative title: "You *need* a headset" (figure 9). As radio loudspeakers were becoming a popular alternative to headphones, Brandes may have felt that it needed to convince readers of *Wireless Age* that they needed headsets; or this may just have been a standard advertising line.[1] Either way, the ad actually bothers to tell readers *why* they need a headset: "You *need* a headset: to tune in with; to get distant stations—both domestic and foreign; to listen-in without disturbing others; to shut out the noise in the room—and get all the radio fun; to get the truest and clearest reception—always."[2] In telling us why we need a headset, Brandes offers a promising entry point into a much longer history of listening.

Headsets were not actually necessary for tuning a radio. Instead, they helped their users to better "DX" or listen for distance—to hear very faint, indistinct, and distant sounds, stretching the existing capacity of their radios and their ears. Picking up faint, faraway stations was one of the holy grails of amateur radio listening (see figure 10). Brandes's ad is clearly aimed at DXing.[3] As the ad also intimates, headphones isolate their users in a private world of sounds. They help create a private acoustic space by shutting out room noise and by keeping the radio sound out of the room. They also help separate the listener from other people in the room. Through this isolation, the headphones can intensify and localize listeners' auditory fields, making it much easier to pay attention to minute sonic details and faint sounds. Brandes's headphones provide the "truest and clearest reception" because of this emphasis on sonic detail through isolation.

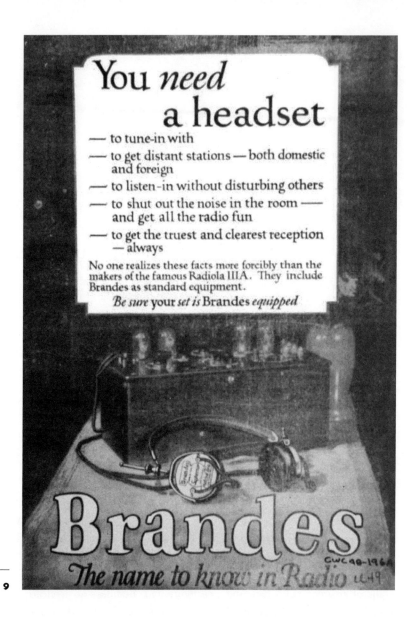

So, according to Brandes, hearing ever more subtle sonic details while isolated in one's own private acoustic space is the way to get "all the radio fun."

Brandes's 1925 headset ad marks a convenient end point for a series of transformations in practical orientations toward listening that began in the 1810s. At that earlier moment, listening was first being articulated to

Figure 9. "You *Need* a Headset"—1925 advertisement for Brandes headphones

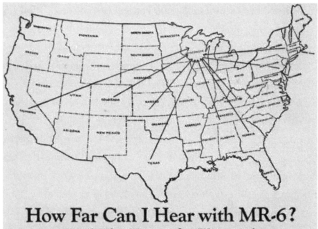

How Far Can I Hear with MR-6?
Read What Happened in Wisconsin!

From Wisconsin alone during one month come reports of De Forest MR-6 Receiving Sets getting California, Colorado, Kansas, Texas, Tennessee, Georgia, Kentucky, Pennsylvania, and New York—distances up to 1500 miles. One man listened across the entire continent, getting Santa Cruz, California and Atlanta, Georgia the same evening. The unsolicited testimonials as to the way this efficient but inexpensive set "listens to the world" are on file in our office and copies will be sent to anyone interested in writing direct to the owners.

Multiply such experiences as these by the thousands of MR-6 sets in use all over the nation—to say nothing of the De Forest Everyman and Radiohome sets—and you get an idea of the way De Forest is serving the nation with the joys of radio.

De Forest manufactures receiving sets all the way from the least expensive to the most elaborate, and laboratory tested high quality parts for those who "build their own". If it's De Forest, it's built in a way worthy to sustain the reputation of that great name.

If you want the best radio has to offer—the songs, the stories, the news of the world—more clearly than you have believed possible and from farther away—you can't go wrong on De Forest!

DE FOREST RADIO TEL. & TEL. CO. JERSEY CITY, N. J.

newly emergent notions of science and rationality through its use in doctors' medical examinations of patients. Over the course of a century, this practical orientation would move from the specialized province of physicians diagnosing their patients to the much larger context of listening to technologically reproduced sound. Brandes could harangue their readers with the imperative "You need a headset" because the headset supplements and crystallizes an orientation to listening that they presumably already had. It presumed a high level of intuitive knowledge about radio listening.

Figure 10. "How Far Can I Hear with the MR-6?"—DeForest advertisement from the early 1920s (courtesy Archives Center, National Museum of American History)

A little over a hundred years earlier, it took more than seven hundred pages for the French physician René-Théophile-Hyacinthe Laennec to make a strikingly similar argument for physicians to use stethoscopes to listen to their patients' bodies. Laennec, who is credited with inventing the stethoscope, published *A Treatise on the Diseases of the Chest and on Mediate Auscultation* in 1819 (a second, somewhat revised edition appeared in 1826).[4] Mediate auscultation is the act of listening to a patient's body through a stethoscope. Laennec's lengthy *Treatise* is a fascinating document because it explains to physicians *why* they would want to listen to patients' bodies, how to listen to patients with the stethoscope properly, and how to interpret the sounds thus heard. This level of explanatory detail was necessary at the time: although physical examination would become the dominant mode of examination in the 1800s, it was still an emergent practice in 1819; mediate auscultation developed at a moment when medicine itself was undergoing a major epistemic shift.[5]

This chapter and the next offer a story about changing meanings of listening. The techniques of listening that became widespread with the diffusion of the telephone, the phonograph, and the radio early in the twentieth century were themselves transposed and elaborated from techniques of listening developed elsewhere in middle-class culture over the course of the nineteenth century. Using the Laennec and Brandes documents as end points, chapters 2 and 3 offer a genealogy of *audile technique,* or *techniques of listening.* By this emphasis on technique I mean to denote a concrete set of limited and related practices of listening and practical orientations toward listening. I follow audile technique through three very different cultural contexts: modern medicine in Western Europe and the United States from the 1760s into the 1900s, American sound telegraphy from the 1840s into the 1900s, and sound-reproduction technologies in Europe and the United States between 1876 and 1930. After introducing the concept of audile technique, this chapter examines the emergence of audile technique in modern medicine. The next chapter considers audile technique in the contexts of emergent sonic media: sound telegraphy, telephony, phonography, and radio.

As should be obvious from the long time span and diverse contexts that I cover, this is not and cannot be an anthropological history of listening practices. It is not meant as a systematic account of how people actually listened, and it certainly does not pretend to exhaust the descriptive possibilities of listening history or catalog all the contexts in which audile technique plays a role.[6] My goal is not to describe what it felt like to listen at any given place or time. Nor do I mean to suggest an evolutionary narrative of listening, where the sense of hearing undergoes a naturalized process of modernization. This is why I use the term *genealogy:* I aim to chart the emergence of a practical orientation in diverse contexts over a long period of time. I am interested in family resemblances among otherwise diverse practices, theoretical or "idealized" constructs of listening, and how those constructs were supposed to be put into practice. In other words, this is a history of "regimes" of listening practices. Even if we acknowledge that many of the programs for conduct considered below were never fully realized, they still tell us a great deal about the construction of the institutions and practices that they sought to organize or explain.[7] To take but one example, despite the fact that physicians were supposed to be virtuoso listeners, at the end of the nineteenth century many America doctors were still poorly trained and haphazardly combined methods of diagnosis and treatment. Yet the medical textbooks and medical education of the time were very much oriented toward turning doctors into rational, scientifically minded, virtuoso listeners. The stethoscope was a symbol of the diagnostic power of the medical profession, even if some doctors were not very good at using it.[8]

My use of the word *technique* in relation to listening is derived from Marcel Mauss's notion of "techniques of the body." "The body is man's first and most natural instrument," writes Mauss: "Or more accurately, not to speak of instruments, man's first and most natural technical object, and at the same time technical means, is his body. . . . Before instrumental techniques there is the ensemble of techniques of the body. . . . The constant adaptation to a physical, mechanical or chemical aim (e.g., when we drink) is pursued in a series of assembled actions, and assembled for the individual not by himself alone but by all his education, by the whole society to which he belongs, in the place he occupies in it."[9] Mauss compiles an extensive list of techniques for investigation: sleep, waking and rest, walking, running, dancing, jumping, climbing, descending, swimming, forceful movements, hygiene, eating, drinking, sexuality, and care of the sick. Although he does not include sensory activities—looking, listening, tasting, smelling, touching—these are certainly implied and even occasionally mentioned in the context of the other techniques.[10] So my argument makes a very short leap from Mauss's list of techniques to a history of techniques of listening in modernity. It is something of an extrapolation: ethnographers can go somewhere to learn about cultural practices through participation and observation; historians and genealogists must reconstruct domains of physical practice from documents and artifacts. But the issue of technique remains salient.

Techniques of the body are constructed through "physical education of all ages and both sexes," and, as we will see, techniques of listening are also the result of physical education, whether this education is institutionalized in professional training or simply accomplished through shared and repeated practice.[11] The term *technique* also conjures up names like Aristotle, Martin Heidegger, and Jacques Ellul. It connotes a connection among practice, technology, and instrumental reason: it is a form of "reasoned production," "a way of revealing," a "means with a set of rules for the game." Under the sign of modernity, technique carries a special value and a special valence—it is connected with rationality. Technique brings mechanics to bear on spontaneity.[12]

This is an incredibly important point for a history of communication technology: after Mauss, the body is the first communication technology, and all the *technologies* of listening that I discuss emerge out of *techniques* of listening. Many authors have conceptualized media and communication technologies as prosthetic senses. If media do, indeed, extend our senses, they do so as crystallized versions and elaborations of people's prior practices—or techniques—of using their senses. So, although *technique* and *technology* are terms that clearly bleed into one another, the distinction is crucial for the history of sound. *Technique* connotes practice, virtuosity, and the possibility of failure and accident, as in a musician's technique with a musical instrument. It is a learned skill, a set of repeatable activities within a limited number of framed contexts.

Listening involves will, both conscious and unconscious—perhaps a better word than *will* would be *disposition* or even *feel*. Orientations toward and styles of listening are part of what sociologists and anthropologists have come to call *the habitus*. Following Pierre Bourdieu, *habitus* denotes a set of dispositions, what he calls *a feel for the game*. The habitus is socially conditioned subjectivity: it combines all those forms of informal knowledge that make up social life. Habitus is a mix of custom, bodily technique, social outlook, style, and orientation. Because habitus is socially conditioned, social position and subjective disposition go together—each influences the development of the other.[13] Industry, bureaucracy, science, rationalism, and the new middle class are all so central to the genealogy of audile technique precisely because techniques of listening represent dispositions articulated within a range of social possibilities.

Modern audile technique combines a relatively stable set of practical orientations toward sound and listening. Although there may be other distinctively modern techniques of listening, the following list represents the orientations common to medicine, telegraphy, and the sound-reproduction technologies considered in this chapter and the next:

1. Listening gets articulated to notions of science, reason, and rationality. Listening becomes a technical skill, a skill that can be developed and used toward instrumental ends. This is hard to describe, and harder to stress, since there are few English words to connote the sonic equivalents of *gazing* or *observing*. We are used to the idea that new orientations toward looking, often thematized as "the gaze," have something to do with changing ways of knowing during and after the Enlightenment. As it was for looking, so it was for listening: audition becomes a site through which modern power relations can be elaborated, managed, and acted out. Starting in a few select contexts, the very meaning of listening drifts toward technical and rational conceptions. Over the long nineteenth century, listening becomes a site of skill and potential virtuosity.

2. In order for listening to become useful as a tool of rationality (and for itself to be rationalized), it had to be constructed as a discrete activity. Chapter 1 introduced the "separation of the senses" and the isolation of hearing in Bell, Müller, and Helmholtz. In the actual practice of audile technique, listening was similarly separated from other sensory activities. As we will see, audile technique is oriented toward a faculty of hearing that is separated from the other senses. Once so separated, it can be intensified, focused, and reconstructed.

3. Concurrent with the separation of hearing from the other senses is a reconstruction of the shape of acoustic space. Audile technique was not simply a representation of acoustic space; it aimed actively to transform acoustic space. The space occupied by sounds becomes something to be formed, molded, oriented, and made useful for the purposes of listening techniques. It can be segmented, made cellular, cut into little pieces, and reassembled. Acoustic space becomes a kind of bourgeois private space. As I will show, even collective conceptions of listening assume that collectivity is entered through this prior, private auditory space.

4. As audile technique problematizes the shape of acoustic space, it also problematizes the content of acoustic space. The previous chapter showed how sound gets constructed as an object in physics, acoustics, and physiology. Whereas voices or music had been privileged instances of sound, now they were merely instances of a more general category of sound. In audile technique, sounds also became meaningful precisely for their sonic characteristics, in a manner parallel to the way in which timbre became a central concern of nineteenth-century acoustics after Helmholtz. On the basis of their sonic character, sounds become signs—they come to mean certain things. Technical notions of listening depend on the establishment of a code for what is heard but exist without an effective metalanguage. A metalanguage of sound would consist of a nonspecialized set of terms that enabled people to describe the details of audile experience in a purely abstract manner. While visual experience has a well-developed metalanguage, sonic experience does not. We have abstract words to describe color, texture, shape, direction, shading, and so forth. Conversely, most of the language used to describe elements of auditory phenomena is metaphoric: aside from specialized languages in musicology, sound engineering, acoustics, and general descriptors such as *loud* or *quiet,* there are very few abstract words in common English for describing the timbre, rhythm, texture, density, amplitude, or spatiality of sounds.[14] Because of the difficulties involved in constructing a metalanguage of sound, audile technique would come to stress listening practice and practical knowledge through listening, rather than formal and abstract descriptions of sounds.
5. Techniques of listening are based in and described through a language of mediation. Audile technique is premised on some form of physical distance and some mediating practice or technology whereby proximal sounds become indices of events otherwise absent to the other senses. This was in part a component of rationalizing listening and turning it into a skill. It was also in part a component of isolating and intensifying hearing as a faculty.
6. Finally, audile technique could come to hold a great deal of symbolic currency: virtuosity at audile technique could be a mark of distinction in modern life. Both doctors and sound telegraphers used representations of listening as part of their professional mystique. The more generalized audile technique associated with sound-

reproduction technologies was widely understood as an index of those technologies' modernity.

Speaking generally, audile technique articulated listening and the ear to logic, analytic thought, industry, professionalism, capitalism, individualism, and mastery—even as it required a good deal of guesswork in practice. The history of audile technique thus offers a counternarrative to Romantic or naturalistic accounts that posit sight as the sense of intellect and hearing as the sense of affect, vision as the precise, localizing sense and hearing as the enveloping sense.[15] Some medical historians have suggested that there is a uniquely modern medical gaze. If this is the case, then modern orientations toward medical listening were a necessary precondition for this gaze as we know it. If, as many media historians have suggested, electric telegraphy heralds the age of modern mass communication, then listening is at the very core of modern media history. If technologies of sound reproduction depended on and actuated versions of audile technique, they drew together a diverse field of practices that had been developing for decades. To capitalize and commodify sound, sound media industries deployed a preexisting notion of sonic space as private property.

Audile technique emerged as a distinctively modern set of practical orientations toward listening. As a way of knowing and interacting with the world, it amounted to the reconstruction of listening in science, medicine, bureaucracy, and industry. It helped constitute these fields. Audile technique was also a distinctly bourgeois form of listening; it corresponded with the emergence of *middle class* as a salient cultural category. Thus, the orientations toward listening that accompanied sound-reproduction technology in the late nineteenth and early twentieth centuries are part of a longer-term historical current. Many writers in the 1920s and 1930s pinned radio's cultural significance on its use of hearing—"a novel sense." Rudolph Arnheim understood radio perception as a kind of blindness, an aesthetics of the audible with the visual component subtracted. For Hadley Cantril and other radio researchers, radio represented a unique psychological phenomenon, where *listening* became synecdochic for all activities of audiencing.[16] These primarily developmental accounts posit the existence of a history of listening and at the same time close it down—radio, film, and sound recording become the agents of acoustic modernity. They treat sound-reproduction technologies as positing a new way of hearing. In contrast, this genealogy of audile technique begins an argument for listening that will be continued throughout the book: over the course of the nineteenth century, audile technique was constructed as a set of related practices. In turn, it was crucial in the construction of sound reproduction as a practice. Thus, my genealogy re-places sound reproduction within the longer flow of sound history.

Audile as an adverb and *listening* as a verb are deliberate choices on my part. Although my use of *audile* is somewhat anachronistic (the word first appeared in the 1880s), its connotations seem especially promising. The word has two primary definitions. As a noun, it refers to a person in whom "auditory images" are predominant over tactile and visual stimuli. An audile is a person in whom auditory knowing is privileged over knowing through sight. As an adverb or adjective, it means "of, pertaining to, or received through the auditory nerves" or "of or pertaining to" the noun sense of *audile*.[17] So the term is useful both because it refers to the physiological, process-based sense of hearing discussed in chapter 1 (as opposed to older terms like *auricular*) and because it references conditions under which hearing is the privileged sense for knowing or experiencing. This is especially important for the practices that I describe below: hearing is not simply one way of knowing or experiencing among others; it is a sense separated and in some cases privileged (as in doctors' uses of stethoscopes) for particular activities. Thus, I use *audile* to connote hearing and listening as developed and specialized practices, rather than inherent capacities.

Activity and practicality are also important for my conception of listening. Too often, hearing and listening are collapsed in discussions of the senses. Certainly, hearing is a necessary precondition for listening, but the two are not at all the same thing. The usual distinction is between hearing as passive and listening as active, but this is not quite right either. As was considered at length in the previous chapter, hearing is a physical and physiological activity, a form of receptivity. Hearing turns a certain range of vibrations into perceptible sound. Over the course of the nineteenth century, hearing was constructed as a set of capacities and mechanisms, and that mechanical, objectified construct of hearing was crucial in the mechanical construction of sound-reproduction technologies.

Audile technique denotes a particular orientation toward listening — but it is not meant to be taken as the only possible orientation toward listening. Other writers have also posited listening as a construct and a cultural process in other contexts. Barry Truax writes that, as a cultural practice, listening can be an active means of gaining knowledge of a physical environment through the apprehension of variations in sonic character, for instance, in the way in which a blind person may use a cane or his or her own footsteps. One of the key means for this facility is through the detection, isolation, and interpretation of subtle variations in a sonic environment, another characteristic of listening that has become highly developed in many cultural contexts. Truax also identifies a range of listener orientations: "listening-in-search," "listening-in-readiness," and "background listening." Each of these practices may be developed differently in different cultural contexts. Steven Feld has discussed the Kaluli ethic of *dulugu ganalan* (or "lift-up-over sounding") as both a structural principle of Kaluli music making and a more general principle in Kaluli approaches to sound. Kaluli listening is highly attuned to direction, timbre, and texture: conversations involve multiple people speaking to multiple audiences at once. Similarly, Michel Foucault's notion of the confessional as he develops it in volume 1 of *The History of Sexuality* could also be read as a particular configuration of listening and speaking practices.[18]

There are a number of extant cultural histories of listening that aim to recount changes in practices of listening. In these histories, technique often appears as a tangent to the main narrative. This is because listening practices are often historicized in a single context. For example, James H. Johnson's *Listening in Paris* documents the gradual silencing of concert audiences in Paris over the course of the nineteenth century. After spending close to three hundred pages documenting the practices of listening and decorum in French concert halls, Johnson posits that musical harmony may have promised a road to collective musical experience in the nineteenth century. For his concertgoers, musical harmony may have been synecdochic for social harmony. But, by the turn of the twentieth century, Johnson argues, nationalist sentiment had effectively undermined Romantic investments in harmony as an expression of cultural collectivity. In other words, nationalist ties were so strong that they could no longer be superseded by the collective contemplation of harmony. Music became resolutely national, and its unifying power lay in composers and performers rather than its formal characteristics.

For all Johnson's attention to narrative continuity, his conclusion posits a radical break between past and present: "In our own atomized, eclectic, post-everything society the metaphor more likely to give voice to our own ineffable impressions and bridge the inner experiences of a fragmented public is the culture of *technique*—the acoustics of the hall, fidelity in recording and reproduction, the perfect sound, all the right notes."[19]

Johnson means that, as listening turned away from the formal melodic and harmonic structure of the music, it turned to the sonic characteristics of the music. Listening for acoustics and fidelity—listening as technique—constructed music as just one more instance of sound. Technique appears as a tangent to Johnson's cultural history precisely because he is interested in the specific practice of concertgoing. But right alongside Johnson's cultural history runs a parallel tale of audile technique—perhaps not the definitive moment of the "modernization of the ear," but certainly a movement that connects Johnson's deceased past with a living present. His conclusion suggests the necessity for the genealogy that I perform here: Where did audile technique come from? How and why did it emerge at the end of the nineteenth century?

Like silent, contemplative concertgoing, audile technique also grew up with the modern bourgeoisie, but it came to music rather late. Audile technique was cultured and developed in relatively circumscribed professional domains of middle-class life. Medicine and telegraphy were two fields where techniques of listening provided professional ethos and prestige. Audile technique functioned as part of the practical and official knowledges of these fields. The use of the stethoscope marked medicine for over a century, and, through its use, hearing surpassed sight as a diagnostic tool. The specificities of listening were themselves able to be applied to the body of the patient; medical listening rendered the interior motions of the human body to medical thought in a new clarity. Sound telegraphy, on the other hand, was the first major electronic medium in the United States, and the development of sound telegraphy represents the instrumentalization of hearing in a bureaucratic-corporate context. Medicine was an upper-middle-class profession, telegraphy a lower-middle-class profession. Both the stethoscope and the telegraphic "sounder" were technologies that crystallized already-extant techniques of listening. As I will discuss in the next chapter, the iconography of listening to early sound-reproduction technologies, especially around ear tubes and headphones, suggests a direct line of descent from the stethoscope and the telegraph to the telephone, phonograph, and radio. Sound-reproduction technologies disseminated and expanded these new technical notions of listening through their own institutionalization.

From roughly 1810 on, audile technique existed in niches at either end of the growing middle class. It would not become a more general feature of middle-class life until the end of the nineteenth century, when sound reproduction became a mechanical possibility and the middle class itself exploded in size and changed in outlook and orientation. In the meantime, audile technique was a well-known but emergent, specialized practice that helped reconstruct the meaning of listening in modern life.

Mediate Auscultation and Medicine's Acoustic Culture

One of the most enduring symbols of modern medicine has been a listening technology: the stethoscope. The stethoscope marks an important point in the history of listening and connects it to the history of medicine's industrialization and professionalization: it marks the articulation of the faculty of hearing to reason, through a combination of spatialization, technology, pedagogy, and ideology. Michel Foucault calls clinical medical experience "that opening up of the concrete individual, for the first time in Western history, to the language of rationality, that major event in the relationship of man to himself and of language to things."[20] If medicine was one of the first sites where the conceptual tools of rationality and empiricism were combined with techniques of investigation to make the human body an object of knowledge, then it turns out that a technique of listening was instrumental to reconstructing the living body as an object of knowledge. Listening was one of the central modalities through which modern medical ways of knowing were developed and enacted. Before the insides of living human bodies could be subjected to the modern gaze, they were subjected to physicians' techniques of listening.

Michel Foucault, Stanley Joel Reiser, and Jacalyn Duffin offer probably the most developed accounts of the use of the stethoscope as a technique of audition, and my analysis here builds on their work. These authors all argue that modern medicine embodies in its development a movement from the theoretical to the perceptual: the rise of empiricism is key here, but more important than the approach is the construction of a new object. These authors chart a new field of knowledge, an arrangement of what can be seen and what can be said.[21] Listening would play a tremendous role in this new medical epistemology.

The history of the stethoscope is not so much about the actual artifact as the technique that it crystallized. As I will use it in this chapter, *mediate auscultation* refers to the practice of listening to movements inside the body with the aid of an instrument (hence *mediate*). Mediate auscultation is the technique of using a stethoscope to diagnose. *Auscultation* is a noun standing for the action of listening or hearkening, and the word's English usage dates from the seventeenth century, although it has a considerably longer history in French and Latin. It picked up specifically medical connotations at the turn of the nineteenth century as the activity of listening to the sound of the movements of organs, air, and fluid in the chest.[22] In fact, auscultation already involved a notion of listening as *active* (vs. passive) hearing. By 1802, the French physician Matthieu-François-Régis Buisson was distinguishing between active and passive listening: "He distinguishes two sorts of hearing, the passive or *audition,* the active or *auscultation,* a division based on equally exact observations, and on which is based the difference between the words, *to hear* and *to listen.*"[23] As I argue above, the physiological notion of hearing as a pure capacity is not quite passive—*receptive* would be a more accurate adjective. But this is still a tremendously significant distinction. At the turn of the nineteenth century, medicine was beginning to construct hearing as a physical and physiological phenomenon, a set of determinate possibilities. Already in 1802 Buisson has a nascent understanding of this receptive, physiological notion of hearing (although further physiological insights into hearing as a unique form of nervous sensation would have to wait for Charles Bell and Johannes Müller in the coming decades).[24] But Buisson's reasoning shows us that, as it emerged, this physiological construct of hearing was accompanied by a practical-social construct of listening.

The phrase *mediate auscultation* was coined by R. T. H. Laennec, who is credited with inventing the stethoscope and the techniques to go with it. He used *mediate* auscultation in opposition to listening to a patient's body without a stethoscope, which he called *immediate* auscultation. I consider the locution to be central here because, for Laennec, *mediate* becomes the normative term, with *immediate* implying an absence of normal mediation. In fact, later writings would simply use *auscultation* to refer to listening to the body through a stethoscope—mediation was always assumed. In other words, although the term *mediate* was dropped from the phrase, it remains the default category for medical listening, down to the present. So a discrete form of listening—as mediated, skilled, and technologized—became centrally important to the construction of modern medical knowledge and its application. It redefined the meaning of listening itself. Techniques of listening and the technology of the stethoscope are at the very core of modern medical practice, even in its later visual forms. As Jacalyn Duffin puts it, mediate auscultation and the stethoscope initiated a whole set of "conceptual and instrumental tools" designed to facilitate the search for the internal causes of external symptoms: "phrenology, pleximeters, specula, ophthalmoscopes, and endoscopes[,] X-ray machines, Crosby capsules, fiber optics, Computerized Axial Tomography (CAT scans), and Magnetic Resonance Imaging (MRI)," all are direct descendants of mediate auscultation and the stethoscope.[25] In other words, an audile technique and technology paved the way for a whole ensemble of visual-medical media.

Like other technological innovations, the stethoscope was designed to operate within the parameters of a set of social relations, and it helped cement and formalize those relations: the doctor-patient relation, the structure of clinical research and pedagogy, and the industrialization, rationalization, and standardization of medicine (along with the improvement of physicians' social status). As Paul Starr writes, medicine became a middle-class profession in part through its acceptance of technology and scientific reasoning and the development of specialized techniques. Over the course of the nineteenth century, the institution and practice of medicine underwent tremendous transformations. With the growth of industrialization and urbanization, people came to depend more on experts and specialists where they had previously depended on themselves. At the same time, medicine itself industrialized: in gaining a more rationalized structure; in taking shape as a self-conscious profession; in a heavier investment in the discourses of science and reason; and, finally, in its adoption of technology. The industrialization and rationalization of medicine is, thus, part of the larger growth of industrial capitalism.[26]

Mediate auscultation and the stethoscope were part of this emergent industrial-rational orientation of medicine. The practice of mediate auscultation developed as a technical response to a social and investigative problem in a clinical setting. Its structure and operation were based on a new set of assumptions about the nature of medical knowledge, treatment, and the patients. Laennec's autobiographical narrative of discovery reflects the importance of these factors to the stethoscope's most basic development. It shows how mediate auscultation and the stethoscope are embodiments of and reactions to this new industrial-professional medical disposition. Laennec wrote that, in 1816, he discovered that a tube of rolled paper applied to the chest of a patient could amplify the sound of her heart. Having failed to get a clear sense of her ailment through application of the hands and percussion of the chest, "on account of the great degree of fatness," he moved on to a more innovative approach:

> I rolled a quire of paper into a kind of cylinder and applied one end of it to the region of the heart and the other to my ear, and was not a little surprised and pleased, to find that I could thereby perceive the action of the heart in a manner much more clear and distinct than I had ever been able to do by the immediate application of the ear. From this moment I imagined that the circumstance might furnish means for enabling us to ascertain the character, not only of the action of the heart, but of every species of sound produced by the motion of all the thoracic viscera.... With this conviction, I forthwith commenced at the Hospital Necker a series of observations, which have been continued to the present time.[27]

The "from this moment I imagined" is key. In his own telling, Laennec gives his scientific training and occupational ideology the weight of immediacy: at the moment of discovery, he abstracted a crude acoustic principle behind the act; from this basic insight, he posited a series of investigations. Every movement of the organs in the human thorax could be tracked by *listening* to the body with the aid of an instrument, and those movements could be rendered meaningful. *This* was Laennec's innovation, not the physical composition of a simple device to accomplish the task. Laennec claimed that he had thus invented the technique of mediate auscultation: listening to the body through a medium at a physical distance.

Although Laennec's own assessment provides us with useful clues to the importance of mediate auscultation for the history of listening, his story is probably a mixture of fiction and fact. "Post hoc" stories of scientific discovery "are never innocent; they are written with full knowledge of the ultimate significance of the event."[28] Laennec was aware of Hippocratic passages on auscultation and had known the basic physics behind mediate auscultation since his youth. Another French physician, François Double, had, in 1817, published a book that recommended direct auscultation of the patient—two years before the first edition of Laennec's *Treatise on Mediate Auscultation*. The British doctor William Hyde Wollaston used a long "stick" to transmit sounds from his foot to his ear, although Laennec claims not to have known about Wollaston's work (which was perhaps derived from Thomas Young's physics experiments published a few years prior).[29] More important, no record exists for the case that Laennec describes as his moment of discovery. Laennec's details are vague, and the medical historian Jacalyn Duffin could find no corresponding record in Laennec's papers. By Duffin's estimate, the case records point to early 1817 as the beginning of research into mediate auscultation and the use of stethoscopes; the earliest published reference to mediate auscultation refers to a case from March

1817, although Laennec may have attempted mediate auscultation as early as the fall of 1816 (which is the time frame for discovery claimed by Laennec and some of his students many years after the fact).[30]

That Laennec's narrative of invention is a bit distorted should surprise nobody. But the context of his invention and the content of his own narrative yield many clues as to what mediate auscultation was designed to do and what it accomplished. Laennec's approach to hearing was technical. Mediate auscultation was a highly structured activity that required practice to perfect. The new practice of listening through a stethoscope was also grounded in the emerging medical epistemology of pathological anatomy. As Duffin has shown, Laennec's work also developed out of and contributed to physiology.[31] Mediate auscultation would help reshape those fields. Its effects lay in three distinct areas:

1. It enacted and embodied a new spatial and social relation between doctor and patient.
2. It reconstructed sound as a field of possible data for medical perception and knowledge.
3. It would, Laennec hoped, elevate the practice to pure science by building a scientific metalanguage of sounds based on a synthesis of the doctor-patient relation and the newly demarcated domain of audible medical signs.

At the same time as they sought to develop a lexicon of acoustic-pathological signs, Laennec and his followers emphasized the importance of practice in listening to patients' bodies. I will develop an analysis of these issues through an extended reading of the third edition of Laennec's *A Treatise on the Diseases of the Chest and on Mediate Auscultation* and some of the other key texts of mediate auscultation in the nineteenth century. Laennec's *Treatise* is a crucial document because it explains in great detail why doctors should use stethoscopes to listen to their patients, how they should use their stethoscopes, and how to understand what they hear through the tube. The *Treatise* consists of a preliminary essay on methods of physical diagnosis and a series of longer essays on the actual diagnosis of different diseases of the thorax, the bronchia, the lungs, the membrane covering the lungs and chest cavity, the circulation, and the heart. Later writings on mediate auscultation would be considerably shorter, essentially taking for granted many of Laennec's arguments.

Mediate Auscultation:
The Social and Philosophical Basis of a Technique

The technology of the stethoscope was simple enough: Laennec's original instrument and those descended from it were monaural ("single-eared") instruments (figures 11–12). Generally cylindrical in shape, they had an earpiece at one end and a hole at the other that would be placed on the patient's body. The hole could be plugged with a stopper for specific applications. Later innovations to the monaural stethoscope included making the middle of the instrument flexible (through the use of rubber tubing) and modifying the tube into two halves so that it could be unscrewed for easy transportation.[32] Laennec had originally intended to call it simply *le cylindre*, but others quickly tried to name the device. It was referred to as a *sonomètre, pectoriloque, pectoriloquie, thoraciloque, coronet médicale,* and *thoraciscope.* Laennec thought all these names improper and instead proposed *stethoscope,* a conjunction of the Greek words for "chest" and "examine" or

---

Figure 11. Diagram of Laennec's stethoscope (courtesy Historical Collections, Health Science Library System, University of Pittsburgh)

"explore."[33] That the word has a visual connotation in *scope* should not be our primary concern: French thought at the turn of the nineteenth century was suffused with metaphors connecting sight, light, and knowledge.[34] Laennec's teacher J. N. Corvisart spoke of physicians having the proper tact and glance (*coup d'oeil,* or "gaze"), so it is no surprise that a device so central to medical knowledge invented in France at this time would have a name with visual overtones. As I argue above, the philosophical privilege of sight is not the same thing as its privilege in practice. This is yet another case of a disjuncture between the aurality of a practice and the ocularcentric language used to describe it.

Physically, the stethoscope was a logical extension of the ear trumpet, which had been in use for centuries. At first, the stethoscope might appear as a kind of reversal of the ear trumpet: instead of the hearing-impaired person listening out into a functional world, the expert physician listens into the diseased body. But the *use* of the stethoscope had some important differences from the use of hearing aids. The stethoscope was not so much

Figure 12. Monaural stethoscopes (courtesy Historical Collections, Health Science Library System, University of Pittsburgh)

the inversion of the hearing aid as the generalization of its principle. Even as it posited the possibility that doctors could become virtuoso listeners, mediate auscultation endowed its practitioners with a functional disability. The unaided ear was not enough: for centuries, the hard-of-hearing had used ear trumpets as hearing aids. Now doctors—whose hearing was ostensibly healthy—could augment their auditory abilities. One early model of the monaural stethoscope, called a *conversion tube,* made the equivalence clear: the stethoscope could also be used as a hearing aid. Like the sound technologies that would appear later in the century, the stethoscope was built on a pedagogy of mediate auscultation that rendered the human ear an insufficient conductor of sound. In point of fact, the ear *was* insufficient for the purposes of internal medicine since the stethoscope was designed to render sounds otherwise imperceptible to the human ear more clearly audible. As it would come to be with the ear phonautograph, the telephone, and the phonograph, so it was with the stethoscope: mediate auscultation fetishized the cultural status and trappings of hearing loss. This is to say that, as far as Laennec was concerned, all of his doctors needed a hearing aid—no matter what condition their hearing was in. Even the doctor's trained ear could never hear enough without the stethoscope.

While empiricism is usually cited as the operative epistemology of early-modern medicine, an epistemology of mediation is equally central to the apprehension of sensory data: the interior of the chest would not yield up its truth to the unaided senses. Doctors had to have the right tools and training to discover interior bodily states through listening. Consider Laennec's objections to listening to the patient's body *without* a stethoscope. In a section in the *Treatise* on immediate auscultation—where physicians would apply their ears directly to the bodies of the sick[35]—Laennec lists six major faults with this technique:

1. Any increased quality of hearing experienced through immediate auscultation as opposed to mediate is a result of the physician's entire face conducting sound when in contact with the patient's body, thus leading to "serious mistakes in cases where pulmonary obstruction is partial and of small extent."
2. It is not physically possible to apply the ear to a number of important regions for diagnosis (such as the angle formed by the clavicle and the head of the humerus in lean persons, the lower region of the sternum when it is depressed, etc.). Moreover, it is not *socially* possible to apply the physician's ear to the body of respectable women:

"In the case of females, exclusively of reasons of decorum, it is impracticable over the whole space occupied by the mammae."

3. The application of the naked ear requires applying more pressure to the patient's chest, thereby further fatiguing the patient.
4. This added pressure can lead to extraneous sounds generated by the patients tightening their muscles or the physician's head rubbing against the patient's clothes, which can be mistaken as respiration.
5. "The uneasy posture which one is frequently forced into, determines the blood to the head and renders the hearing dull. This circumstance, and the repugnance which every one must feel to apply the ear to a patient that is dirty or whose chest is bathed in perspiration, must always prevent the habitual or frequent use of this method." Since auscultation is most advantageous for the early detection of disease before it presents any visual signs, the physician's presumed reluctance to use it in every single case may prevent effective diagnosis.
6. Finally, the stethoscope adds to the naked ear and the sounds of the patient's body its own acoustic properties, which aid in the detection of certain physical properties in the patient.[36]

Here, in nascent form, we find the basic tenets of audile technique laid out decades before it would be articulated to technologies of sound reproduction. To offer an anachronistic paraphrase: "Physicians! Auscultators! You *need* a headset! To tune in the sound of the body, to hear interior states that you otherwise wouldn't, to eliminate the noise of the room and your own body, and to always get the truest and clearest reception of gasses and liquids inside the chest." A century before Brandes headphones, it is all there in the *Treatise:* the separation of hearing from the other senses, the production of noises otherwise inaudible to the naked ear, the demarcation of interior and exterior sounds, and the drive for true fidelity. But, because Laennec explains *why* physicians need headsets, his work helps us understand the philosophical and social roots of audile technique. There are two main issues at work in Laennec's objections to immediate auscultation: he is very much concerned with an epistemology of mediation (and mediation through the use of technology), but there is also an important relation between physical distance between doctor and patient and social distance. I will consider each issue in turn.

The *Treatise* is in part a guide for physicians to get the perceptible world into their senses. It aimed to teach doctors how to listen but also how to think about the act of listening to patients. Mediate auscultation is the physical, spatial configuration of a particular way of knowing. That is to say, it really encompasses two kind of knowledge. Mediate auscultation requires what *knowledge* commonly means in English: "textbook" learning, "cognitive" knowledge. To be an effective auscultator, a doctor must know the different internal organs, their possible pathologies, and the relation between those problems and sounds. Inasmuch as the new medical techniques of listening relied on and developed the larger field of pathological anatomy and physiology, we could say that they are part of a larger "epistemic" shift.[37] But mediate auscultation also requires that experience and disposition be socially organized, that they be conditioned, yielding a feel for the activity, a habitus. An epistemology of mediation works at this level of practical knowledge or habitus. Inasmuch as it offers formal medical knowledge, it does so on the basis of this practical knowledge—the doctor must have mastered a technique and developed a feel for and disposition toward listening. As Malcolm Nicolson has argued, "The successful adoption of percussion and stethoscopy was dependent not only upon the diffusion of academic knowledge of the techniques but also upon an experiential, learning process being undergone by prospective exponents."[38] This practical orientation toward medical listening was built on mediation.

Laennec's objections to immediate auscultation also draw a clear line between the stethoscope—an instrument for listening—and the human body (also conceived as an instrument for listening). In so doing, the *Treatise* can be read as offering insight into the changing meaning of listening itself. For Laennec, the stethoscope adds its own acoustic properties, which aid doctors in listening to patients. It can be placed against the patient's body without causing changes in the flows of fluids or gases in the patient. Applying the naked ear to the patient, meanwhile, results in physical discomfort for both doctor and patient. It results in bone conduction and other extraneous effects—blood flow to the doctor's head, the patient's muscles tightening, etc.—that may lead to misdiagnosis. Laennec even makes a plea for physics, arguing that the stethoscope can go where the ear cannot. So, for Laennec, the very qualities of the unaided ear render it unsuitable for the task of listening that he has set for it. In order to hear the interior of the human body properly, the doctor's ear requires technological supplementation; it requires a mediator. One might be tempted to read this as a form of dehumanization, a delegitimation of the human body in the face of inhuman technology. But the technology of the stethoscope is precisely a human enterprise. Laennec is not dehumanizing listening so much as he is seeking to add to, qualify, and modify the meaning of listening itself.

Mediate auscultation carries with it a variety of contextual demands: a framing of the listening event, a structuring of the doctor-patient relation according to clear physical and social roles, and a particular preference for instruments. Consider this range of provisions found in textbook instructions for the use of stethoscopes:[39]

1. The role of the stethoscope is to be primarily *instrumental;* it is to be viewed as a means to an end, an enhancement of medical perception rather than its substance.
2. The stethoscope must be of good quality: it must fit the listener's ears comfortably; the end should not exceed a certain diameter (usually of about 1 inch); and its edges should be rounded so as not to dig into the patient's skin. Austin Flint, an American popularizer, adds that the tubes of the stethoscope should not be obstructed, nor should they be at all stiff or produce any sound of their own.
3. The patient's body should not be covered in heavy or loose coverings. Laennec is more willing to use the stethoscope to listen through clothing than are his followers later in the century.
4. The stethoscope must be applied carefully to the surface of the patient's skin so as to leave no gap between the skin and the end of the instrument; however, excessive pressure should be avoided so as not to cause the patient discomfort.
5. The position of the patient's body varies according to the area being examined. Physicians should avoid a position that would require too much bending or stooping on their part. Patients may be positioned sitting upright, leaning forward, or in other specified positions depending on the nature of the examination.
6. The examination should cease if the patient is in any way excessively excited or nervous, as this will have an effect on respiration.
7. Any examination of the patient will be effective only if the physician's ear is already trained and thoroughly practiced.

These fundamentals serve as a kind of frame for the examination: establishing a number of constants for the detection of sonic variables. The physical positions of doctor and patient are prescribed, as are the relations among doctor, patient, and instrument. The relation between doctor and instrument is particularly important here. The textbook authors describe listening through the stethoscope as a kind of concentration, an isolation of the sense of hearing from the other senses. This isolation was an advantage of mediate auscultation over immediate auscultation; it was also an essential component of stethoscope pedagogy. The proper execution of mediate auscultation depended on the proper separation of hearing from other sensations and also the proper orientation toward the sounds heard through the stethoscope.

The development of the stethoscope and mediate auscultation coincided with the development of new theories of sense perception based on a "separation of the senses." As discussed in chapter 1, each sense "gets its due," so to speak, in Enlightenment science. From this point on, it becomes possible—at least in the pages of a medical textbook—to think of each of the senses as ideally and totally isolated from one another at a fundamental level. If the senses were, before the late eighteenth century, conceived as a kind of complex whole, they now became an accumulation of parts—a tool kit. So, at roughly the same time that the physiologists Charles Bell and Johannes Müller were positing the physiological separation of the senses, Laennec posited a practical separation of the senses. According to Laennec's objections to immediate auscultation, not only does the use of the stethoscope make up for some of the insufficiencies of the human ear, but it also isolates the faculty of hearing from the other senses and renders aberrant any conduction of audible vibration by body parts *other* than the ear. Hearing is to be separated from touch (and especially bone conduction)—anything else is an aberration. This is the substance of his first objection to immediate auscultation: bones outside the ear can conduct vibration and, therefore, sound. The physician's ear, like the physician's eye, thus becomes a separated sense, divided from the others, whose specificity was to be preserved and intensified through the proper use of instrumentation. The stethoscope rendered audible phenomena otherwise unavailable to doctors' senses, thereby increasing their powers of investigation. "The prohibition of physical contact makes it possible to fix the virtual image of [i.e., *to listen* to] what is occurring well below the visible area," writes Foucault.[40] *Mediate* auscultation displaces its "immediate" counterpart as the default status for medical listening in Laennec's account. A simple instrument marks and helps solidify a whole medical epistemology of mediation.

As hearing became an object of medical knowledge through physiology and otology, listening becomes a route to medical knowledge.[41] This new understanding of perception also has significance for the practice of perception: since each of the senses are, ideally, autonomous, one of the purposes of technique is to develop them toward that ideal state. It makes perfect sense for Laennec to suggest that the ear must not be supplemented, in the first instance, with touch through the hands or the combined senses of touch and hearing (via bone conduction) enabled by placing the face directly against a patient's body. In order to get the truest possible sense data (reception), for the doctor to *really* listen, hearing must be separated from the other senses. Once so separated, hearing can be supplemented by techniques and technologies especially designed for it. As the ear comes to be conceived as a technology, an apparatus to register a piece of the vibrating world inside doctors' heads, the ear becomes particularly amenable to other technologies. Laennec's stethoscope helped accomplish this, and later stethoscopes would improve on this principle of isolation and intensification.

In addition to separating hearing from the other senses, mediate auscultation demanded a particular kind of framing of sound. It put a frame around some of the sounds audible through the stethoscope, rendering some sounds as interior sound and others as exterior noise. Only sounds inside the frame were to be analyzed or considered for diagnosis. The sounds of the apparatus itself, and the other sounds accompanying auscultation, were to be ignored:

The ability to abstract the mind from thoughts and other sounds than those to which the attention is to be directed, is essential to success in auscultation. All persons do not possess equally this ability, and herein is an explanation in part of the fact that all are not alike successful. To develop and cultivate by practice the power of concentration, is an object which the student should keep in view.

Generally, at first, complete stillness in the room is indispensable for the study of auscultatory sound; with practice, however, in concentrating the attention, this becomes less and less essential.[42]

This combination of abstraction and framing was soon embodied in a modification of the stethoscope: the binaural stethoscope. Although ideas for binaural instruments go back to at least 1829, the first widely used binaural stethoscope was designed by Arthur Leared in 1851. It consisted of a small chest piece that connected with two gutta percha pipes for the physician's ears.[43] The binaural model quickly found favor for several reasons: by providing sound to both ears, it further helped isolate physicians from other sounds and concentrate the sound in their auditory fields. It also held itself on the physician's head, thereby freeing both hands for use on the patient. Many users also claimed that it provided a better quality of sound,

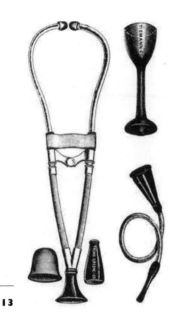

in terms of both volume and clarity.[44] George Cammann published plans for a binaural stethoscope in 1855, and his model became the accepted standard of stethoscope design for the rest of the nineteenth century: thirty years later the *Journal of the American Medical Association* would declare that "Cammann's binaural stethoscope just as he left it, is really the best instrument . . . for auscultatory purposes that we have" (figures 13–14).[45]

While the binaural stethoscope helped crystallize in physical form the isolation and intensification of hearing central to the method of mediate auscultation, it still required proficiency in listening technique: the instrument itself produced sound outside the proper listening frame. Of Cammann's binaural stethoscope, Flint writes, "The advantages, however, . . . are not appreciated until after some practice. At first, a humming sound is heard which divides the attention and thus obscures the intrathoracic sounds. After a little practice this humming sound is not heeded, and it ceases to be any obstacle."[46] As a part of the entire procedure, the character of the instrument itself must be erased from consciousness during mediate auscultation. In classic technological deterministic fashion, the tool stands in for a whole process from which it erases itself. Mediate

---

Figure 13. Cammann's stethoscope (courtesy Historical Collections, Health Science Library System, University of Pittsburgh)

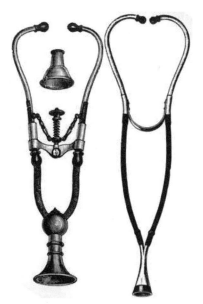

auscultation was thus a "license to forget,"[47] a kind of reification. The forgetting associated with technology was the forgetting that all learners do as they achieve mastery—technique moves from a conscious effort to a kind of second nature, a disposition, a feel for the game. Mediate auscultation may have been the social basis of a new medical hermeneutic, but it also marked a new kind of medical habitus. Not only did doctors develop a "feel for" listening to patients' bodies through the stethoscope, but they also developed a whole set of dispositions: toward the spatial aspects of the doctor-patient relation; toward use of the stethoscope as a buffer between doctor and patient; and toward the status of medicine as a profession practiced in hospitals.

Mediate auscultation thus encapsulated a whole set of meanings associated with hearing, listening, listening with technology, and the sounds heard via listening with technology. Social position and disposition go together. Mediate auscultation is the artifact of a particular habitus, a particular subjective view of the social world.[48] This is obvious in Laennec's most basic argument for mediate auscultation over immediate auscultation: "Independently of its deficiencies, there are other objections to its use:

Figure 14. Binaural stethoscopes (courtesy Historical Collections, Health Science Library System, University of Pittsburgh)

it is always inconvenient both to the physician and patient; in the case of females it is not only indelicate but often impracticable; and in that class of persons found in hospitals it is disgusting."[49] Long before the germ theory of disease would have suggested a need for distance between doctor and patient, Laennec was worried about the propriety of male doctors touching women's breasts ("mammae"), the disgust that he and others might feel at various illnesses and conditions of patients, and the class difference between doctors and their hospital patients. As a technique of and disposition toward listening, mediate auscultation located its practical philosophy of sensory mediation with a practice of social mediation. Laennec's objections to the use of the naked ear in medicine are instructive because they show the dual basis of an audile/medical procedure: in an emergent organization of sensation itself and in the organization of social differences.

Mediate auscultation is a response to the analogy of physical and social distance. Laennec does not mention in his justifications of mediate auscultation the danger of contracting disease. He presents the need for distance between doctor and patient as emphatically social. While, from today's ultrasanitized standpoint, Laennec's disgust at his smelly, sweaty, and sickly patients might seem perfectly reasonable, we should not be so quick to accept it simply as a transhistorical fact. To begin with, Laennec was smelly, sweaty, and sickly a good deal of the time himself. He is known as "one of the great invalids of tuberculosis history," sometimes so breathless that he "could not tolerate the weight of his clothing and would have to strip to the waist in order to keep working at his desk."[50] Even the Paris of Laennec's time was fairly odiferous. But before we veer too far into psychobiographical matters, we should consider the threshold of disgust as itself a historical problem. As Norbert Elias has argued, an ever-receding threshold of disgust, shame, embarrassment, aggression, and perception is a subjective condition of European modernity: "Social life ceased to be a danger zone in which feasting, dancing, and noisy pleasure frequently and suddenly give way to rage, blows, and murder, and becomes a different kind of danger zone if individuals cannot sufficiently restrain themselves, if they touch sensitive spots, their own shame-frontier or the embarrassment-threshold of others. In a sense, the danger-zone now passes through the self of every individual. Thus people become, in this respect too, sensitive to distinctions which previously scarcely entered consciousness."[51] In other words, as social life becomes "safer" and more rationalized, as people put some distance between their everyday life and their perceptible mortality, the messiness of the natural world becomes more difficult to bear. Hence, people are more easily disgusted or frightened; hence, people perceive more and more subtle action as aggressive or improper, and, hence, they become more sensitive to ever-smaller details of the data provided by their senses. As an instrument, the stethoscope is an artifact of these tendencies. As a practice, mediate auscultation codifies a whole set of bourgeois sensibilities into medical knowledge. Laennec's objections to immediate auscultation clearly touch on shame frontiers and embarrassment thresholds: the stethoscope can even hear through a patient's clothes and relieves male doctors of the awkwardness of touching women's breasts. Bourgeois decorum can be upheld even among the sick and their healers. So Laennec's disgust at his patients is an eminently cultured or cultivated disgust. But, as the stethoscope codifies disgust, it is also about ever-increasing sensitivities: subtle changes in liquid or gaseous motions; the tiny, barely audible details of a patient's breathing or heartbeat. Where the audible distinctions among patients' bodily sounds barely entered the physicians' consciousness, they would soon preoccupy it, providing the necessary clues to the true state of the patients' bodies.

Of course, distinctions between sounds were not the only salient distinctions in medicine at the turn of the nineteenth century. Mediate auscultation was very much about mediating class distinctions. In the clinical setting at this time, the physician was almost invariably of a higher class status than the patient, and Laennec was no exception. Of 770 patients described in Laennec's records, there are occupations listed for 662. None of them are doctors, lawyers, politicians, aristocrats, or musicians. Only 10 of 234 women listed were housewives. Laennec's female patients were day laborers, maids, lingerie makers, porters, masons, dressmakers, and so forth—in other words, they had to work for a living. A quarter of his male patients worked in the construction industry, with most of the rest divided among textiles, equipment and maintenance jobs, and unskilled labor.[52] This information tells us that, although Laennec was not particularly wealthy, he would most often have been of a higher class position than were his patients, men or women. Only with the professionalization of medicine and increasing standards of cleanliness, along with urbanization and the gradual specialization of medical knowledge, would the middle and upper classes venture into hospitals more frequently.[53]

Doctors like Laennec thus occupied a strange position. An aspiring middle-class profession found itself conducting work in one of the culture's most reviled spaces. Laennec, like his contemporaries, thought that work in a hospital was essential to learning the methods of mediate auscultation.[54]

No doubt this was linked with the aspirations toward middle-class respectability of the medical profession in general. These aspirations also called for the application of middle-class decorum to the bodies of poor women as well, which blended male physicians' self-understanding with their attitudes toward poor women. The physical distance between doctor and patient at the moment of examination was, thus, a reassertion of social distance even when the two inhabited a common space. Even the very language used to describe these procedures renders social difference as spatial. Mediation can be read in terms of distance as well as instrumentation: *immediate* auscultation (as opposed to just *auscultation*) was listening to the body without distance between physician and patient, whereas *mediate* auscultation, listening to the body from a distance—with a physical medium or barrier between doctor and patient—becomes the default term in medical description.

Laennec's stethoscope research allowed for some interesting variations on this formula. In so doing, it may have helped promote the cause of medicine among elites. During the first decades of the nineteenth century, when Laennec was conducting his clinical research, experimental techniques and student procedures were most often tried out on the poor before they were applied to the rich. In contrast, Laennec happily used his experimental instrument and techniques on patients: "Modestly performed through the clothing, auscultation was painless, harmless, and singularly noninvasive."[55] In other words, mediate auscultation allowed for the preservation of decorum among elites as well as across social classes. In one particularly famous case from 1817, Laennec examined Mme Germaine de Staël, daughter of the Neckars who founded the hospital where he worked. Although de Staël's doctor was resistant to Laennec's methods, and although they did not provide any useful insight into her illness and treatment, her doctor's account of Laennec's technique provided him with useful publicity.[56]

Laennec's disposition was connected with his position. Not only was he an aspiring bourgeois physician, but he was also politically reactionary, even penning a royalist pamphlet himself.[57] It is perhaps accidental that the stethoscope and mediate auscultation were developed by a royalist in Restoration France, but it is at least a historically interesting coincidence. Foucault calls the stethoscope "the measure of a prohibition transformed into disgust, and a material obstacle," and for good reason.[58] What better tool to maintain the potentially fluid class distinctions between doctor and patient—in whichever direction they may run? Disgust rises in social importance when physicians and patients are crowded into common spaces and social distance must be maintained in spite of physical proximity. Disgust becomes more of an issue when the poor, already repugnant to the physician's sensibilities, are subjected to crowded and often squalid hospital conditions. Disgust is a key component of the bourgeois sensibility to which the newly professionalized doctor aspired. Techniques of listening were social forms of bodily treatment.

So mediate auscultation mediated in several ways: the stethoscope mediated and shaped the auditory relation between a listening doctor and the sounds of a patient's body. The stethoscope mediated between doctor and patient as social beings by keeping intact the physical distance implied by social distinctions. At the same time, it embodied the ever-receding threshold of disgust and kept the listening doctor at some physical remove from sickness itself. In constituting a new kind of social-spatial relation between doctor and patient, mediate auscultation opened up a new kind of acoustic space for physicians. It required not only a new orientation toward listening but also a new orientation toward what was heard: a new way of organizing sounds in medicine.

The Audible Becomes the Knowable: Changing Regimes of Medical Knowledge

One of the reasons for the apparent breakthroughs associated with the stethoscope was that, in Laennec's own opinion, it made use of hearing—a "novel sense" in diagnosis.[59] The novelty of hearing was not in its presence, however, but in its application: mediate auscultation is not so much a shift *to* listening in medical practice as a shift *in* listening. This new form of listening constituted a whole field of sounds as potentially meaningful and relevant for diagnosis. In other words, audile diagnosis shifted from a basis in intersubjective speech between doctor and patient to the objectification of patients' sounds—in mediate ascultation, patients' voices existed in relation to other sounds made by their bodies, rather than in a privileged relation to them.[60] Speaking patients with mute bodies gave way to speaking patients with sounding bodies. This marks a significant shift in medical epistemology and heralds the rising importance of physiology in medical knowledge.

Diagnosis in the seventeenth century and for most of the eighteenth century was based on a combination of the patient's own narrative testimony and the physician's own visual examination of the patient. Doctors relied heavily on the patients' subjective accounts of illness, their personalities and manners of expression, and less on their own perceptions. Physical examination was usually limited to the taking of the pulse and occasionally to viewing the body.[61] Beyond the patient's speaking voice, the sounds of the body were totally disregarded: one physician was so shocked that he could hear a patient's heartbeat from the bedside that he called it "almost incredible." Another, when asked by his own patient the meaning of a "blubbering sound" in the chest, could offer only a meager guess.[62] Without a hermeneutic for hearing patients' bodies, the nonverbal sounds that they made were meaningless. Even the voice itself was not considered to be a physical phenomenon. For doctors, what the patient said was essential for diagnosis, but the sound of the patient's speech was almost wholly irrelevant.

The growth of physical examination in the eighteenth century coincided with a change in orientation toward the body as a whole. Accompanying the changes in diagnostic methods was a change in theories of disease, from an understanding of disease as simply an imbalance of the bodily humors to the idea that different symptoms might relate to entirely different diseases. As the eighteenth century wore on, efforts to classify diseases on the basis of observations of patients (and less and less on the basis of personal narrative) gave way to autopsies to better track the marks left by disease on internal organs: "The practice of dissecting bodies to find physical evidence of disease began to transform some eighteenth-century physicians from word-oriented, theory-bound scholastics to touch-oriented, observation-bound scientists."[63] This kind of physical observation brought with it the requirement for new methods. Thus, visual, tactile, and audile techniques of examination increased in importance as medical empiricism gained currency: physicians had to revitalize old techniques of examination and develop new ones.[64]

As with Laennec's disgust at his patients, this new sensory sensitivity to small and detailed phenomena represents one of the hallmarks of European-bourgeois "modern" subjectivity. Doctors were becoming "sensitive to distinctions which previously scarcely entered consciousness."[65] The sound of a patient's heartbeat or the blubbering sound in a patient's chest moved from being the curious epiphenomena of an illness to indices of its exact nature, state, and case. As doctors oriented their senses more toward details and minutiae, these classes of phenomena became increasingly significant as signs. Sounds always present to the senses became meaningful in new ways, and doctors sought modes of listening that would give them full access to that new, meaningful audible world.

This is not to say that a physical examination that made use of the doctor's senses was invented in the eighteenth century. One can find examples of auscultation and other acoustic methods of examination throughout medical history. Hippocrates is usually cited as the first written example of immediate auscultation, with a long list of followers.[66] But the privileging of sensory data and the explicitly empiricist orientation to medical knowledge was a new development. This shift is apparent when we consider a kind of "in-between" document: Leopold Auenbrugger's monograph entitled *Inventum Novum*, published in 1761. Auenbrugger's book was perhaps the most significant example of audile-tactile examination prior to Laennec's. Auenbrugger's treatise is notable because it shifts responsibility for the apprehension of disease from the speech and appearance of the patient to the doctor's senses. Auenbrugger distrusted the accounts of his patients and wanted to make diagnoses by "the testimony of my own senses," thereby displacing the centrality of patients' own narrative accounts of their illnesses.[67] The technique that he advocated was called *percussion*, the striking of the body to get a sense of its interior composition. His technique was likely derived from watching his father tap on casks of wine or beer to check their level; his audile orientation probably came from his interest in music. But Auenbrugger's treatise was short, vague, and incomplete. He neither systematized his observations nor provided a clear explication of his practice. In fact, some physicians confused his procedure with an older practice called *succussion*, which involved listening for fluid in the chest by physically shaking the patient.[68] In other words, Auenbrugger had the empiricist bias, but he did not carry it through toward a systematic approach characteristic of the emergent scientific worldview.

Percussion did not find much favor in the medical profession until Laennec's time. A variety of conditions mitigated against Auenbrugger's ideas immediately catching on. In the 1760s, there was still a widespread prejudice among doctors against engaging in physical activities in diagnosis. While surgeons used their hands and instruments, physicians still shied away from the trappings of physical labor on behalf of their patients. Even at the Spanish Hospital in Vienna where he worked, most of the responses to Auenbrugger's research were initially negative.[69] Auenbrugger did little to promote his work, and his treatise circulated little outside German-speaking areas until the turn of the nineteenth century: Rozière de la

Chassagne was his first translator into French in 1770 (although Auenbrugger was not widely read in France until J. N. Corvisart discovered and retranslated *Inventum Novum* three decades later), and John Forbes brought Auenbrugger to an English-speaking audience at the same time as he translated Laennec's *Treatise*.

There were clearly institutional and practical reasons why Auenbrugger's work did not immediately catch on. But *Inventum Novum* is interesting for our purposes because it fit a paradigm that had not yet come into favor within the medical establishment. Percussion required the interpretation of auditory symptoms and was based on a theory of disease as a localized phenomenon. While this construct of disease would prevail in pathological anatomy and physiology, it was still not a popular view among doctors in the mid-eighteenth century. At that time, diseases of the chest were known by names for symptoms: fever, difficulty breathing, cough, wheezing, spitting blood, pain, and palpitations. As long as doctors were primarily interested in symptoms, they would have no reason to probe the interior of a patient's body for deeper causes of an illness. Percussion was essentially meaningless without a system of medical knowledge based on a physiological model of disease (itself made possible through the gathering of anatomical and physiological data through dissection). Without the larger ideological edifices of empiricism, pathological anatomy, and physiology, physicians of the late eighteenth century found listening to the interior of the body to have no practical, informative purpose.

These same issues are relevant for our understanding of Laennec's work. For mediate auscultation to make sense, a paradigmatic shift in medical epistemology was necessary. It was only when the body came to be understood as an assembly of related organs and functions that percussion—and very shortly thereafter mediate auscultation—would take on such a primary role in medical diagnosis that Laennec's work would be hailed by Forbes as "one of the greatest benefits" ever bestowed on medicine. In the United States, this is for the very good reason that Forbes's was the first English translation of *Inventum Novum* and he coupled it with a digest of Laennec's *Treatise* for general consumption. Following Forbes's example, auscultation and percussion were usually grouped together in English-language medical textbooks.[70]

Mediate auscultation is also clearly an innovation of the technique of percussion. Laennec saw his work as completing and improving on the work carried out by Auenbrugger: "If percussion furnishes us only with indications which are circumscribed and often doubtful, it becomes most valuable when combined with mediate auscultation; and we shall find hereafter that the pathognomonic signs of several important diseases, and among others of pneumothorax, emphysema of the lungs, and the accumulation of unsoftened tubercles in the upper lobes, are derived from the contemporaneous use of these two methods."[71] Laennec's orientation toward the body as a drum (and wind instrument) was no doubt fertilized by his teacher Corvisart, who was the first French doctor to put Auenbrugger's ideas to work in French hospitals. By percussing the chest of a patient, Corvisart could predict postmortem findings in autopsies before patients died; this was unprecedented.[72] As medicine became more empirical, techniques like percussion and auscultation would more easily find favor. While still a student, Laennec made extensive use of autopsies and this new empiricst orientation to the senses. His own research was very much shaped by the empiricist paradigm.

As a mode of empirical verification, autopsy was unsurpassed in popularity in nineteenth-century medicine: it offered a means of checking diagnoses for ailments that were not cured (a quite frequent event).[73] Developed from the work of Giovanni Battista Morgagni, François-Xavier Bichat, and others, early-nineteenth-century anatomy took the autopsy as its primary site of knowledge, the moment when the body would give up its truth. As medical thought moved from recording patients' accounts of symptoms to the localization of disease, it required a means to transcend the subjectivity of the patient; doctors had to verify the condition of interior organs and bodily states. Autopsy served exactly that function: through the organization of hospitals and clinics that provided facilities both for handling the sick and for medical research, it became possible for a patient's death to become a "spontaneous experimental situation" since autopsy could commence almost immediately after death.[74]

Laennec initially demonstrated the effectiveness of his diagnoses through autopsies of his patients. Each of the forty-eight cases discussed in *Treatise on Diseases of the Chest* concludes with Laennec's subsequent findings in an autopsy performed within about a day of the patient's death. In fact, it was those findings that retroactively confirmed the diagnoses of mediate auscultation. It could thus be said that, in those first few years of diagnosis by stethoscope, patients' bodies were made to speak, but only retroactively. The appearance of lesions on the organs, the sight of the tissues, and fluids confirmed the auditory diagnosis sometimes made but a day earlier.[75]

As a novel technique of diagnosis through audition, mediate auscultation initially required visual proof for its legitimation. Once established, it would take on a life of its own. Mediate auscultation enabled the movement of the primary site of knowledge in pathological anatomy back from the dead to the living. Although Laennec's mode of analysis was grounded in pathological anatomy, it was a significant departure from its more orthodox applications because he mixed anatomic and physiological approaches. He was less concerned with the status of the organs themselves than their movements, relations, and functions. Mediate auscultation was a hydraulic, physiological hermeneutics, charting the motions of liquids and gases through the body and relating those movements to issues of function. In the age of auscultators, hearing surpassed sight in diagnostic precision. Only in the patient's death could vision again take hold as the primary sense used by doctors for diagnosis (although the invention of anesthetic allowed for a more visually contemplative orientation toward surgery). This primacy of audile diagnosis would continue into the twentieth century.[76]

Techniques of listening are thus central to modern medicine as we know it. In the founding moments of modern medicine, listening moved from an incidental modality of intersubjective communication to a privileged technique of empirical examination. It offered a way of constructing knowledge of patients independent of patients' knowledge of themselves or what they might say about themselves. The truth of a patient's body became audible to the listener at the other end of the stethoscope.

As with Laennec's understanding of auscultation as an acutely active form of listening, his approach to patients' voices was foreshadowed by his teacher Buisson. Buisson distinguished among three forms of voice: the voice "strictly speaking"; song; and the spoken word. In the form of a doctor-patient exchange, speech was still essential for Buisson. But his notion of the voice "strictly speaking" was a physiological function, the voice as sound. In other words, the voice was useful for diagnosis, not only because of what patients said, but because it could function as a timbral index of interior states of the body. Buisson's notion of the voice "strictly speaking" was therefore a functional understanding of speech. The voice was a physiological process, "invested with intrinsic and extrinsic elements reflecting the mind and the biology of its owner."[77] Following Auenbrugger and Buisson, Laennec found listening to the patient's body a means of diagnosis superior to listening to the patient's narrative account of the illness. The sounds of the patient's body were independent of the patient's free will: patients could not "conceal, exaggerate or lessen" the sounds that their bodies yielded on examination by mediate auscultation.[78] Like Auenbrugger,

Laennec trusted his own senses above anything the patient might say or do. Patients' speech might deceive, but their bodies would yield up their own truth on examination. As Stanley Reiser argues,

> The stethoscope focused the attention of physicians on a new class of disease signs—the sounds produced by defective structures in the body. . . . A model of disease, deduced from these sounds and from the assorted lesions in the body found at autopsy, largely replaced the model constructed from the patients' subjective impressions and the physicians' own visual observations of the patient. The physician's withdrawal from such person-centered signs of illness was increased by the fact that the auscultatory process required the physician to isolate himself in a world of sounds, inaudible to the patient. Moreover, the growing success with which disease could be diagnosed through auscultation encouraged the physician to favor techniques that would yield data independent of the opinions and appearance of the patient.[79]

The technique of mediate auscultation (and not the stethoscope per se, as might be inferred from Reiser's language) was predicated on a relativization of the human voice. In diagnosis, the voice became one sound among many contending for the physician's attention in the audible world. Frequencies were frequencies.

The pedagogy of mediate auscultation facilitated a general shift in diagnosis from a privileging of the mouth (via voice and speech) as the most important sonic location on the body to a diffusion of bodily sounds to be apprehended and sorted out in the ear. The history of audile diagnosis thus runs parallel to the shift from automata to tympanic machines. The voice became one contender among many for the trained auditor's attention. In any case, the voice was the only sound capable of speaking an untruth in Auenbrugger's and Laennec's hermeneutics.[80] While other sounds could conceivably mislead a doctor and lead to misdiagnosis, they did not actively deceive—interpretation was simply a form of physician's error. Speech, on the other hand, could be understood perfectly and still mislead a person.

This displacement of the voice and the patient's own subjectivity goes even further since, until the discovery of X-rays at the end of the nineteenth century, auscultation was the only available method for apprehending the interiority of patients' bodies without physically cutting them up. In fact, Reiser goes so far as to say that the physician could, "in a sense, autopsy the patient while still alive,"[81] a statement that makes sense only given the privileged status of autopsy in the acquisition of medical knowledge: while the dead patients lay forever muted, their bodies could yield up immutable truth through the empiricist's skillful use of the scalpel. The body of the patient was a "whole network of anatomo-pathological mappings . . . the dotted line of a future autopsy."[82]

Laennec's understanding of the voice as simply one possible sonic aspect of the body is well illustrated in a small subsection of his book dedicated to the auscultation of the voice in the different regions of the lungs and throat. Building on Buisson, he postulated a series of acoustic states of the voice, each connected with a set of interior physical conditions: broncophonism, aegophonism, and pectoriloquy are each moments when the voice is conducted in a particular way through an area of the chest. In this account, the whole thorax becomes a kind of resonating chamber; the lungs especially become like the interior of a musical instrument:

> the loose texture of the lungs, rendered still more rare by its intermixture with air, is a bad conductor of sound; and the softness of the bronchial branches, after they cease to be cartilaginous, renders them very unfit for its production; while the smallness of the calibre must render whatever sound is produced more acute and weaker in them than in the larger trunks. But if any one of these adverse conditions is removed, and yet more, if several of them are so at the same time, the sound of the voice may become perceptible in the smaller bronchial tubes.[83]

Here, the voice is interesting purely as sound: voice becomes vocalization or, in Buisson's terminology, "voice as such." The pure voice becomes a kind of sound effect—a container of timbre and an index of the states that shaped it. Laennec's living body is not that different from the automata of eighteenth- and nineteenth-century inventors; he honors a French tradition of the body as machine descending from Descartes.[84] The voice becomes a physiological mechanism and a wind instrument. It may reveal the truth of the patient's condition, but only if the physician listens well after extensive practice.

Laennec's discussion of the voice comes in the middle of a larger methodological section of the *Treatise;* it is preceded by a discussion of the mediate auscultation of respiration and followed by discussions of the cough, "rattle," and "metallic tinkling." Laennec found the different modulations of the cough and respiration to be similar to those of the voice. However, the rattling and tinkling were perceptible only through the stethoscope. These are, for Laennec, entirely new sounds. They could be described only by analogy, although their definition was simple enough. The rattle consists of "all the sounds, besides those of health, which the act of respiration gives rise to, from the passage of air through fluids in the bronchia or lungs, or by its transmission through any of the air passages partially contracted." The metallic tinkling sounds like the object of its analogy and is perceived through the stethoscope during speaking, breathing, or coughing.[85] In addition to these basic divisions, Laennec proposed a whole set of pathological subcategories: the moist crepitous rattle; the mucous, or gurgling, rattle; the dry sonorous rattle; the dry sibilous rattle; the dry crepitous rattle with large bubbles or crackling; utricular buzzing; amphoric resonance. In essence, he defined and classified new sounds that were hitherto inaudible: "Laennec had to invent language to express the sounds and their anatomical significance, and once he possessed the words, the ideas became objects to be sought."[86] At times, the *Treatise* is an acoustic lexicon of motion in the body. Laennec's object is a body that lives and moves from the inside out; physiology and pathological anatomy come together in the *Treatise*. Each sound becomes a chart of the space through which it moves: Is there fluid in the lungs? Is there an inflamed membrane? Has an area become hard or more porous? Even the simple directional movement of sound could provide important cues. In a widely cited case from the section on the voice, Laennec listens to the voice through the body of a patient displaying no outward symptoms of tuberculosis:

In the very earliest period of my research on mediate auscultation, I attempted to ascertain the differences which the sound of the voice within the chest might occasion. . . . In the case of a woman, affected with a slight bilious fever, and a recent cough having the character of a pulmonary catarrh, on applying the cylinder [stethoscope] below the middle of the right clavicle, while she was speaking, her voice seemed to come directly from the chest, and to reach the ear through the central canal of the instrument. This peculiar phenomenon was confined to a space about an inch square, and was not discoverable in any other part of her chest. Being ignorant of the cause of this singularity, I examined with the view to its elucidation, the greater number of the patients in the hospital, and I found it in about twenty. Almost all these were consumptive cases in an advanced stage of the disease. . . . Two or three, like the woman above-mentioned, had no symptom of this disease, and their robustness seemed to put all fears of it out of the question. Notwithstanding this I began immediately to suspect that this phenomenon might be occasioned by tuberculosis excavations in the lungs. . . . The subsequent death, in the hospital, of the greater number of the individuals who had exhibited this phenomenon, enabled me to ascertain the correctness of my supposition: in every case I found excavations in the lungs, of various sizes, the consequence of the dissolution of tubercles, and all communicating with the bronchia by openings of different diameter.[87]

Laennec called this phenomenon *pectoriloquy* and proceeded to chart its perceptibility depending on the location of the lesion in the lung. *Pectoriloquy* (translated by John Forbes as *pectoriloquism*) means "the chest speaks"—easily reminding us of *ventriloquy*, which means "the stomach speaks." But the difference is significant and illustrates the changing status of speech itself: whereas ventriloquism is a skill perfected by the speaker to fool the listener, pectoriloquy is a vocal sound effect determined by the speaker's physiological condition and is audible only to virtuoso listeners using stethoscopes. The speech is different as well: the ventriloquist wants speech to be heard as a message, yet none of Laennec's many passages on pectoriloquy give even the slightest hint of what his patients said when they spoke. For the purposes of mediate auscultation, the voice was both an instrument and a collection of sounds. Laennec's method treated the body as a collection of objects and flows all related to one another; a series of critical coincidences that could be mapped out and verified.

Pectoriloquy was a significant discovery because it allowed doctors to detect symptoms of disease in patients before patients themselves could detect them. In fact, it was the discovery of pectoriloquy that led Laennec to appreciate the full potential of mediate auscultation as a mode of diagnosis. Since Laennec had not yet coined the term *stethoscope* in 1817, some commentators labeled his cylinder the *pectoriloque*. Giorgio Baglivi's seventeenth-century lament—"Oh how difficult it is to disgnose diseases of the lung"—became a standard epigraph of textbooks on mediate auscultation.[88] Laennec himself cited it in the hope that his work would put it to rest. Whereas, in clinical medicine previous to the rise of pathological anatomy, the most significant signs of illness (which are any perceptible effects that the illness would have on the body) are patients' symptoms (fever, e.g.), mediate auscultation aided physicians in their search for signs of illness beyond the horizon of the patient's perception. Mediate auscultation bore "the imprint of a new medical era," as Laennec treated the illness primarily as an objective, physical phenomenon to be discerned through physical examination. Here, empiricism, pathological anatomy, and physiology came together in new techniques of listening and a system of signs yielded by these techniques.[89]

Medical listening built a bridge in the 1810s and 1820s between pathological anatomy, a science of form, and physiology, a science of animation.

At the level of clinical practice, techniques of listening did more than any other medical innovation to render the interior of the human body as a dynamic field of action. The shift to audile examination opened up the possibility for treatment of the live subject: in many cases, diagnostic knowledge preceded any notion of a cure: "A central point of contention was the value of precise diagnosis, given the still impoverished state of medical therapy. Auscultators claimed that by discovering the disease in its earliest stages, by converting suspicion into certainty, illness could be treated with a vigor and success not possible when doctors depended on the traditional, inexact means of diagnosis."[90] Audition was a key modality in perceiving states of patients' bodies.

Listening is, therefore, crucially important for diagnosis: doctors could hear what they could not see. When Laennec concluded his February 1818 address to the Académie des Sciences with the claim that pectoriloquy was the only certain sign of phthisis (tuberculosis) present before symptoms "could raise suspicions," few doctors before him had been able to make such a direct assertion about an internal organic condition in the absence of other symptoms.[91] Listening could yield certain medical diagnosis. Yet some authors have erroneously attributed medical empiricism to an emergent primacy of vision in the form of "the gaze." Certainly, techniques of observation were central to the development of modern medicine. But, in repeating the dogma that vision gave us modernity, we miss both the centrality of listening to a modern form of knowledge and a distinctively modern form of listening. Foucault performs some interpretive gymnastics in order to locate mediate auscultation as a subspecies of the gaze. In his discussion of pathological anatomy as an epistemic shift, "the ear and hand are merely temporary, substitute organs" for the eye, an assertion based on the prior assertion that "spatial data . . . belong by right of origin to the gaze."[92] To say that spatial data belong "by right of origin" to the gaze is to make an essentially theological argument about the origins and purposes of the senses. The audiovisual litany is a powerful ideological frame for the history of the senses, but it is not an accurate description of that history. For Foucault's explanation to work, vision must be a rational, technical, and spatial sense, and nothing more, while hearing must be a temporal and nontechnical sense, and nothing more. Yet Laennec's lengthy passage on pectoriloquy shows us that it was perfectly possible to create and develop new purposes for hearing in the early nineteenth century. As Laennec's pedagogical language makes abundantly clear, the senses themselves could be trained and shaped to the needs of reason. It is not that hearing becomes more like the gaze in mediate auscultation; it is that both senses become tools of reason and rationality in nineteenth-century medical examination.

In rendering the interiority of the body available to the physician's ear, mediate auscultation was geared toward the spatial decomposition of the body and its surfaces. In providing a means of assessing the motions of the chest—the thoracic cavity—it simultaneously rendered the body as active and provided an instrument for the identification of actions as *pathological*. Mediate auscultation was the technique whereby the dead body of pathological anatomy first came back to life. Decades before medical films, mediate auscultation offered a "distinctly modernist mode of representation in Western scientific and public culture—a mode geared to the temporal and spatial decomposition and reconfiguration of bodies as dynamic fields of action in need of regulation and control."[93] Mediate auscultation rendered the body in motion, beyond the patient's own perceptions, as a field of signs to be heard and interpreted.

Sounds as Signs

To recap the argument so far, mediate auscultation actualized a philosophy of mediation in a medical practice of listening. As they learned to use the stethoscope, doctors learned to restructure their auditory space. Mediate auscultation articulated both a physical and a social distance between doctor and patient, enacting a distinctly modern sensibility about bodily presence and distance. All these factors organizing space helped create and frame a sonic event where sounds were grouped into "interior" sounds, which had diagnostic meaning, and "exterior" sounds, which were to be ignored. Both interior and exterior sounds were considered as sounds "strictly speaking" in Buisson's sense: they were considered for their sonic characteristics, rather than for the meanings that they might have in other contexts. The most striking example of this new approach to medical listening was in the voice itself: whereas doctors had been interested in patients' voices for what they were saying, they now listened to the voice for its sonic content alone.

All these framing devices and audile-interpretive practices in turn set the stage for a new medical semiotics. If sounds were, indeed, signs of interior states, then it logically followed that the sounds and their meanings could be cataloged. Indeed, Laennec's *Treatise* reads like a sonic lexicon at times. This was, in part, an extension of a long pedagogical tradition going back to Hippocrates, whose teachings were organized as a series of "if-then" statements. Some of Laennec's own teachers had used this method as well. As an attempt to systematize the sounds heard through the stethoscope into a codified set of diagnostic signs, Laennec's *Treatise* was one of the first attempts to develop a metalanguage of sound, a set of descriptions for the shape and texture of sounds that was independent of subjective experience (i.e., independently verifiable). This is the major thrust of Laennec's *Treatise*. As we will see, it is also Laennec's most spectacular failure (as judged by physicians later in the century). Ultimately, the promise of an acoustic lexicon went unfulfilled: the attempt to create a metalanguage of sound quickly became nothing more than a set of metaphors and unverifiable (and therefore unscientific) observations. Laennec's lexical dreams served only to reiterate the primacy of audile-diagnostic technique: method over fixed knowledge. Pedagogy and practice became all-important.

The desire for an identifiable order in the resonant world of patients' bodies, an acoustic-thoracic hermeneutic of perceived sounds, drove the early development of mediate auscultation. The transformation of clinical experience into diagnostic knowledge was the purpose of most of Laennec's *Treatise*; 640 of a little over 700 pages are devoted to detailed discussion of different illnesses in the thoracic region. Throughout, Laennec integrates the use of mediate auscultation into a comprehensive diagnostic method. The result is a lexicon of the body: since it has been established that each sound corresponds with a physical condition, the physician's task is to learn to discern the sounds well enough to be able to diagnose the physical condition. Articulated to and through the stethoscope, mediate auscultation became the mark of a new age in clinical diagnosis.

Laennec explained the physical characteristics of each condition in some detail, accompanied by a discussion of symptoms that may be found in the patient and signs that the physician may discover on examination. Although some space is devoted to treatment, these are the shortest sections in the *Treatise*. Take, for instance, this excerpt from the discussion of emphysema of the lungs: "In the case of one lung being principally affected, the augmented sonorousness and increased size of this side, will discriminate the disease from all others, except pneumo-thorax, from which likewise, as will be shown when we come to treat of that disease, it can be readily distinguished." Surrounding that is a description of the different noises yielded up by emphysema of the lungs: in this case, Laennec treats the sound as a variation of the crepicious rattle with large bubbles.[94] Thus, the illness becomes identifiable through sound (as well as other observable phenomena). There had to be a correspondence between sound and sign.

In building his semiology, Laennec had two governing ideals: specificity and sensitivity. *Specificity* meant that, in order to be useful as a sign, a sound had to correspond to a single condition and be connected to that condition in every instance. Laennec sought to avoid "false positives" (like pectoriloquy without tuberculosis). *Sensitivity* meant that, in order to be useful as a sign, a sound had to occur in every instance of the lesion. Too little sensitivity would result in many false negatives. Laennec was less concerned with this criterion, although it remained relevant for him. To be effective as a diagnostic sign, the sound that Laennec heard had to be a positive indication of an interior lesion in almost every case. Otherwise, it would not be a useful measure for diagnosis. Still, Laennec was careful to point out that lesions were not themselves the causes of disease; they were simply effects. He wrote that clinicians should be prepared for false positives and false negatives: a patient with lesions could be healthy, and lesions could be absent in a very sick individual. In short, he tempered pathological anatomy with physiology.[95]

The *Treatise* and those works that followed it—and, more broadly, the method of mediate auscultation itself—can be read as attempts to set up a system of signs for the purposes of diagnosis. The sounds are signs; they must indicate something: "The sounds which constitute signs represent certain physical conditions pertaining to the chest. The normal or healthy signs represent physical conditions existing when the organs are not affected by disease; the abnormal or morbid signs represent physical conditions which are deviations from those of health, being incident to the various diseases of the chest. The physical conditions represented by signs may be distinguished as normal or healthy, and abnormal or morbid conditions."[96] If Laennec and his followers sought to assemble a semiotics of the body, they did so at a particular level. To use the language of Charles Sanders Peirce, Laennec sought to posit *indexical* connections between sonic signs and illnesses. *Indexical* signs accompany their object in experience; an index is a sign "which refers to its object not so much because of any similarity or analogy with it, nor because it is associated with the general characters which that object happens to possess, as because it is in dynamical (including spatial) connection both with the individual object, on the one hand, and with the senses or memory of the person for whom it serves as a sign, on the other."[97] Pectoriloquy designates a very specific condition of the lungs as well as the location of that condition. Cases where there was no direct correspondence between sound-sign and a referent were a source of some frustration for Laennec. In his exploration of the heart and arteries, Laennec came on the bellows sound, the name of which derived "from the circumstance of its exactly resembling the noise produced by this instrument when used to blow the fire." Yet he could not fix the bellows sound in relation to a lesion in the body. For a few pages, he explored the variety of circumstances in which he encountered it, but each time his hypotheses were led astray; he was forced to a vague conclusion: "for various reasons I consider this particular sound as owing to a real spasmodic contraction of the heart or arteries." The uncertainty here led to further experiments and speculation, but no conclusive position was reached; the section ends with a call for further research and a caution against mistaking certain phenomena for the bellows sound.[98]

The attempt to create a codified set of sounds was one of the reasons for Laennec's success where others before him, like Auenbrugger, had failed: in addition to inventing a new technique, he provided a complete guide for its use.[99] In fact, the rationalization, codification, and instrumentalization of the sounds produced through mediate auscultation are key elements of its modernity as a technique of listening. Sound operates in the service of science. Yet it is also the least credible aspect of Laennec's method. Laennec built his sonic semiology without consulting new developments in acoustics or physics, mathematics, or statistics that might have been useful for his attempt to construct a system of signs. A few decades later, Helmholtz's theory of upper partials would show that a single sound could be created by a range of different sources, so long as they produced the right overtones. After Helmholtz, sounds alone could not be direct indices of internal lesions.

Laennec's definitions of sounds aimed for strict referentiality, where a particular sign would refer to a single condition and that condition alone. In naming characteristics of sounds, he strove for a sonic metalanguage. The various rattles, pectoriloquy, bronchiophonism, and so forth were all attempts to develop terms to elevate indexical signs to symbols, where abstract qualities of sounds could be apprehended, described, and discussed. But, here, Laennec's ambitions collided with the acoustic properties of sound: any single sound could be caused by many different things. Even his scientific-sounding names for sounds were not purely abstract descriptions of the characteristics of the sound; they were simply names designating a set of common experiences. Doctors would have to wait for others to develop a purely analytic language of sound.[100]

By the late nineteenth century, many of Laennec's ideas about sounds as indices of internal lesions came under fire. His typology of sounds was deemed inaccurate, and his notion of the correspondence between diseases and specific sounds was difficult to prove. The Czech physician Josef Skoda attempted to reproduce many of Laennec's results and often found that the precision sought by Laennec did not exist. Skoda argued that each sound heard through auscultation and percussion could be traced to a physical alteration of the texture of the body but that each alternation could have been produced by one of several causes. Rather than being acoustic signs of pathology, each sound indicated nothing more and nothing less than a physical condition of the body. The physician's task was, then, to interpret the acoustic signs along with others to produce a proper diagnosis. By the late nineteenth century, Skoda's notion of diagnosis had eclipsed Laennec's.[101] Similarly, Austin Flint would write that "very few signs are directly diagnostic of any particular disease. They represent conditions not peculiar to one but common to several diseases."[102]

If mediate auscultation could not yield a proper sonic lexicon, it did yield a whole set of iconic and indexical signs—signs that are grounded by coincidence in lived experience rather than arbitrary relations.[103] Sonic signs produced through mediate auscultation were indexical in that they were produced by some interior condition in the body, even if that condition could not be linked to a specific lesion or disease, as Laennec originally hoped. Laennec's descriptions of the sounds themselves are similes: *like* a metallic rattling, *like* a bellows blowing on the fire, *like* a musical tune containing these notes. It would be for innovators and popularizers like Austin Flint to attempt to better codify the sounds themselves. Flint began his textbook by borrowing Helmholtz's triad for distinguishing among sounds: pitch, intensity, and quality. Using music as the analogy, he defined *pitch* as musical pitch, *intensity* as the volume or perceived degree of force, and *quality* as, essentially, the sound's timbre (his analogy is to two different instruments playing the same note). Yet even this apparently more scientific language for the discussion of sound quickly degenerates back into analogy: "There are some other points of difference; namely, the duration of certain sounds, their continuousness or otherwise, their apparent nearness to or distance from the ear, and their strong resemblance to particular sounds, such as the bleating of the goat, the chirping of birds, etc."[104] Although Flint claims that these additional sounds are of "lesser importance" in diagnosis, even his more abstract classifying scheme, in the last instance, resorts back to analogy for the most exact description. Having set up an analytic system, Flint retreats somewhat, insisting that sound can be described *only* by analogy.

Other medical textbooks offer a more direct analogical approach, for instance, suggesting that a most accurate imitation of the heartbeat is accomplished by

> pronouncing in succession the syllables *lupp, dupp*. The first of these sounds, which is dull, deep and more prolonged than the second, coincides with the shock of the apex of the heart against the thorax, and immediately precedes the radial pulse. . . . The second sound, which is sharper, shorter, and more superficial, has its maximum intensity nearly on a level with the third rib. . . . These sounds, therefore, in addition to the terms first and second, have also been called inferior and superior, long and short, dull and sharp, systolic and diastolic—which expressions, so far as giving a name is concerned, are synonymous.[105]

One could dismiss the authors' obvious difficulties with description, arguing that all language is fundamentally metaphoric, so the analogy comes as no surprise. Yet this would conceal a double process in the pedagogy of mediate auscultation and listening to the body in general. Auscultators were trained to recognize the specific qualities of a sound in the stethoscope and to use that specific sound to recognize conditions in the patient's body. The language of analogy is a language of iconicity, where ostensibly different sounds (bleating goats, internal lesions) resemble one another: "Anything whatever, be it quality, existent individual, or law, is an Icon of anything, in so far as it is like that thing and used as a sign of it."[106] Of course, the physicians are not listening for goats in their patients' bodies, but, at the same time, they need some kind of language to describe what it is they *are* hearing.

The main effect of this analogical description in textbooks was that mediate auscultation could never be fully abstracted from experience; a full and complete sonic lexicon could not be written. Thus, another discourse accompanies and quickly overtakes the lexicography of mediate auscultation: a discourse of clinical experience and refinement of technique. Throughout the *Treatise,* Laennec chides the "inexperienced observer"; pitfalls in diagnosis can largely be overcome through clinical experience. Forbes, in his introduction to the work, summarizes the position, a common one in medical pedagogy:

> It is only by long and painful trials, (*inter toedia et labores,* as Auenbrugger says of his congenerous discovery) that any useful practical knowledge of it can be

acquired. When, therefore, we hear, as we sometimes do, that certain persons have *tried the stethoscope,* and abandoned it upon finding it useless or deceptive; and when we learn, on inquiry, that *the trial* has extended merely to the hurried examination of a few cases, within the period of a few days or weeks, we can only regret that such students should have been so misdirected, or should have so misunderstood the fundamental principles of the method.[107]

Nothing will substitute for the experience of extensive clinical practice and training; nothing will substitute for the sustained experience of hearing through the stethoscope. Later writers would concur: "I have to suppose that you have made your ears familiar with these sounds." "Of the peculiar quality of any particular sound, one can form no definite idea otherwise than by direct observation."[108] Clinical experience was institutionalized in medical pedagogy as a way of guaranteeing a kind of common experience, a certain practice of practical knowledge.[109] The goal of clinical experience, then, was to render medical knowledge more true and more present through its immediate perception—hearing the rattle, seeing the lesion on the lung—and at the same time transform abstract knowledge into a very specific kind of practical knowledge.

The sign created by clinical medicine aspired to be an *index* of that state, an absolute accompaniment. Instrumentalized, the sounds "discovered" through mediate auscultation are connected with the interior states of the body—in the experience of the well-trained physician, they become indices. Flint, for instance, argues against any kind of abstract formalization beyond the experience of cultivating a clinical technique:

The study of different sounds furnished by percussion and auscultation, with reference to distinctive characters relating especially to intensity, pitch and quality, distinct signs being determined from points of difference as regards these characters, may be distinguished as the analytical method. It may be so distinguished in contrast with the determination of signs by deductively taking as a standpoint either the physical conditions incident to diseases or the sounds. If we undertake to decide, *a priori,* that certain sounds must be produced by percussion and auscultation when certain conditions are present, we shall be led into error; and so, equally, if we undertake to conclude from the nature of the sounds that they represent certain conditions. The only reliable method is to analyze the sounds with reference to differences relating especially to intensity, pitch, and quality, and to determine different signs by these differences, the import of each of the signs being then established by the constancy of association with physical conditions. *It is by this analytical method only that the distinctive characters of signs can be accurately and clearly ascertained* (emphasis added).[110]

For the sounds produced by mediate auscultation to signify properly—that is to say, for them to signify as indices of internal conditions—the auscultating doctor must have achieved a certain level of technical virtuosity in listening. It is not a simple matter of a lexical or formal correspondence; one must learn the feel of a set of coincidences, learn which events coexist with which other events: "Preconceived notions frequently oppose themselves to the reception of the truth, and have to be got rid of before the real state of matters can be ascertained. Hence the great importance of deriving your first impressions of the sounds to be heard by auscultation, not from books or lectures, but from the living body itself."[111] Both Laennec and Skoda understood that the sounds perceived through mediate auscultation were themselves produced by the conditions that they indexed, even if Skoda was right to distinguish between a condition, an illness, and a lesion.

As mediate auscultation became institutionalized, as it became a regular practice in medicine, instruction in listening moved from attention to the pathological to attention to the normal. While Laennec's earlier *Treatise* deals entirely with pathological signs, Austin Flint's innovation was to begin medical education with the healthy body, to construct a set of normal positions from which the diseased body deviated.[112] Indeed, Flint and others argued that such knowledge of the healthy body was a necessary precursor to diagnosis of disease. Without it, the physician ran a serious risk of misdiagnosis. Once again, this position was based in the privileging of clinical experience over a system of abstracted and, therefore, objectifiable signs.[113]

One of the early complaints against Laennec and his method was that there were simply too many fine gradations of sound for any single person to master, that Laennec's own claims for the stethoscope were far too grand. As Skoda and others later demonstrated, they were. But the difficulty of learning proper auscultation contributed to its value, made it a mark of initiation, a form of virtuosity: "This difficulty in attaining a complete practical knowledge of *Auscultation* is one of the greatest drawbacks to its value; as it will ever prevent the indolent and careless from making themselves masters of it. But I will venture to add, that no one who has once mastered its difficulties, and who cultivates his profession in that spirit which its high importance and dignity demand, will ever regret the pains taken to overcome them, or willingly forego the great advantages which he has thereby acquired."[114] Certainly, as this passage might suggest, the technique of mediate auscultation should be considered as part of the larger experiential approach to medical education at this time, especially as it related to the professionalization of medicine. But, if it rendered virtuosity a proper skill of the virtuous professional, it also depended on the more basic assertion that audition was a skill to be cultivated as well as refined for scientific purposes. A skilled doctor had to be well practiced at listening.

Over the course of the nineteenth century, the stethoscope became the hallmark of medical modernity. "Within the nineteenth century," writes Audrey Davis, "the instrument had been applied to every cavity and organ in the body."[115] The development of mediate auscultation applied medical and scientific reason to listening, just as a particular practice of hearing the body became integral to the everyday functioning of medicine. Part of physicians' elevated cultural status at the end of the century was based on the valuation of the skills specific to their profession. They were virtuoso listeners; they could hear the body in ways inaccessible to laypeople: mediate "auscultation helped to create the objective physician, who could move away from involvement with the patient's experiences and sensation, to a more detached relation, less with the patient but with the sounds from within the body."[116] Mediate auscultation was an artifact of a new approach to the work of sensation, in which listening too moved away from ideals of an intersubjective exchange between doctor and patient into the quiet, rhythmic, sonorous clarity of reason and rationality.

3 Audile Technique and Media

Mediate auscultation was the first site where modern techniques of listening were developed and used, but audile technique would proliferate across cultural contexts in the second half of the nineteenth century. This chapter explores the expansion of audile technique in media contexts: first, sound telegraphy and, later, sound-reproduction technologies like telephony, sound recording, and radio. In all these contexts, listening carried with it a great deal of cultural currency. It became a symbol of modernity, sophistication, skill, and engagement.

Telegraphy helped popularize technical notions of listening even as it constructed audile technique very differently from medicine. While medicine was a relatively elite practice, telegraph operators were at the lower end of the middle-class spectrum in both income and prestige. While mediate auscultation was about listening to the human body, sound telegraphy was about listening to a network that linked people separated by distance. Mediate auscultation had to create a physical distance between doctor and patient as participants; sound telegraphy presumed great distances between operators. Mediate auscultation was linked to scientific reason; sound telegraphy was linked to bureaucratic reason. If mediate auscultation is significant because of doctors' systematic attempts to elaborate a hermeneutics and pedagogy of listening, sound telegraphy both further generalized a notion of technicized listening and brought it for the first time into the realm of mediated communication, mass culture, and everyday life. Doctors went through years of training to become virtuoso listeners, but the telegrapher was a self-made auditor. Sound telegraphy itself was not handed down through textbooks and institutionalized training; rather, it developed as a result of workers' changing orientations to the machines that they used. The telegrapher's auditory skill drove the acceleration of telegraphic communication, and hearing became a hallmark of its efficiency—a synecdoche for the effectiveness of the network itself.

By the time technologies for reproducing sound became commercially available, there were already well-established and well-known repertoires of audile techniques. While some authors argued that sound-reproduction technologies made novel use of hearing, this chapter will demonstrate that their novelty was in the innovation of longer-standing cultural practice rather than in creating new modes of listening from scratch. The construct of a private, individual acoustic space is especially important for commodifying sound-reproduction technologies and sound itself since commodity exchange presupposes private property. The audile techniques articulated to these new technologies emerged out of smaller domains of middle-class life to encapsulate a larger middle-class sensibility. Through audile technique, people could inhabit their own private acoustic space and still come together in the same room or even across long distances. They could listen alone and listen together at the same time.

Even as the specific techniques of listening varied across contexts, the basic outlines of audile technique remained fairly constant. Techniques of listening articulated listening to reason and rationality. They separated hearing from the other senses so that it could be extended, amplified, and otherwise modified; listening became a discrete skill. Audile technique reconstructed acoustic space as a private, interior phenomenon belonging to a single individual. It problematized sound and constructed an auditory field with "interior" and "exterior" sounds. Techniques of listening instrumentalized and promoted physical distance and epistemological and social mediation. The long history of audile technique thus stands as a crucial component of sound's history in the modern era. Many of the meanings that we commonly attribute to listening—along with a few that scholars have forgotten—were articulated and elaborated over the long nineteenth century.

Telegraphy: "Ancient and Modern"

A cartoon from the 1870s (figure 15a–b) bearing the same title as this section depicts two telegraph offices. The first office, which is clearly meant to

Telegraphy—Ancient and Modern.

represent the "ancient" way of doing things, portrays a beleaguered telegraph operator sitting at a table in an open room, amid a mess of telegraph tape, trying to read it as messages come off the wire. To his side, a man and three smiling boys look on from the window; one of the children points and either laughs or speaks. The second office shows no public at all—suggesting a greater level of organization. The telegraph operators are now kept separate from the public. The door to the office remains closed. Inside

---

Figure 15a–b. Telegraphy—ancient and modern (courtesy University of Illinois Libraries)

the room, the picture simply depicts two telegraph operators seated across from one another at a table divided into cubicles. They appear comfortable in their chairs, and each is taking messages neatly on a pad while listening to the sound of the telegraphic messages coming in. One operator has two notes neatly hung in his cubicle. Both wear the visors that had become the distinctive mark of a professional telegrapher. There is a general impression of calm and of organization. As he listens to the sound of his own machine, each telegrapher has his own private space.

This is a professionalization narrative: the changing characteristics of the office, changes in dress, and changes in telegraphic technique all coincide to valorize the modern, professional telegrapher. But, for our purposes, the message of the cartoon is even more basic: visual or written telegraphy is ancient and outdated, while sound telegraphy is modern, clean, efficient, and even somewhat professional. The message is unremarkable except when read in the contexts of telegraph history and media history more generally. In these broader terms, this simple cartoon is suggestive of a historical shift from vision to hearing widely acknowledged in accounts of telegraphy but rarely considered at any length as having a significance of its own. The following pages thus consider the history of sound telegraphy and its significance for the history of listening.

Many of the key accounts of telegraph history place it as the first major electronic medium in American history and often as a precursor of the modern mass media. Harold Innis considered the telegraph a major turning point in media history. Although he was mainly concerned with the control of knowledge, he considered the telegraph unrivaled in its power to strengthen or weaken organizations' control over the flow of news. Menahem Blondheim follows Innis's lead to cast telegraphy as a turning point in the history of news and information, arguing that the telegraph helped destroy old monopolies of knowledge but promoted new ones in the guise of the Associated Press and Western Union.[1] Daniel Czitrom connects the rise of American telegraphy with the birth of a kind of media consciousness and shifting attitudes toward technology. Along with James Carey, Czitrom sees the telegraph as the first medium to separate the social facts of transportation and communication; he emphasizes that, through its telegraph business, Western Union became the first major corporate monopoly. Each author uses the telegraph to "stand metaphorically for all the innovations that ushered in the modern phase of history and determined, even to this day, the major lines of development of American communication"—to quote Carey.² In addition to the narratives of the telegraph's own development, Carey and Czitrom consider its importance in the rationalization and reorganization of news production and dissemination and even in transformations of cultural sensibilities around language and time. More recent writings from other perspectives have challenged the notion of the telegraph's "revolutionary" nature and its foundational role in American media history.³ Yet the telegraph retains its importance in media histories for many reasons: its connection with the commodification of news; the solidification of a newspaper elite; the promotion of the tendency toward monopoly in local markets; as well as the mythical status of the telegraph as the first medium to use electricity for long-distance communication.

Before I continue, a qualification is in order. What we now commonly call *telegraphy* is really *electric telegraphy*, which is a comparatively recent development in a longer history of telegraphy. An older form of telegraphy, now called *semaphoric telegraphy*, can be traced back to the Greeks and the Old Testament. A semaphoric telegraph uses lines of sight and relay stations to convey messages over a distance. With the aid of hills or towers, fires or flags, and a system of agreed-on signs, simple messages can be quickly relayed over very long distances. At the beginning of Aeschylus' *Agamemnon*, for instance, the watchman is depicted as hopelessly bored, waiting for the fire from a distant hill, "the sign of the beacon," so that he can report that Troy has fallen.⁴ This basic system of semaphoric telegraphy would remain in place for about two thousand years. The first major improvement was proposed by the British natural philosopher Robert Hooke in 1664, who suggested using telescopes, thereby greatly increasing the possible distance between relay points. Although mechanical telegraphs were first proposed in ancient Rome, one was not built until 1794 in France. Devised by Claude Chappe, the French mechanical telegraph used a system of bars and levers that allowed for 92 possible positions (actually, it allowed for 192, but, for reasons of clarity, the French used only 92). Each position corresponded to a numbered word in one of three books, so the mechanical telegraph had a vocabulary of 25,392 words.⁵

Semaphoric and mechanical-semaphoric telegraphs were visual-tactile media: they relied on the sense of sight for the transmission of information over a distance. The electric telegraph, however, was another story altogether. From the outset, the electric telegraph allowed for an interchangability among the senses: electric telegraphy could produce visual or sonic data. For the purposes of this chapter, I will use *sound telegraphy* to refer to a specific set of practices involved in telegraph operators listening to the Morse-based electric system. There were a number of experiments with more properly "sonic" telegraph devices over the course of the nineteenth century. The most well-known is the harmonic telegraph, which used multiple pitches to send multiple messages down a single line. Hermann Helmholtz also made some efforts to connect his tuning-fork apparatus with a telegraph. But these are not my main concern here.

As I will discuss below, the interchangability among senses in the electric Morse system is actually central to the practical development of telegraphy in the United States. Yet the electric telegraph has been relatively neglected in the history of the senses, and the history of the senses has been neglected in accounts of telegraphy. Redressing this absence is not simply a matter of completeness or further inclusion. Rather, unlike many other media, the electric telegraph spent time as both an apparently visual medium and an apparently acoustic medium. Historians of the senses often tend to think in terms of binary logics: thinking, practice, or technology is either visual or auditory. Rick Altman has called this the *ontological fallacy,* where scholars extrapolate from historically specific practices to make transhistorical claims about the nature of a medium.[6] The history of sound telegraphy requires a shift in focus from the essential sensory characteristics of a particular technology to the history of its deployment. It requires a shift in focus from the sensory classification of media to the history of the deployment of the senses through and around media. It also shifts in focus from the essential or natural aspects of listening to those that were historically contingent. Telegraphic listening actually consisted of many learned practices that developed over time. Precisely because of the electric telegraph's sensory interchangeability (or at least complementarity), we should consider telegraphy and listening to the telegraph from the standpoint of *technique.* The electrical telegraph could be configured to be apprehended by either eye or ear. As we will see, the choice of one or the other was a practical question. While the senses are technically interchangeable in telegraphy, vision and hearing play very distinct roles in its cultural and industrial history.

Widely regarded as the first intimation of electric telegraphy, an anonymous letter (from one C.M.) to the *Scots' Magazine* in 1753 entitled "An Expeditious Method of Conveying Intelligence" outlined an electric-telegraphic system that consisted of one wire for each different character that it would transmit. The wires would run between the two points to be put into communication, and the apparatus worked by connecting the wires to an electric machine, in the order of the characters to be conveyed. At the receiving end, the electric pulse would lift a piece of paper labeled with the appropriate character, and an operator at the receiving end would write down the message. But the author thought this method to be interchangeable with an acoustic method, which would involve replacing the paper with bells decreasing in size from A to Z, each wire being connected to a bell "and the electrical spark, breaking on bells of different size, . . . [informing] the correspondent, by the sound, what wires have been touched. And thus, by some practice, they may come to understand the language of chimes in whole words, without being put to the trouble of noting down every letter."[7] At its very outset, there is a sensory interchangeability in electric telegraphy: sheets of paper or tuned bells produce the same effect as far as the author is concerned. This exchangeability is based on an instrumentality of perception. Electricity here takes the form of "pure" information that is conveyed from one node of the network to another and rendered intelligible through the route that it takes. Sensation and action occur at either end of the network. The route is the fixed thing, the perception the variable.

C.M.'s theory of telegraphy is a nascent cybernetics. This is no surprise since cybernetics itself was a communication theory developed on the basis of technical issues in the communication network that replaced telegraphy in the United States: the telephone system. Still, the anachronism is tempting. The whole history can be read backward, with telegraphy as an instance of the cybernetic model of communication. In telegraphy, people's sense organs and muscles become extensions of machines that convey messages over a network: "When I give an order to a machine, the situation is not essentially different from that which arises when I give an order to a person. . . . Information [that man] receives is coordinated through his brain and nervous system until, after the proper process of storage, collation, and selection, it emerges through the effector organs, generally his muscles."[8] While this logic may have been present from the outset, the actual development of telegraphy was based on a series of sensory preferences, rather than a preference for interchangeability itself.

No other major examples of early electric telegraphy operated on the acoustic principle. When it finally broke in America (and in England as well), the electric telegraph was understood primarily as a visual medium.[9]

This is probably because of the legacy of semaphoric and mechanical telegraphs: Chappe's telegraph was in use in France by 1800 and was well known and imitated throughout the world. Two electric telegraph systems were invented at roughly the same time. In 1837, William Cook and Charles Wheatstone (Cook was the entrepeneur, Wheatstone the inventor) devised an electric telegraph that moved a needle to convey information. In the same year, the American Samuel Morse devised an electric telegraph—and it is the history of the Morse telegraph that I will discuss here. As did his contemporaries, Morse understood telegraphy as an essentially visual medium. In fact, his original patent application was for a specifically visual telegraph; writing in 1837, he claimed:

> About five years ago, on my voyage home from Europe, the electrical experiment of Franklin, upon a wire some four miles in length was casually recalled to my mind in a conversation with one of the passengers, in which experiment it was ascertained that the electricity traveled through the whole circuit in a time not appreciable, but apparently instantaneous. *It immediately occurred to me, that if the presence of electricity could be made* VISIBLE *in any desired part of this circuit, it would not be difficult to construct a* SYSTEM OF SIGNS *by which intelligence could be instantaneously transmitted.*[10]

Morse's original telegraph worked through a relatively simple process: pressing down the transmitter key completed a circuit, and a receiver on the other end would use a stylus to make an indentation on a piece of paper, which would be drawn beneath the stylus by a clockwork mechanism set into motion when the circuit was completed. Vail's meticulous description of the early telegraph's working suggests that the visuality of the technology had now been hardwired in. Any sounds that the machine made at this point were incidental, mere epiphenomena of its making a visible recording (figure 16):

> If, then, the hammer is brought in sudden contact with the anvil, and permitted as quickly as possible to break its contact by the action of the spring, and resume its former position, the galvanic fluid [electricity], generated at the battery, flies its round upon the circuit, no matter how quick the contact has been made and broken. It has made the iron of the electro magnet a magnet; which has attracted to it the armature of the pen lever; the pen lever, by its steel pen points, has indented the paper, and the pen lever has, also, by the connecting wire with the break, taken it from the friction wheel; this has released the clock work, which, through the agency of the weight, has commenced running, and the two rollers

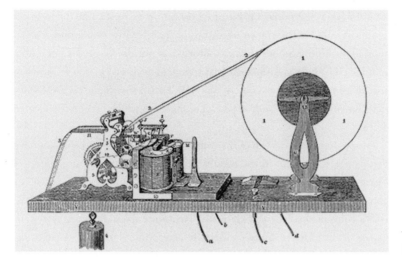

16

have supplied the pen with paper. But, as only one touch of the key has been made, the clock work soon stops again, if no other touches are made, by the action of the break upon the friction wheel.[11]

If, as Friedrich Kittler suggests, nineteenth-century media transmogrified a holy trinity of image, sound, and writing, telegraphy would commingle all three.[12] The Morse telegraph was a machine for writing at a distance; how that writing would be perceived was an open question. Initially, Morse and Vail's machine produced a script to be viewed. The length of the indentation would depend on how long the operator at the transmitting end kept the key down. Thus, the code appeared as "dots and dashes" because those were the two different indentations made by receivers. Morse's famous "first" (i.e., public) message, "What hath God wrought?" was accomplished through this means.[13] However, these technical facilities were only half of the Morse telegraph's innovation. The other was Morse code.

Morse's code was originally a simple cipher code, where a series of dots and dashes would arbitrarily stand for a number that corresponded to a word. It was improved by Alfred Vail in 1837 to a version of the modern Morse code alphabet (which was finalized in 1844) in which each letter was represented by a series of dots and dashes. As a result, a telegraph operator

Figure 16. Diagram of printing telegraph. The noise of the apparatus based on this model eventually became the basis for sound telegraphy.

would have to memorize only a limited number of series, one for each letter and symbol to be transmitted, rather than having to memorize a potentially infinite number of words. Morse was not the first to invent a telegraphic code; he was simply the first to invent such a code that found wide use. In 1851, a European conference made Austro-Germanic code the standard for international and land telegraphy everywhere except the United States.[14]

Morse and others spoke of his code in terms of signs, and the language here is not accidental. Morse code is a set of symbols. It is purely a set of conventions, a series of signals to be perceived by an operator, each series corresponding to a specific letter in the English alphabet. Because of the code's conventionality, it obviates the need for C.M.'s elaborate system of papers or tuned bells. Thus, as with the development of mediate auscultation, it is not simply the apparatus but the technique of perception and its codification that were the significant innovations in the early history of telegraphy. Telegraphy was the combination of a physical technology, a system of signs, and a technique of its use.

Although Morse held a patent for both sound and visual telegraphy, he initially ignored sound telegraphy altogether. Like Laennec's autopsies, the recorded (written) telegraphic messages provided a kind of proof that the event happened—after the fact. The visual verification of a telegraphic message was used to prove the system's accuracy to audiences. Morse and Vail's early public demonstrations were clearly based on this reasoning, where physical proof of a message's content, along with a notated time of reception, could be lined up against the content of the original transmission. The demonstrations in 1838, and the later demonstrations in 1844, were based on this performance technique.[15] Contemporary accounts emphasized that belief in the telegraph was at least in part founded on seeing the telegraph at work. Although the telegraph would be considered an extension of the postal system early in its history, these early demonstrations predate any large-scale attempt to put it to practical use. A Rochester city newspaper offered this comment on the telegraph's arrival there in 1846: "The actual realization of the astonishing fact, that instantaneous personal conversation can be held between persons hundreds of miles apart, can only be fully attained by *witnessing* the wonderful fact itself." Around the same time, a Philadelphia paper effused, "It is difficult to realize, at first, the importance of a result so wholly unlike anything with which we have been familiar; and the revolution to be effected by the annihilation of time . . . will not be appreciated until it is felt and *seen*" (emphasis added).[16] For these writers, the telegraph effected an annihilation of space and time that had to be seen to be understood. Yet there is already a taste of telegraphic synesthesia in these accounts: as a modification of an age-old phenomenon that was at least partially sonic, telegraphic conversation had to be seen to be believed.

The telegraph may have profited from its initial visibility, yet a different set of values quickly displaced and supplemented telegraphic investments in visualism. By the 1850s, listening to the telegraph became the favored method among operators—although the printing receiver continued to be used for a variety of specialized tasks. The germinal forces of bureaucratic capitalism and the values of efficiency and accuracy would belong in telegraphy, not to the eye, as might be supposed by conventional critical wisdom, but to the ear. The telegraph was a machine for writing at a distance: operators could decipher the code by reading the script; or, instead, they could decipher the message by listening to the process of inscription on the receiving end. Once telegraphers started listening to their receivers, the receiver's script itself became a vanishing mediator: operators could simply listen to the machine, decode the message as they heard it, and then discard the tape with the dots and dashes on it. It became more efficient simply to listen to the machine, and this efficiency was essential to an ever-growing telegraphic bureaucracy that handled more messages with each passing day.

This shift from visual to sound telegraphy began at the level of practice; only slowly did managers and companies become interested. While Czitrom credits Alfred Vail with "working out" a sounder,[17] a different story can be pieced together from other accounts. Very quickly, telegraph operators learned that they could discern messages more clearly and with greater speed by listening to the machine than by reading its output. The noise that began as a by-product of the machine's printing process became over time its most important aspect.[18] It is both impossible and irrelevant to establish proper credit for the "invention" of sound telegraphy. The idea of sound telegraphy existed before any kind of electric telegraphy had been accomplished; the actual practice was probably discovered all over the country by creative telegraph operators in the late 1840s. No single person can claim to have invented sound telegraphy; that people did anyway says more about patent battles than about the nature of the discovery.

We do know that sound telegraphy developed almost as quickly as telegraph lines were set up for operation in the United States and operators were put to work. The story of James Francis Leonard, a telegraph operator from Kentucky, is an instructive example. By June 1849, Leonard had dispensed with the "cumbrous paper" of the telegraph printer and took messages entirely by ear. Leonard's biographer, John Townsend, claims that Leonard was the first practical reader of the Morse alphabet, although many telegraph operators had probably caught snippets of words and phrases by sound from their machines. While Townsend does not present convincing evidence for Leonard's "firstness," suffice it to say that Leonard was one of the first practical sound readers.[19] Operators all over the country were experimenting with listening around the same time as Leonard. For instance, a Milwaukee company was using sound telegraphy almost exclusively from 1851 on.

Initially, the Morse company fought hard to prevent sound telegraphy, even to the point of getting laws passed to prohibit its practice. This may have initially been in response to concerns about patent rights and equipment, although Morse did hold a patent for sound telegraphy. Yet the interest in sound telegraphy quickly spread. By July 1849, Leonard was wired to come to Louisville for a trial of his method, even though the Louisville manager, James Reid, saw sound telegraphy as a potential "danger" to the profession. Leonard quickly proved his worth and became a fixture in the Louisville office. In 1851, he declined an offer from P. T. Barnum to take his telegraphy skills on the road (which was probably a wise business decision, given that sound telegraphy was becoming quite widespread at this point). In 1855, Leonard achieved wider notoriety when Samuel Morse—who had apparently changed his mind regarding the need for speed in telegraphy—requested that all telegraph office employees take speed tests so that the fastest could be shown at the first International Exposition in Paris, later that year. Leonard was able to average about fifty words a minute and got up to fifty-five. Morse had Leonard's fifty-five-word message copied down, and then took it with him to Europe for the exposition.[20]

Sound telegraphy spread elsewhere in the country as well, and, in each place, it was hailed for its efficiency, speed, and accuracy. An 1853 account suggests that similar approaches to Leonard's were gaining popularity throughout the country:

> It was at first supposed that it would be found in practice very advantageous to have signs of letters permanently made on paper. But we find that the advantage of recording mere signs is not universally acknowledged, and that even parties who have purchased from Morse the right to use his system have discontinued the recording of dots and dashes. . . .

The company owning the telegraph line running from Buffalo to Milwaukee, called the Erie and Michigan Telegraph Company, working under Morse's patent, have for some time past discontinued the practice of recording the signs produced by the process above mentioned, and have instead thereof received their messages by sound. This they have done for the last two years without interruption, having found that they could receive three messages by sound in the same time which would be occupied in receiving two under the other system; and, moreover, that in receiving by sound they made fewer mistakes than they were liable to in the use of the dots and dashes, and also dispensed with some of the operators.

The mode of receiving messages by sound is very simple, and one operator is sufficient. The operator sits by his table in any part of the room where the message is received, and writes it down as the sounds are produced. The different sounds are made by the striking of the pen lever upon a piece of brass; thus, three raps in rapid succession are made for the letter A, two raps, an interval, and then two raps more, are made for B, and so forth.[21]

Here, the assumed preference for visual data comes under direct critique. Of course, this very criticism is made possible in part by Morse's earlier public demonstrations, which established the telegraph's functionality and utility for large audiences and for which printouts were doubtless of great utility. Now that the telegraph was established, however, and becoming the technical side of a vast and complex corporate apparatus, the values of efficiency, accuracy, and speed quickly eclipsed any assumed preference for seeing the messages; their verity could be assumed. This greater efficiency also means that the telegraph's value as a form of fixed capital was enhanced for the firm, which could therefore let go of some of the operators previously thought indispensable. Indeed, the "talking telegraph" almost entirely displaced the printing telegraph on these very terms: "Many—perhaps we might correctly say most—telegraphers can 'read' more or less readily by sounds, and therefore *that* is not the subject of our wonder. But we were surprised to see the *whole* business of extended lines, at a junction so important as Cleveland, transacted *exclusively by sound,* without any use of recording apparatus—transacted satisfactorily, too, amid the apparent confusion incident to the clicking of so many instruments in such close proximity."[22]

Here, we see sound telegraphy emerging as both an aid to efficiency and a *skill*. The telegrapher's fabled ability to tune sound in and out at will was not in and of itself a new thing. Sound had already been framed as "interior" and "exterior" in mediate auscultation, and directed listening, "listening in search," has a long history. Like mediate auscultation, telegraphy yielded sounds that were indices of states otherwise imperceptible to the listener: the doctor heard the patient's insides through the stethoscope, the telegrapher heard distant messages through the receiver. Proximal sounds had become effects of relations at a distance. Moreover, sound telegraphy required that sound be framed into foreground and background, inside and outside: the operators at Cleveland, in a room full of clicking instruments, knew full well that each click was linked with events at a great distance from that room, and they could focus their attention on their instruments alone. The operator focusing attention on his or her (although more likely his) instrument was not simply listening for a particular event in a confusing environment but listening for an event in the environment that corresponded with an event in Milwaukee, or Baltimore, or Chicago, or wherever. The noise in that Cleveland office was the noise of a network, the concentration of signals at a nodal point. Even as they were hearing physically proximal sounds—after all, sound did not travel through telegraph wires—the operators were listening to other points on the network. Indexical sounds brought the distant world near.

By the end of the 1860s, sound telegraphy was the rule, rather than the exception, although it never achieved complete dominance in the field:

On the American lines the system most commonly employed is that of an acoustic telegraph known as a "Sounder." A continued practice with Morse apparatus leads the employees involuntarily to recognize the signals by ear, and when they have once attained proficiency in this way of reading, they seldom or never return to that more fatiguing one of reading by sight....

The great drawback of this system is the want of a record, which is so necessary for the justification of the employees and the administrations. On this account the Morse, or some other recording system, will continue to be employed upon all lines on which telegrams of importance are transmitted.[23]

Although this account is equivocal on the question of hearing versus seeing, it acknowledges the central place of sound in telegraphic communication while still suggesting that an operator's written translation of a sound transmission may not be sufficient for evidentiary purposes.

Like mediate auscultation before it, sound telegraphy's organization of listening was intimately tied to the larger process of the separation of the senses and the construction of sounds as indexical signs. Like mediate auscultation, sound telegraphy was essentially a technique of listening, this time articulated to a technique of writing. Sound telegraphy was perhaps the first media site where proximal sounds directly corresponded with distant events. While photography would provide a visual record of distant times and places, telegraphy offered an audible trace of contemporaneous distant events. Still, from the perspective of subsequent media history, the acoustic data of telegraphy were extremely limited. The telegraph had no transducer; it did not mechanically transform sounds into electricity and then "reproduce" sound at the other end. The Morse sounder (or even a printing telegraph) simply made a noise to correspond to a distant event. To follow McLuhan for just a moment, telegraphy could have intense affective and ideological significance precisely because of this relative paucity of sensory data. The experience of telegraphy required a great deal of involvement from its users.[24] Mediate auscultation crystallized the physical distance between doctor and patient; it turned that distance into an epistemological and aesthetic virtue. Sound telegraphy used techniques of listening to overcome distance between stations, and this distance could be both metaphoric and real. Sound telegraphy simultaneously became a focus of fascination and fear and a medium through which people could become invested in distant events and locales.

Carolyn Marvin tells of the fear and interest surrounding telegraphic weddings, where women were conned into having a telegraphic marriage ceremony with a suitor whom they had previously known only through written correspondence. In one case, this turned out to be a fraud, costing a Milwaukee widow $3,000 as well as a good deal of embarrassment; in another case, a white woman married a man whom she thought she liked a good deal, until she discovered that he was of African descent.[25] This kind of deception—what Marvin terms "crimes of confidence"—would continue with other media down to the present day. This is possible because of the phenomenon of *presence availability,* to use a term of Anthony Giddens's. Building off the work of Erving Goffman and Edward Hall, Giddens argues that social availability is structured into front and back spaces, where the front space is the locus of social information that is available to others and back space is the locus of social information that is hidden.[26] Giddens and John Thompson both argue that the rise of the mass media has coincided with the growth of forms of communication that entail very small front spaces (relatively little available information) in relation to relatively large back spaces (lots of unknown factors).[27] The telegraph is a good example of this phenomenon since the clicks of the sounder are the only information available about what is happening at the other end of the line.

Thus, the possibility for deception is high in these contexts, as is the possibility for misunderstanding. Yet this lack of information about a correspondence did not necessarily diminish the feelings of intimacy regarding communication at a distance. Like mediate auscultation, telegraphy stratified the sounds into meaningless "exterior" sounds and intensely meaningful "interior" sounds. In the technique of sound telegraphy, the sounder and the meanings attributed to the sounds that it made came to dominate the auditory field.

The idea of telegraphic intimacy thus had a certain amount of currency, and men and women were sometimes represented as successful in using the telegraph to these ends. Marvin also tells of a successful telegraphic wedding that was conducted when an operator was required to work on his wedding day. All this is to say that women were used as a symbol to represent the emotional capacity of telegraphy. At the same time, real women were largely excluded from the ranks of professional telegraph operators, and much of the professional and technical literature represented women as unfit for the job, unable to acquire the requisite technical knowledge. Popular narrative, meanwhile, was considerably more flexible in the matter of women's technological competence than the professional literature.[28] The story "Kate: An Electro-Mechanical Romance" provides a literary rendition of this level of intimacy between a city woman and a country woman communicating by telegraph. Having gotten her father to leave the telegraph office, the main character, Kate, wires another operator with whom she's become good friends, after making sure that the intervening stations are off the line:

Mary replied instantly, and at once the two girl friends were in close conversation with one hundred miles of land and water between them. The conversation was by sound in a series of long and short notes—nervous and staccato for the bright one in the little station; smooth, legato and placid for the city girl. . . .

[T]he two friends, one in her deserted and lonely station in the far country, and the other in the fifth story of a city block, held close converse . . . for an hour or more, and then they bid each other good night, and the wires were at rest for a time.[29]

Intimacy here is managed purely through expression, which carries such a burden that the personality traits of the characters enter into the very sounds of the apparatus. From even this short passage, one could imagine a whole musicology of the "singing wires" active in the literary imagination. Although a telegrapher's touch was as distinctive as a signature, the notes of the telegraph in "Kate" clearly mark the differences of class, region, and sophistication between two women. These sounds are forerunners of the voice of the telephone operator that would be presented to elite patrons as evidence of "a smooth and knowledgeable broker of social relations between middle-class households, . . . at the same time that she was only a servant, not truly a member of the class to whose secrets she had access."[30] But as with the telephone operators—who were under constant suspicion of eavesdropping and entering into all sorts of "intimacies" with middle-class men—the investment in the sound coming over the wires was presented as a vestige of the body (the voice, the movement of a hand) that had squeezed through the grain of the apparatus itself.

So, although it presented a tiny front space by the estimations of Goffman, Giddens, and Thompson, sound telegraphy was invested with the possibility of a depth of feeling and communication that was hitherto reserved for face-to-face and written interaction. If people had access to the medium and the skills with which to use it, they could experience telegraphy as listening to events at possible distances of hundreds of miles away, a listening that was held up as separate from any other sense data or even their absence. In sound telegraphy, distant events became audible purely through a sonic trace, even if this audition had to carry a huge burden of social knowledge with it. Interestingly, this is a moment when my history could gel well with psychobiological explanations of auditory experience based on the audiovisual litany: the "spherical" field of auditory perception, as opposed to the forward directionality of vision, would logically better lend itself to new kinds of spatial relations. The placement of a sound (produced at a distance through sound telegraphy) in a listener's auditory field absent any other sensory data would be more like other, more familiar forms of auditory experience. In other words, people are used to treating things that they can hear but cannot see, smell, touch, or taste as "present," and, therefore, it would make sense that the first sense of a kind of intimate, distant immediacy would be accomplished aurally. But this kind of reasoning offers an incomplete explanation: it discounts the kinds of affective and imaginative investments that people put into writing before telegraphy and into visual telegraphy before sound telegraphy. Since the copresence is imagined, the psychobiological organization of a sensorium is not necessarily determinate.

The cultural and technical dominance of sound telegraphy is probably the first major example of listening in a media context in American history. Sound telegraphy does not reproduce sound so much as link it, enmesh it in a relation of correspondences, and organize it according to the logic of an indexical code—a particular kind of signification. The skill involved in telegraphy was part of the mystique of the profession. Being able to listen to the machine and decipher the code was an acquired skill. It is true enough that even a child could learn Morse code with a little practice; but sound telegraphy was nevertheless an object of practical mastery and even virtuosity—mostly for men—as illustrated by the role of telegraphers in popular fiction and the role of the telegraph itself in the relaying of news. Although, as Carolyn Marvin points out, telegraphers had even less occupational prestige than other kinds of electric workers,[31] by the mid-nineteenth century sound telegraphy—and the attendant image of the able telegraph operator—was embedded in American popular culture and everyday life.[32]

Audile Technique Disseminated; or, A Short History of Headset Culture

Audile technique developed over the course of the nineteenth century in a variety of contexts. The two considered thus far, mediate auscultation and sound telegraphy, share important characteristics: their articulation to scientific and instrumental reason in clearly defined and framed institutional settings; their practical separation of hearing from the other senses; their reconstruction of acoustic space; their construction of sounds as a primary object of audition; and their powerful symbolic currency in American culture. These were the same orientations toward listening that would come to be articulated to sound-reproduction technologies, almost from the moment of their emergence. The audile/experiential characteristics attributed to early telephony, phonography, and radio—the problematization of sound and the construction of an auditory field with "interior" and "exterior" sounds, the "extension" and separation of hearing from the other senses, the focus on technique, and the connection between sound, listening, and rationality—were, thus, refinements and extensions of existing cultural practices.

The growth of the early sound-reproduction technologies would be better characterized as further disseminating previously localized practices than as "revolutionizing" hearing as such.[33] This section considers that process of dissemination by briefly exploring the diffusion of constructs of audile technique in discourses about listening: especially in images of listening to reproduced sound ranging from 1880 through 1925. Three aspects of audile technique are especially salient for the images that I consider below: the separation and idealization of technicized hearing; the construction of a private acoustic space; and the subsequent commodification and collectivization of individuated listening.

Both mediate auscultation and sound telegraphy relied on the construction of an individualized acoustic space around the listener. The binaural stethoscope crystallized this orientation toward acoustic space in an artifact; we will see the same process at work in headphones and telephone booths. It is true that people often listened together to sound recordings and, later, to radio shows. Yet even these collective modes of listening already assumed a preexisting "privatized" acoustic space that could then be brought back to a collective realm. As we will see, the construction of acoustic space as private space is in fact a precondition for the commodification of sound. This is because commodity exchange presupposes private property. Acoustic space had to be "ownable" before its contents could be bought and sold.[34]

These principles worked together, and we can see their coevolution in the history of binaural or even stereophonic listening technologies. Audile technique worked by separating hearing from the other senses so that listening could be more easily directed and manipulated. The individuation of acoustic space and the stratification of sounds occurred along with the separation of hearing from the other senses. This is the basic principle behind Laennec's stethoscope—a device that physicians used to concentrate their hearing in a particular place and deemphasize other sounds in the room. As part of this longer history of technique, Cammann's innovations with the binaural stethoscope primarily involved delegating more of the work of framing to the technology: where physicians had been framing the auditory field through the conventionalized use of a monaural stethoscope in a single ear, they could now frame it more fully simply by putting tubes in both ears. With tubes in both ears, a new, more malleable auditory field becomes possible. The origins of "binaural" (and, in some cases, stereo) devices lay in this development. Two ear tubes opened up the possibility for two different sound sources. By connecting each ear with a slightly different sound source, the listener would get a three-dimensional sense of the auditory field: what is now called the *stereo image*.

The earliest experiments with stereo audition were with the differential stethoscope. A binaural instrument with two chest pieces connected to two tubes, one for each ear, the differential stethoscope made it possible to listen to two separate points on the patient's body and compare the sounds.[35]

The physician George Carrick wrote of the differential stethoscope almost as if it were a prosthetic device: "By allowing us to place each ear [*sic*] on a different part of the chest at the same time, it enables us to differentiate sound easily, i.e., to recognize the stronger from the weaker."[36] The differential stethoscope provided a weak stereo effect; by placing the two chest pieces in different places on the patient's body, the user created the illusion of a three-dimensional auditory field. This version of the stethoscope was promoted as assisting in the location of sounds within the body and also as helping physicians adjudicate the differences between two sounds. Although the differential stethoscope never found common use, the possibility of shaping the auditory field in order to produce a specific sonic frame and directional sense of hearing was a necessary presupposition in its construction. In other words, it was one of the first instruments built that took for granted Laennec's ethos of hearing; it extends the stethoscopic principle to include the possibility of rudimentary echolocation.

These practical notions of stereo audition would later be developed in telephony and elsewhere. Alexander Graham Bell, for instance, took up a series of experiments with "stereophonic" telephony in the late 1870s, on the grounds that "there seems to be a one-sidedness about sounds received through a single ear, as there is about objects perceived by one eye." Bell's experiments were simply part of a long line of experimentation in combinations of sound technology and audile technique. But already they demonstrated that acoustic space is not simply reproduced or *represented* through audile technique, that the technique instead precisely affects the shape, contour, and texture of the listener's acoustic space. A simple experiment moved from the attempt to imitate "direct" audition to the construction of directed audition: going into the experiments, Bell's hypothesis was that a stereophonic telephone would provide for directional hearing more like "that experienced by direct audition." Bell discovered precisely the opposite. Telephony, even when in stereo, transformed the directionality of hearing—listeners could detect the relative latitudinal position of a sound (right or left) but not its longitudinal position. Even "if the sound be caused to move in an irregular or serpentine path—the sensation at C, D [the stereo receiver] is as though the sound had been moving in a straight line—horizontally in front of the observer from left to right or *vice versa*."[37] The combination of audile technique and sound technology worked to reshape acoustic space.

That the technicized auditory field had certain characteristics setting it apart from "direct audition" is central to understanding the development of listening in the age of technological production. Foremost among these characteristics is the emphasis on directionality and detail against a "holistic" perception of the auditory environment. After all, microphones and transmitters are instruments sensitive to the "feeblest sounds."[38] This was also the basis of technique in mediate auscultation and telegraphy—an emphasis on the minutiae of rhythms and the slight shift in timbre emanating from the chest, or the staccato of the Morse sounder. How different in their organization, in their framing, were these sounds from the faint voice at the other end of the early telephone—a voice the upper partials of which allowed its listener to imagine the more fundamental tones—or from the scratchy sounds in the grooves of the phonograph cylinder? Certainly, there are some differences. While medicine and telegraphy professionalized audile technique, sound reproduction popularized it. But, across the decades, these practices share the focus on detail, the notion of listening as a "separated" sense, where hearing did not have to correspond with other sensory phenomena. Most important, this type of listening shifted from attention to the sound qua speech or music or some other phenomenon to the sound qua sound. Judging by the increasing emphasis on detail in audile aesthetics, we can draw a direct line of descent. When Horkheimer and Adorno wrote in 1944 of an aesthetic of the detail coming to preeminence in music and mass culture, they traced its lineage through art music. Yet the rise of audile technique no doubt contributed to the emergence of this aesthetic as much as the Romantics' rebellion against organization.

In a fashion similar to James Johnson's history of listening (discussed at the beginning of chapter 2), technique appears suddenly in Horkheimer and Adorno's description of musical aesthetics. This is simply because they are not primarily concerned with the history of technique. Yet their work provides some interesting insight into the historical and aesthetic importance of one artifact of audile technique: the emphasis on sonic detail. The aesthetic of the detail is at the center of their very perceptive analysis of the *form* of mass culture.[39]

Adorno and Horkheimer's analysis of mass culture has been much maligned in the past few years as elitist, but a serious reading of their work shows their attention to many of the issues now dominating the analysis of mass culture—the increasing concentration of media ownership and the commodity status of entertainment—as well as their attention to the aesthetic dimensions of mass-cultural experience. Although they stridently argued for the need for structure and a sense of totality in musical listening (as did Adorno in his other writings), their analysis of the increasing emphasis on the detail in music was essentially correct: the dominance of African-American blues and the descendants of that music (ragtime, jazz, rhythm and blues, rock and roll, country, etc.) have strongly conventionalized song structures that allow for improvisation, subtle variation, and an emphasis on rhythm and timbre. That Horkheimer and Adorno thought this a bad thing (and that some of their readers, myself included, think this a good thing) does not diminish the essential correctness of their description of the central aesthetic features of mass-mediated music. The history of the collision between a new emphasis on sonic details in predominantly white spheres of cultural practice and white interest in African American musical forms has yet to be written. But it is clear that the detail resonates at the very core of modern American practices of musical listening.[40]

Beyond its privileging of sonic details, audile technique is based on the *individuation* of the listener. The auditory field produced through technicized listening (whether by convention or prosthesis) becomes a kind of personal space. The individual with headphones is perhaps the most obvious example of this phenomenon, and this is why I have chosen to concentrate on images of headphones here. Consider a typical valorization of headphone listening (figure 17). A telegraph operator sits in the corner of a train car, shut off from his surroundings by his physical seclusion and the headset that connects him to the railway telegraph. His head is turned down in concentration, and the apparatus sits in his lap, yet his posture is one of repose: the slight slouch, the crossed legs. His comfort represents a facility with the machine, a telegraphic literacy. His bodily disposition also represents a separation from the environment: he sits alone in the back corner of the car, his eyes averted from his own environment, his ears covered and linked through the train to the telegraphic network. From the bedside physician in the hospital to the railway telegraph operator in the passenger car: through technology and technique listeners could transcend the "immediate" acoustic environment to participate in another, "mediated" linkage.

This isolation and localization of sound and specifically of hearing would be taken to further extremes. Not only was hearing to be separated from the proximal auditory environment, but the act of communication itself was to be separated from the surrounding physical environment. The American Telephone Booth Company hoped to facilitate this framing through physical separation of the telephone from the rest of the office through the use of an indoor phone booth. Here, it is not only the separation of hearing but also the other end of the medium—the isolation of

Figure 17. Train telegraph operator, 1890

telephonic speech from the rest of the office—that is at stake: "In order to do away with the noise incident to telephonic service, the Telephone is often put in an out-of-the-way place. These troubles are overcome by using a sound-proof Booth."[41] The "trouble" in this case is the intermixture of the sound of telephony with the sound of office work. The booth here serves as a framing device. It reduces the front space of telephonic communication appropriate to that demanded by the conventions of the medium. We can conclude, however, from the fact that telephone booths for offices never caught on, that this particular physical supplementation to the ideological framing of telephony was not necessary to the proper functioning of the medium. On the other hand, the prevalence of telephone booths on streets, in airports, in shopping malls, and in other public places suggests that this construct found other, more fertile niches.[42]

In the phone booth image and the ones that will follow, there is a great deal of attention to (to borrow Reiser's phrase) the *isolation in a world of sounds* so central to the functioning of sound technology. Audile technique requires the sonic equivalent of private property. This suggests that the diffusion of audile technique is also the dissemination of a specific kind of bourgeois sensibility *about* hearing and acoustic space over the course of one hundred years. It is no accident that, at each stage of this history, it is an emergent crest of the middle class where one finds these ideas about hearing: doctors seeking middle-class legitimacy in the early and mid-nineteenth century, telegraphers seeking middle-class legitimacy in the mid-nineteenth century, and, finally, the growing consumerist middle class at the turn of the twentieth century and into the 1920s, a group of people learning to believe in connections between consumption and individuation. As a bourgeois form of listening, audile technique was rooted in a practice of individuation: listeners could *own* their own acoustic spaces through owning the material component of a technique of producing that auditory space—the "medium" that stands in for a whole set of framed practices.[43] The space of the auditory field became a form of private property, a space for the individual to inhabit alone.

This is not a universal way of portraying listening or even privacy. For instance, in *The Sight of Sound,* Richard Leppert analyzes a series of seventeenth-century paintings that represent various forms of collective listening. Obviously, paintings of artistocrats and peasants differ in important ways from drawings and engravings of middle-class people in the late nineteenth century. But there are several differences especially relevant for our purposes: the seventeenth-century paintings emphasize place and landscape—they give a sense of *where* they are located. Sonorial activity was presented outdoors, suggesting the expansiveness of sound and, along with distinctions of class and gender, distinctions of species, genus, and kingdom. For instance, Leppert notes that Jan van Kessel the Elder's *Bird Concert* carefully separated socially significant sound like music from the chirping of birds, submitting the latter to the former, thereby affirming humans' continued dominion over avian species.[44]

As Leppert argues later in the book, it is only in the nineteenth century that an obsession with privacy and domesticity emerges. By this time, the concern is no longer keeping the birds in their proper place on the great chain of being but negotiating social difference among individuals. Painters moved from negotiating differences across species or class to negotiating the spaces between individuals. This coincides with the rise of the bourgeoisie and the codification of art music. Even public spaces become more and more private. Where opera and concert audiences had been noisy and unruly, quieting down only for their favorite passages, they gradually became silent—individually contemplating the music that they had enshrined as autonomous art. We can see a similar trend with the gradual silencing of later audiences for vaudeville and film: as a form of expression becomes more legitimate and more prestigious, its audience quiets down.[45] This quieting has the effect of atomizing an audience into individual listeners. As we are told today every time we go to the movie theater, in "observing silence" we respect other people's "right" to enjoy the film without being bothered by noisy fellow audience members. The premise behind the custom is that, in movie theaters (and a variety of other places), people are entitled to their private acoustic space and that others are not entitled to violate it.[46]

By the 1920s, the possibilities for collective listening to sound-reproduction technologies presumed a prior individuation and segmentation of acoustic space. Acoustic space could be individuated through any number of techniques, all creating an acoustic inside and outside, all shaping the auditory field and emphasizing the detail over the whole. The whole process of technicization operated according to a logic somewhat analogous to a bastardized version of "social contract" mythology: economic, social, and cultural forces produce property-owning individuals who then perceive themselves as voluntarily entering into a collective and later participating in a "general will."[47] The iconography of collective listening embodied this kind of reasoning: the individuated listener comes before the collective sonic experience.

Collective versions of technical listening were designed to allow many people to hear the same thing at once while still putting the sound directly in their ears. In other words, the technicized, individuated auditory field could be experienced *collectively*. The instructional stethoscope was the first technology developed on this principle. It allowed several students to hear the same sounds at once: it attached a single chest piece to many listening tubes. The first such model was designed in 1841, and instructional stethoscopes were in use throughout the nineteenth century.[48] Instructional stethoscopes were doubly useful to medical pedagogy: they modeled not only the character of the sound to be heard, but also the proper techniques of listening. The instructional stethoscope facilitated listening in a collective yet individuated manner. It is an interesting twist on Reiser's discussion of the physician "isolat[ing] himself in a world of sounds":[49] here, the isolation is *collective;* each student would hear the same things as all the others while still within an enclosed sonic field.

By the early twentieth century, instructional stethoscopes were replaced by electrical stethoscopes that would make the sounds of a patient's body

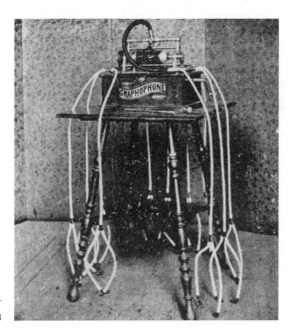

18

audible throughout a room and sound recordings of amplified heartbeats.[50] But this notion of collective yet individuated listening persisted elsewhere—both in sound recording and, later, in radio. "Hearing tubes" were a common alternative to horns on early phonographs (figure 18). They provided a way of increasing the volume of relatively quiet mechanical instruments and also a means of private listening. This mode of listening caught on in private homes, but it was essential for the first context in which the phonograph industry turned a profit: phonograph parlors that used coin-in-the-slot machines. As I discuss in the next chapter, phonograph parlors or arcades were a place where commuters (perhaps awaiting a train or a trolley) could stop in for a short time, drop a coin in the slot of a phonograph, and listen to a short tune or sketch. Limiting the sound to one listener at a time helped increase onlookers' curiosity and maximize sales. The phonograph parlor also handily demonstrates the connection between the construction of a private auditory space and the commodification of sound itself.

Put simply, acoustic space modeled on the form of private property allows for the commodification of sound. There needs to be a form of private

Figure 18. Hearing tubes in Edison catalog, ca. 1902 (courtesy Archives Center, National Museum of American History)

162   THE AUDIBLE PAST

property before there can be a commodity form—people must be able to own something before it can be bought and sold. Hearing tubes and audile technique construct an individuated, localized sound space, allowing the experience to be sold to a single individual. The patron at the phonograph parlor paid for a certain amount of time in a certain kind of acoustic space. Of course, the practice of sharing ear tubes was quite common in late-nineteenth-century phonograph parlors—each person could put one tube in one ear. But even this sharing was predicated on a prior individualization. Like any commodity, two or more people could pitch in, purchase it, and share its use. Decades later, an acoustics firm selling its talents to film theater designers would summarize this emergent mind-set: "Sound is like other commodities in that there are different qualities—hence different values. Sound for sales purposes . . . costs money to produce. It must be sold at a profit. If the sound merchant does not know how to measure and weigh it, he is out of luck and his profit and loss figures will show up red."[51] The physical, practical, and metaphoric privatization of acoustic space and auditory experience allowed for sound—the thing in that space—to become a commodity.[52]

Private acoustic space was, thus, a centrally important theme in early representations of sound-reproduction technologies. An advertisement for the Berliner gramophone from the late 1890s basically updates the imagery and practice of the instructional stethoscope (figure 19). Pictured in a domestic setting, the gramophone is described as an entertainer in the Victorian middle-class household: "Sings every song with expression, plays the piano. . . ." But the pictures present it as a surrogate for the entertainer: the top photograph depicts a man attending to the reactions of the listeners; the bottom photograph has a woman attending to the machine. In both pictures, the listeners are at attention: their hands at their ears, their faces turned down in concentration (or, in one case, up in delight). These audiences are immersed—"alone together," to use William Kenney's phrase—in a world of sound.[53] The message is one of mediation: listeners isolate themselves in order to have a collective experience *through* the gramophone.

In his history of recorded music, Kenney writes that listening to sound recording was both an individualized and a collective experience. He conceives of it as "large numbers of individuals around the country and indeed the world, 'alone together,' actively using their phonographs to replay as they wished commercially mediated musical messages." Phonographs, "far from promoting only 'ceremonies of the solitary,' paradoxically encouraged widely shared patterns of popular behavior, thought, emotion, and

A Talking Machine for the Family at so low a price that it is brought within the purchasing power of everybody, is one of the latest achievements of scientific invention.

The Berliner Gram-o-phone

Talks distinctly, sings every song with expression, plays the piano, cornet, banjo, and in fact every musical instrument with precision and pleasing effect. The plate called "The Morning on the Farm" gives a perfect reproduction of the lowing of cattle, crowing of the rooster, the call of the hawk, the neigh of the horses, the bleating of the sheep, and in fact every sound which is familiar to the farmyard. The records are endless in variety, including nearly every song you are acquainted with.

Accompanying illustration (above) shows the machine operated with hearing tubes for three people. Tubes for **two** people go with each machine. Extra hearing tubes, so that any number of people may hear, are furnished at 75 cents extra for each person. Two records are included with every machine. Extra records 60 cents each, $6.00 per dozen.

ALL FOR $10.00.

OUTFIT. The Outfit includes Talking Machine, Style 7½, provided with revolving table covered with felt, fly-wheel so balanced as to turn evenly and arm which holds the sound box with reproducing diaphragm, rubber tubes as described above. Box of 100 needles. All nicely packed in a box and sent express prepaid to any point in the United States upon receipt of price.

Send Money by Postal Note, Express Money-Order or New York Draft.
SPECIAL OFFER. With each Machine ordered before Nov. 10th, we will include an *Amplifying Horn.*

FOR SALE by all MUSIC DEALERS.   Send for Catalogue. Free of Course . .

NATIONAL GRAMOPHONE COMPANY, 874 Broadway, NEW YORK CITY.

Figure 19. Advertisement for the Berliner gramophone (courtesy Archives Center, National Museum of American History)

Figure 20.

sensibility."[54] Kenney is clearly referring to a geographically and temporally dispersed audience within the United States—individual people who listened to the same record in New York, Los Angeles, Duluth, Urbana, and San Antonio and thereby entered into an imagined community of shared musical experience. But his suggestive locution is also applicable to the space *around* the sound-recording devices in the home. With their ear tubes and postures, the listeners in the Berliner ad are also listening "alone together" to recorded music. Their shared auditory experience is based on a prior segmentation of sound space into auditory private property.

The gramophone may have been a picture of domestic sociability, but the same kind of collectivity could be organized for instructional purposes. Student operators at a Marconi Wireless School (figure 20) are organized in a cellular fashion. Each has his own partition on the desk, and each wears a headset to hear and write down signals. The experience is highly individuated, standardized, yet also collectivized. These Marconi operators are also listening "alone together." One assumes that they all hear the same thing, but even that is not necessary for the experience to be collective since they are all inhabiting their private acoustic spaces in their headphones. The

Figure 20. Students at the Marconi Wireless School (courtesy Archives Center, National Museum of American History)

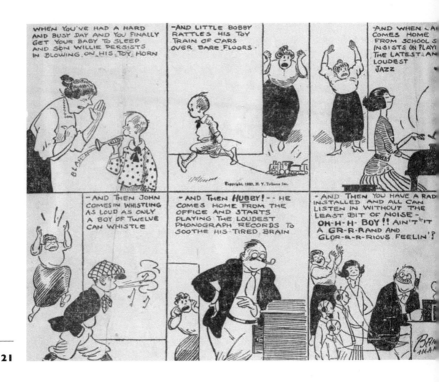

21

framing of the shot focuses on listening to the apparatus: operators turn their faces down, contemplating the signals that they hear and writing them down. The photograph emphasizes the role of training in audition—from the seriousness of the students, to the explanations of code on the board, to the teacher watching over, listening is clearly a skill to be nurtured and developed.

The collectivized isolation of listeners could also become a proper strategy for containing noise and bringing acoustic order to a chaotic milieu. A 1923 cartoon (figure 21) dramatically represents the problem of the containment and organization of noise in a domestic setting: a housewife exhausted by the noise of the day's household activity finally has peace and quiet when the rest of the family put on their headsets and plug into the radio. This cartoon offers a vision of radio as a pacifier similar to the account of radio offered by social critics of the time: the characters in the cartoon make all sorts of noise until they are quieted—alone together—by the ra-

---

Figure 21. Cartoon of frustrated housewife, 1923 (courtesy Archives Center, National Museum of American History)

dio set in the living room. The crowd becomes the mass right before our eyes in the Sunday paper.

This image is also interesting because it depicts a form of family togetherness and communal listening that begins from cellular acoustic space. As with the gramophone advertisement, people listen together by using headphones together. But there is a question as to whether the headphones are even necessary in the image here. Within a few years, images of family togetherness around the radio would use loudspeakers instead of headphones. But audile technique would remain. Even in this and other collectivized settings, technique could be the governing mode of listening. As James Johnson said of concert audiences, it was technique that was most likely to "bridge the inner experiences of a fragmented public."[55] Even Johnson's locution evokes the prior division of acoustic space into private property: since inner experience is fundamentally private and, therefore, in need of bridging. A history of group listening to phonographs or radios is obviously a step beyond the history offered in this chapter, but I would predict that, even in these moments of collectivity and togetherness, people's practical techniques of listening involved a certain prior individuation of acoustic space. They entered this audile collective like the mythical individuals who would enter into a social contract: first free and separate and then together. Private acoustic property and the commodity form of sound emerged together.

In the images considered thus far, listening is emphasized through its iconographic relation to looking. Sensory separation and framing of the medium are represented visually through representation of physical posture and downward or otherwise averted or undirected gaze. In each case, the gaze (as it is pictured) is averted, elided, in an effort to represent hearing. Yet even this hearing is a subject of gazing—the viewer is implied, not only by the positioning of people's bodies, but also within the frame of the stethoscope, the gramophone, the Marconi school, and the cartoon. Each picture presents a spectator within the frame that could easily be read as an analogue of the spectator outside the frame: in the pedagogical situations, a competent, socially superior observer ensures the solemnity of the situation and guides the ears of those in training. In the domestic scenes, it is a question of attending (in both senses of the word) to others' pleasures through maintaining the operation of the machine and at the same time observing and enjoying others' enjoyment. The collective activities in these events are possible only after the listeners have been individuated— their separation effected through bodily disposition, the mix of prohibition and exhortation enacted through social convention and the ideology of the universal bourgeois individual.

The collectivity represented in this mode is more of an interconnection than anything else—listeners are linked with the network and, through the network, can reach one another. This was Kenney's original point: standardized, commodified music allowed people separated by expanses of time and space to hear the same thing. The same principle works for telephony. The N. W. Ayer Agency, which had a major advertising contract with AT&T through the 1910s and 1920s, cast telephony in terms of intensification of the expressive and perceptive capacities in hearing. Two advertisements from the agency's vast output can illustrate this tendency (figures 22–23). Telephony not only increased personal agency as a kind of fixed capital— "the multiplication of power"—but also provided a kind of audile immediacy at a distance previously reserved for the telegraph. The sounds heard on the telephone corresponded to sounds possibly thousands of miles away; proximal sound was linked to distant activity: "Each answer is made *instantaneous* by the Bell telephone service" (figure 22). Even in advertisements for the telephone, however, the medium quickly disappears:

A generation ago, the horizon of speech was very limited. When your grandfather was a young man, his voice could be heard on a still day for perhaps a mile. Even though he used a speaking trumpet, he could not be heard nearly so far as he could be seen.

Today all this has been changed. The telephone has vastly extended the horizon of speech.

Talking two thousand miles is an everyday occurrence, while in order to see this distance, you would need to mount your telescope on a platform approximately 560 miles high.[56]

This ad copy renders the telephone as a necessary supplement for the deficiencies of hearing or, rather, its "limited" horizons. Laennec would be proud. As with telegraphy, for a moment ads for the telephone could claim it as the marvel of the modern age, and the ability to *hear* in new ways was a hallmark of progress and modernity. A variation on this theme can be found in Ayer's "The Man in the Multitude" (figure 23). Here, a man with a telephone sits in a chair above and to the side of a panoramic view of a great mass of people. He looks down and away from the crowd, like the earlier telegraph operator, his legs are crossed—he is comfortable, at ease, and in command. His hand is to his ear, and his eyes are free to do as they

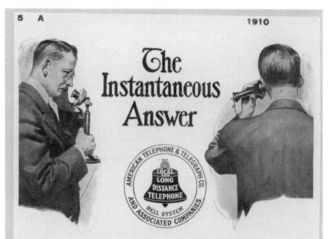

Figure 22. "The Instantaneous Answer"—an N. W. Ayer advertisement for AT&T's phone service, 1910 (courtesy Archives Center, National Museum of American History)

The Man in the Multitude

That the human voice may be transmitted across our continent by telephone is the marvel of this age of wonders. Yet the full significance of the achievement is not realized if it is considered strictly as a coast-to-coast connection.

The Transcontinental Line not only bridges the country from east to west, but, by having finally overcome the great barrier of distance, it has removed the last limitation of telephone communication between all the people of the nation.

This means that the voice can be sent not only from New York to San Francisco, but from *anywhere* to *anywhere*—even from *any one* to *any one*—in the United States.

Wherever you are, it is possible to reach any one of our hundred million population. You can single out from this vast throng any particular individual with whom you desire to speak.

To bring this about, the Bell System has spent years and millions, extending its lines everywhere, anticipating the ultimate triumph. It has had the foresight and the courage to unite this great country, community by community, into one telephone neighborhood.

With success achieved by the Transcontinental Line, the established Bell highways make you, wherever you are, the near neighbor of your farthest-away fellow citizen.

AMERICAN TELEPHONE AND TELEGRAPH COMPANY
AND ASSOCIATED COMPANIES

One Policy   *One System*   *Universal Service*

Figure 23. "The Man in the Multitude"—another N. W. Ayer advertisement for AT&T's phone service, 1915 (courtesy Archives Center, National Museum of American History)

24

please. In the distance, the picture shows another man with a telephone; a telescopic line connects the two: "You can single out from this vast throng any particular individual with whom you desire to speak." Ayer is selling the directionality of audile technique and sound technology as a form of agency. What's good for Cammann and his binaural stethoscope is good for AT&T—listeners can better direct, orient, and focus the auditory field through its supplementation. Of course, this kind of agency, supplementation, and immediacy was phrased in terms of personal business success. But the logic behind it goes beyond advertising aimed at businessmen. Ultimately, sound-reproduction technologies were portrayed as providing intimacy (erotic, familial, personal) as well as immediacy.

"Her voice alluring draws him on" is the caption for the cover of the *Telephone Review,* published by the New York Telephone Company (figure 24). Spilling out of the frame are three young women, the one in the center with a telephone in her hands, her eyes looking directly at the

Figure 24. "Her Voice Alluring Draws Him On"—cover of the July 1913 *Telephone Review* (New York) (courtesy Archives Center, National Museum of American History)

observer—all three are cast in vaguely suggestive poses. In the background is a stormy sea, and in the upper-right-hand corner of the picture reside three sirens perched on an island. This image of telephonic erotics is no doubt still cast with male eyes in mind, but its implication is quite different—here the possibility of telephonic audition is itself rendered as a kind of excess, both sexual and sensual in nature—an intensification, supplementation, and possible supersession of the immediacy of unaided hearing. Again, this image points to a reconstruction of acoustic space: in the space of telephony, the operator's eroticized voice draws in male ears as it comes down the line. Meanwhile, the sirens in the distance look on, silently. Once able to induce shipwreck through merely grazing the ears of male sailors with their voices, they have been overcome by the controlled erotics of modern sound technology. The unruly eros of antiquity, where the male subject could lose control at any moment, has been replaced by a regime of segmentation, distance, self-control, and modulated desire.

The New York Telephone Company anticipates Norbert Elias's riff on Freud and Weber: repression and mediation give us modern life. "Restraint of the emotions" wrote Elias, became especially important as "outward differences of rank had been party levelled." As gradations of social difference became ever finer and more contingent, physical distance and decorum, which had previously been issues only among social equals, became a more general concern. Technologies of sound reproduction did not cause this transformation, but they did crystallize it—"once, in conjunction with a general transformation of human relations, a reshaping of human needs was set in motion, the development of a technical apparatus corresponding to the changed standard consolidated the changed habits to an extraordinary degree. This apparatus served both the constant reproduction of the standard and its dissemination."[57] The New York Telephone Company's illustration thus indexes an entire middle-class habitus of listening.

This notion of direct interconnection extended to other, less exoticized, although equally fantastic, scenarios. A 1923 cartoon from the *Syracuse Telegram* (figure 25) pictures a female-male couple wearing headphones plugged into two heart-shaped radios, communicating with one another through the ether. In the distance, a broadcasting station called the "Permanent Love Wave Co." is bordered by a sign warning, "Philanderers keep off." The cartoon thus acknowledges the anxiety around radio communication even as it paints a romantic image of modern love: both the woman and the man hold their hands to their ears, their eyes averted from any kind of direct gaze, their bodies poised in anticipation but still at ease with the

Figure 25. "Tuning In"—cartoon from the 6 September 1923 *Syracuse Telegram* (courtesy Archives Center, National Museum of American History)

medium. The message was clear: the ether was a fertile medium in which to culture the intensity of romantic love, and this intensity could be felt by listening alone together. Even the most intimate social-sonic relations could presuppose the segmentation of acoustic space into private space prior to its collectivization. The warning sign is particularly interesting: it is clearly of a piece with the stories about telegraphic "crimes of confidence" discussed above, but, in the context of the picture, it also carries another meaning. It suggests that the ethereal love enabled by radio is a spiritual connection: those who would confuse this deeper intimacy with physical attraction between human bodies should stay away. Once again, the distance between bodies is essential to the conception of private acoustic space and audile technique.

By the early 1920s, a visual vocabulary of auditory immediacy had been established. Headphones could appear in almost any situation, as much a symbol of connection to a common commodity culture and of that culture's integration into both domestic and public life as of anything specific about listening (figures 26–27). Families, children, adults, and pets all appeared in the ubiquitous headphones. Cartoonists, meanwhile, lampooned these conventions, either by mocking onlookers' impressions of technicized listening (figure 28) or by making a joke of the "isolation in a world of sounds" that other media representations cast as both necessary and desirable. These images might be attributed to headphones' status as an easily recognizable trapping of radio technology in general, a status that made them useful to the press depicting a "radio boom." But the *meaning* of the headset was not simply as an index of radio. It was an extension, modification, and refinement of one hundred years' worth of techniques of listening. The private experience of radio listening was certainly different from the private experience of the physician listening to a patient's body, but the two practices bear a striking morphological similarity to one another. Audile technique organized, framed, and conditioned both those experiences.

Over the span of a century, technicized listening moved from an esoteric medical technique that required extended training and explanation to a common motif in the "radio boom" of the early 1920s. Beginning at the margins of middle-class culture, audile technique first came to be a distinguishing feature of medicine and telegraphy. Rather than "revolutionizing" hearing, sound-reproduction technologies would expand on and further disseminate constructs of listening based on audile technique. Audile technique moved from Laennec's *Treatise of Diseases of the Chest* to advertisements for telephony, sound recording, and radio. What took hundreds of pages to explain in the 1820s could by the 1920s be accomplished within the space of a single page or even the single frame of a comic strip.

Claude Fischer, Susan Douglas, and William Kenney have all persuasively argued that technologies of sound reproduction embodied a diverse range of practices and engendered a diverse range of responses. As Kenney writes, accounts of the diversity of public attitudes toward and responses to a technology represent "an important corrective to overly simplified, unitary interpretations of 'the' influence" of a technology.[58] It would be too much to claim that audile technique was the single modality through which sound-reproduction technologies were interpreted or used. At the same time, the history of audile technique connects sound-reproduction technologies with the longer history of listening in modernity. The salient features of audile technique considered here—the connection of listening

Figure 26. "Ecstatic Interference"—artist's drawing, 1922 (courtesy Archives Center, National Museum of American History)

Figure 27. Victorian woman with headset and radio (courtesy Archives Center, National Museum of American History)

Figure 28. Cartoons from the April 1923 *Wireless Age* (courtesy Archives Center, National Museum of American History)

and rationality; the separation of the senses; the segmentation of acoustic space; the construction of sound as a carrier of meaning in itself; and the emphasis on physical, social, and epistemological mediation—are all fundamental to the ways in which people listened to and with sound-reproduction technologies, and all these practices of listening have a long history over the course of the nineteenth century.

For all this, techniques of listening do not simply turn sound technologies into sound media. As Kenney points out, it takes both a shared cultural sensibility and a standardized, industrialized record business to get the same recording to the different people in different places so that they could listen alone together. As the Ayer Agency pointed out on AT&T's behalf, telephony derived its power and significance from its status as an industrial and technical network, linking together distant people and places so that they could listen alone together. For sound technologies to become sound media, they would have to be articulated together in networks through the organization of new media industries and new middle-class practices.

4 Plastic Aurality: Technologies into Media

[*Trill.*] There are more things in heaven and earth Horatio, than are dreamed of in our philosophy. [*Trill.*] I am a Graphophone and my mother was a Phonograph.
—**MESSAGE RECORDED ON A GRAPHOPHONE** deposited at the Smithsonian Institution in a sealed container, 1881

In the summer of 1885, a relatively brief meeting took place between Charles Sumner Tainter and Edward Johnson. As Thomas Edison's business representative, Johnson sat across the table from an improved version of the Edison phonograph, more than five years after Edison's invention had faded from fame and favor to obscurity. Tainter, an employee of Alexander Graham Bell's, represented the interests of the Volta Laboratory, which had been working on improving sound-reproduction technologies since 1881. He brought with him an improved version of Thomas Edison's phonograph called the *graphophone*. The graphophone improved on Edison's original tinfoil phonograph by using a wax cylinder as the recording medium and by engraving in the surface of the cylinder rather than indenting it. Volta was seeking a collaboration with Edison for obvious reasons—Edison's name and fame as an inventor would be an invaluable marketing tool, and Edison himself had abandoned the phonograph in 1880 to pursue other, more profitable ventures in incandescent lighting. From Volta's perspective, it was reasonable to suppose that a collaboration could benefit both parties.

Johnson, however, was demure during the meeting, which he subsequently reported to Edison. According to Andre Millard, Edison was despondent on finding out that someone else had improved the phonograph—an invention to which he referred as his "favorite" and his "child."[1] Covering for his boss, Johnson sent Bell a letter of explanation claiming that he could not endorse Edison's entry into a collaboration with Volta until he was shown something "more practical than the last machine shown me." "While I am sanguine in a measure—I am not sufficiently so to stake my reputation for safety as a guide in such matters on the issue."[2] Yet Johnson's optimism was Edison's research plan. Although the Edison/Volta collaboration never materialized, the new Edison phonograph did a few months later. Like Tainter's graphophone, it used a wax cylinder and recorded by engraving instead of indenting.

Male Birth and Baby Machines

While Edison's feelings in this matter (as reported by Millard and others) appear simply to add color to an otherwise standard scenario of nineteenth-century-style entrepreneurial competition, deeper analysis connects them with a whole way of thinking about new technologies. Edison was not alone in feeling that inventions were like children to men: "Mr. Morse claims to be the inventor of the first marking Telegraph. He states in a letter dated December, 1852, published in the American Telegraph Magazine, that in November, 1835, he made a telegraph in the New York City University, and he contends that at the time 'the child was born, and breathed and spoke.'"[3] The metaphor of birth has a strangely persistent presence in the history of technology—a history of condensations, connections, collaborations, and recombinations. Yet the analogue of technological creation in this metaphor would have to be asexual reproduction. Here are the inventors themselves—almost all men—declaring that they have finally, proudly, given birth. Alexander Graham Bell wrote the following letter to his father after finally arriving at a working model of the photophone, a machine that transmitted sound through the use of light. The language is so outrageous and yet so banal that the sources almost speak for themselves:

Dear Papa,

I have just written to Mamma about Mabel's baby and I now write you about my own! Only think!—Two babies in one week! The first born at 904 14th Street . . . , the other at my laboratory on the nineteenth.

Both strong, vigorous healthy young things and both destined I trust to grow into something great in the future.

Mabel's baby was light enough at birth but mine was LIGHT ITSELF! Mabel's baby screamed inarticulately but mine spoke with distinct enunciation from the first.

I have heard articulate speech produced by sunlight! I have heard a ray of the sun laugh and cough and sing! The dream of the past year has become a reality—the *"photophone"* is an accomplished fact.[4]

Clearly, the metaphor of birth is about ownership and authorship of an artifact; one cannot claim to be the author of a work that entered the world incomplete.[5] Yet Bell's emphatic letter to his father goes beyond simply metaphorizing the logic of patent rights. His investment in the machine as his child is made clear through the devaluation of his real child.[6] "Mabel's baby screamed inarticulately but mine spoke with distinct enunciation from the first"; the machine appears in the world finished, fully functional, working perfectly. Bell seems aware that the birth metaphor itself does not quite work since human babies depend on their parents for some time after birth. But birth is a convenient metaphor because it appears to fix origins in absolute terms. Existence becomes a binary operation; the only moment between the final formation of a machine and its nonexistence is the inventor's giving birth. The metaphor is less than a step shy of technological determinism's conventional wisdom: thus born into the world, a new generation of machines shall inherit the world and make it their own.

Although the male-birth model of technological history makes for good news copy and better patent applications, historians of technologies and media have long since moved away from it. The odd figure of male birth allows for an historical sleight of hand. In addition to fitting well with the vanities of patent law then and now, it allows for a kind of naturalization, a mystification of technology and the institutions and practices that surround it. Once machines come into the world through singular moments of birth, they can have "impacts." They take on a little bit of humanity by becoming autonomous agents coming from outside the world of human activity to affect it, even as the birth metaphor deprives them of their greater humanity as products of collective human endeavor. The critical task, as always, is to restore to these machines their greater humanity, their entanglement with the human even at their most mechanical moments.[7] This chapter charts some of the larger institutional and cultural contexts through which sound technologies became media.

Each of the preceding chapters has considered a certain "plasticity" of sound—the malleability of sound itself and the malleability of practices of hearing and acoustic space. This chapter goes further to argue that the distinguishing characteristic of sound media is again this plasticity or malleability—this time on a larger social scale. As with the form of sound and the function of hearing, modernity marks a new level of plasticity in the social organization, formation, and movement of sound. These moments of plasticity, where the social organization of sound can and does change, are perhaps the defining characteristic of the modern sound media. Almost every major social or cultural history of telephony, sound recording, and radio attempts to account for the form that these media would come to take. When we view media history as a domain within social and cultural history, this makes good sense. For a history of sound, however, it is precisely the moments prior to this crystallization that are most interesting—it is the mutability as opposed to the eventuality of form that is at stake.

I am therefore less concerned with the particular forms that sound media would eventually come to take than with the malleability of form itself. A medium is a recurring set of contingent social relations and social practices, and contingency is the key here.[8] As the larger fields of economic and cultural relations around a technology or technique extend, repeat, and mutate, they become recognizable to users as a medium. A medium is therefore the social basis that allows a set of technologies to stand out as a unified thing with clearly defined functions. Elsewhere, I have argued that this aspect of media history—where technologies, institutions, and practices come together in a recognizable and repeatable form—is well described in Georg Lukács's notion of reification. To use Lukács's language, social relations take on a "phantom objectivity"; over time, they become associated with technology itself in the minds and practices of users.[9] This is readily apparent today, to offer an oversimplified illustration: casual users associate sound recording with music and entertainment, radio with broadcasting, and telephony with point-to-point communication. We know, for instance, to call the various kinds of wireless telephones (cellular, PCS, etc.) *phones* instead of *radios* because they are associated with the institutions and practices of the phone system, despite the fact that they are themselves wireless transmitters (which would theoretically, at least, make them radios). *Phone* is really a linguistic shorthand for a whole set of related institutions, technologies, people, and practices that are conveniently (and perhaps necessarily) forgotten when we place our calls.

As is already well documented by media historians, connections among function, practice, institution, and network—call them point to point, broadcast, and archival—were not clearly assigned to a single technology over another. Early users did not necessarily differentiate between telephony, sound recording, and radio in the way in which we are disposed to do so today. Sound technologies had to be differentiated from one another and connected with differing social practices and contexts to become media. This is a story of *articulation,* the process by which different phenomena with no necessary relation to one another are made into a social unity: "The concept of articulation provides a useful starting point for describing the process of forging connections between practices and effects, as well as of enabling practices to have different, often unpredicted effects. Articulation is the production of identity on top of difference, of unities out of fragments, of structures across practices."[10] Each machine embodied a whole set of articulations; in turn, it was articulated to larger economic, technical, and social functions and relations among many other possible and actual uses. In other words, radio had to become broadcasting; telephony had to become a network of point-to-point connections; phonography had to become an archival medium. While, at the mechanical level, the machines apparently had the potential for each of these functions, their social and cultural history explains the terms of their use.

The features of modern sound culture explored in this book—plasticity, contingency, objecthood, supplementation—make sense only in the larger contexts of industrial capitalism, middle-class culture, enlightenment science, and colonialism from which they emerged. They develop in the context of a large industrial-capitalist society, where people are used to the presence of many complex machines in their everyday lives; where the commodity form and the cult of reason and technology combine to make abstraction a common mode of thought and experience; where work and leisure, production, distribution, and consumption, are separated nominally (if not always clearly or effectively); where large networks of transportation and communication congeal into an infrastructure for the interconnection of distant experience; and where the work and resources of the many are appropriated through a system of unequal exchange to benefit the few.

Sound technologies emerged from a changing context of research, innovation, and development; they grew in the spaces of a transforming middle class, and they were nourished with surplus materials and labor generated by industrial capitalism. This chapter examines the immediate social field within which the telephone, phonograph, radio, and other related sound-reproduction technologies were invented and then considers their development in the United States in the context of the changing middle class at the turn of the twentieth century. Had the conditions of their emergence been different, the shape of the sound media would no doubt have been different. Sound reproduction emerged alongside magazine reading, photography, motion pictures, and a host of other new mass-cultural practices, but it is, I argue, a distinctive field of practices, related technologically, practically, and institutionally. My point is not to catalog all the concurrent developments in order to establish a hierarchy of importance, but rather to establish the conditions under which a given set of practices and possibilities has emerged.

Yet this is not simply a narrative of the capitalization or commodification of sound. *That* story would require a closer and more constant attention to economics, resulting in an examination of copyright and intellectual property, the connections between the commodification of musical and spoken performance and the later buying and selling of actual sounds, the politics of urban industrial noise, and the whole spectrum of class relations, positions, and experiences as they shape the practices of listening, speaking, sounding, and music making.[11]

Instead, this chapter has much more modest goals: it argues that the various sound media should be understood as institutionally and economically related (since chapters 2 and 3 establish some ways in which sound-reproduction technologies are related to one another in form and practice). Although these facts are well documented in the existing media histories, telephony, sound recording, and radio are still largely treated as separate social and cultural phenomena. No doubt, this is because historical questions are formed and interests driven by the current state of affairs (i.e., given the present state of sound recording or telephony, how did we get where we are today?). Yet all these media obviously had technical and practical similarities. This chapter explores their institutional and economic connections to argue that the sound media emerged from a shared cultural and industrial context. They descend from a common cultural origin, as parts of an initially cohesive social and cultural field.

The plasticity of media forms, the possibilities for connecting telephony to broadcasting and sound recording to point-to-point communication, emerge from this field of connected industrial, technical, and cultural phenomena. After establishing the research and innovation field from which the various sound-media technologies and industries emerged, this chapter turns to the plasticity enabled by the changing contours of middle-class life. Beyond the sedimented media forms with which we are familiar today, the very possibilities for articulation were shaped by the conjunctures of modern sound culture, industrial capitalism, and the emergence of a new middle class.

Sound Becomes an Industrial Problem: The Research and Innovation Field

The technology to reproduce sound was invented at a time when the status of invention and development was itself changing. Western Union had become the first major and, according to some writers, the first modern corporation in the United States. The telegraph was also the first of the electronic media, and, although not exactly a "mass" medium, like the newspaper it was woven into American everyday life and imagination in countless ways. The concerns of Western Union drove the vast majority of the research in communication technology at the time.[12] Telegraphy was, thus, one of the major contexts for the development of sound-reproduction media,[13] and the telegraph industry provided both an income and a research program for the people who would go on to invest in the new sound technologies.

When considered from the perspective of invention and initial development, sound-reproduction technologies emerged from a relatively small, elite group of people, many of whom were in contact with one another. That is to say, the research and development of sound-reproduction technologies emerged from a fairly coherent and consistent social field. Consider two of the most well-known figures in this history: Alexander Graham Bell and Thomas Edison. Bell and Edison both came to experiment in sound reproduction through their work on telegraphy. Both at one time or another worked on the harmonic telegraph—a means to send multiple telegraphic messages over a single line by varying the pitch of the telegraphic tone. Both had read widely in the European literature on sound and sound reproduction. Both were familiar with the work of Helmholtz, Reis, König, Scott, Henry, Tyndall, and other key writers. While they were of different backgrounds, both benefited from the relative lack of institutionalized engineering and invention, neither having formal training in electricity or invention more generally.[14] Both read and published in the popular scientific and technical journals of the time, like *Scientific American* and the *Electrical Review*. Both popularized their discoveries through well-established scientific and technical (and, later, popular) lecture circuits as well as through newspapers and magazines.[15]

The economics of their operations bears some similarity and interconnection as well, although Edison was clearly the more ambitious capitalist of the two and Bell and his benefactors were out of the telephone business by the early 1880s. Both started up financing their own experimentation with extra money from work that they were doing for others. Both, when given the opportunity, established a laboratory for carrying out their work and hired assistants. During the period of development, invention moved more and more from an "artisanal" to an "industrial" mode of production, and both men worked toward rationalizing the invention process. Edison and Bell both hired many assistants and eventually set up multiple research labs and whole research complexes—what was *invention* in the 1870s became *research and development* by the 1920s. Edison used the profits from his patents on the quadruplex telegraph to found his laboratory in Menlo Park in 1876. Widely regarded as the first independent research and development laboratory in the United States, Menlo Park was founded to "produce a stream of useful innovations, practical innovations that would meet a well defined need in commerce or industry."[16] In other words, the purpose of Menlo Park was to develop technologies that would generate more capital (through their sale to industry) to develop even more technologies. Edison the person—but, more important, Edison the entire company—helped transform invention and innovation into an industry in its own right.[17]

Although Edison is generally better known for developing a research and development operation, Bell also lent his name at least as much as his expertise to a laboratory. When awarded the Volta Prize (which included $20,000 in cash) by a French panel in 1880, he used the capital to set up the Volta Laboratory in Washington, D.C., to pursue his various research projects. Although Volta was less commercially successful and is not as widely noted in histories, it demonstrates a morphological similarity between Edison's and Bell's approach to invention. The Volta Laboratory provided facilities and income for two assistants as well as funds to hire machinists to build apparatus and to provide occasional facilities for other guests. Bell's two main employees were Charles Sumner Tainter, who essentially was Thomas Watson's replacement, and Chichester Alexander Bell, a favorite cousin of Alexander Graham's. The lab also employed a number of craftsmen and maintained a workshop separate from the laboratory because of the "need for secrecy," all the involved parties being well aware of possibilities for industrial espionage and patent challenges.[18] Although Bell was present for some of the experiments, most notably his work with Tainter on the photophone, much of the work on sound recording was carried out in his absence. Bell, like Edison, became as much the manager of a research and development apparatus as he was a researcher himself. A letter from his frustrated cousin shows the extent to which Bell was an absent manager:

> I called at your house this evening to see you but found you out.
>
> When Tainter and I went up to Manchester (with the usual result when you have been asked to take the least trouble about our affairs) it was arranged that if the association were allowed to continue until May next, with certain modifications in the understanding between us, you would on your return here in October, go over our inventions, decide on what patents were to be taken out, and have specifications drawn up. As you said yourself, that was to be *the* work for the winter.
>
> The end of the year has now come and you have neither looked into our affairs nor shown the slightest inclination to do so.
>
> This is the last appeal I shall make to you to carry out your agreement. As you have broken faith with us, it is quite open to me to declare the Association ended at once. And as everything connected with it has become distasteful to me, I shall not hesitate to do so.[19]

Though Bell evidently responded to his frustrated cousin's letter and the association was not dissolved by Chichester or anyone else, the letter indicates very clearly the extent to which Bell expected the lab to carry on its work without him—to the point where his employees had to travel in order to seek him out and get him to do his part. The lab's work and day-to-day operations were thus shaped as much by Bell's absence as by his involvement.

The lab was also as much a commercial as a scientific endeavor. Bell wrote the following summary of discussions from the spring of 1881:

> We fully decided before the publication of my paper on the photophone to postpone purely scientific work upon Radiophony or any other subject and to devote our time to something that would pay. My intention being to put my share of profits into the Laboratory itself so that I might have a self-supporting laboratory.
>
> Upon looking over the ground, Dr. C. A. Bell, Mr. Tainter and I decided that the most promising field for *joint* work—would be to perfect the "Phonograph" or "Graphophone" or whatever we may decide to call it. To perfect a means of reproduction sound from a record. Dr. C.A.B. has also some special chemical work which promises preliminary results.[20]

The notes, which essentially amount to a diary entry for Bell, demonstrate that Bell's thinking was not all that far from Edison's. Although he still professes a scientific interest in invention, this is mitigated by his desire to produce a self-sufficient lab through developing the phonograph into a commercially viable machine. Already, he is concerned with naming issues in part because he will have to distinguish his machine from Edison's.

In addition to contributing to the industrialization of invention, both men had to seek outside financing. Not only did Bell and Edison use similar financing techniques, but the sources of their money were in some cases identical. While Bell evolved into a manager in his own right, his early experiments were financed in large part by his father-in-law, Gardiner Hubbard, a major telegraph investor. When the telephone made Bell a millionaire in the late 1870s, he used his telephone fortune to finance further experiments.[21] Edison also financed his research and development operation through the sale of stock and patent rights.[22] More important, the two men's financial resources were very closely intertwined. Three of the five major investors in the original Edison Speaking Phonograph Company were major investors in the telephone. Charles Cheever and Hillbourne Roosevelt already owned the New York City telephone rights when they got into the Edison company; Gardiner Hubbard was the third Edison board member with telephone connections. In fact, one account suggests that it was Hubbard who encouraged Bell and his colleagues to begin work on innovating the phonograph.[23] Although Hubbard departed from Bell Telephone not long after his son-in-law, his last act was to bring in Theodore Vail, who led the company into the twentieth century and helped organize AT&T's telephone monopoly. In fact, for several years, Bell Telephone and Edison Phonograph shared offices at 203 Broadway in New York City. Bell and Edison themselves corresponded about business and technical matters; long before the 1885 meeting between Tainter and Johnson, Bell was proposing a secret industrial summit—he did not want any photographs of the meeting to be published.[24]

Both Bell and Edison's names were lent to corporations in an effort to cash in on their public personae. Although Bell got out of the telephone business in the 1880s, AT&T repeatedly used his name and presence throughout his life to promote its interests. Edison retained an interest in the musical phonograph industry until 1929 and an interest in dictation machines even longer; his name and face graced a whole line of hardware and software through the 1920s.[25] As Susan Douglas has suggested, the entire period can in fact be characterized by its cultural and commercial fixation on inventors and the cult of invention; the personalities of the inventors thus became important currency for promoting the inventions.[26] In civic, commercial, and journalistic discourse, Bell and Edison came to be associated with the telephone and the phonograph, respectively, although the operations run by each (and many others) made important contributions to the invention attributed to the other.

There was a great deal of technical cross-fertilization among technologies as well. Developments in one area often were directly connected with developments in another. In some cases, a single development applied to all areas at once. Edison held a patent on a carbon transmitter that greatly increased the volume and distinctness of Bell's first telephone; Emile Berliner provided Bell with a variable-resistance transmitter, the principle of which is still in use in telephones today, and a way around the Edison patent. Bell's employment, in turn, provided Berliner with the capital to begin his researches into the gramophone. Later, Berliner assistants would benefit from their initial experience in the Volta Laboratory before going to work for Berliner.[27] Bell's collaborators at the Volta Laboratory were acutely aware of the technical interrelations among sound-reproduction devices: "There may be some difficulty in drawing the line between telephonic and non-telephonic inventions; as some telephone apparatus may be required or advantageous in our present-work."[28] Advances such as Lee De Forest's audion, or vacuum tube, made possible speech and music over radio, the electric amplification of sound recordings, and long-distance telephony.[29]

As the industries grew, these connections moved from the personal and the technical to the industrial. As sound technologies became media, and as industries grew up as parts of those media, relations among different industries and practices of sound reproduction were frequently in a state of flux. AT&T's research and development in the 1910s and 1920s led to major changes in the way in which music was recorded and reproduced, and telephone lines facilitated the creation of national radio networks in the late 1920s.[30] When George O. Squier patented the transmission of music over telephone lines in 1922, he opened up the possibilities for the hybrid of sound recording and telephone transmission that eventually became Muzak—a company that first presented itself as an alternative to jukeboxes and would later gain fame for its Taylorization of hearing in factories, elevators, and other commercial spaces.[31] Citing rising sales of radios and declining sales of phonographs in 1922, an article in *Radio Broadcast* pondered whether radio would eventually replace sound recording as the provider of music and other audio entertainment. The author's conclusion was that the two could coexist, yet he has no clear scenario for their working together, except to note that phonograph recordings are repeatable and radio broadcasts are not.[32] A decade later, national radio networks would provide a means for the recorded-music industry both to promote and to disseminate its recordings, first through live performances, later through the broadcasting of recorded music.[33]

The relation between the emerging sound media and other forms of industrialized expressive modes of sound culture, such as the sheet-music industry and live performance, was also characterized by periodic transformations. Music publishers treated live and recorded performance as a form of promotion. Publishers hounded sound-recording companies to record their songs and vaudeville acts to perform their latest releases to help stimulate sheet-music sales. Although it was difficult to get radio stations and live performers to pay royalties, this was apparently seen as a separate problem from promotion until the early 1920s, when royalties from the sales of records began to decline. The American Society of Composers, Authors, and Publishers (ASCAP) went after radio stations and by 1924 had won over executives like RCA's David Sarnoff, who said, "Radio must pay its own way."[34]

Working musicians had similar problems: until unions could fix record company wages for recorded performance, earnings for recording musicians varied widely. Both cinema and radio initially provided new and promising live performance venues. While movie theaters first boomed as a source of income and then dried up with the popularization of sound films from 1927 on, radio slowly became a source of performance revenue. Early radio operators could often not afford to pay invited musicians to perform. Often, the musicians were so enthusiastic that they would perform for free, seeing radio as a new form of promotion and potential exposure. When ASCAP went after radio stations for not paying royalties to publishers, musicians benefited as a side effect (although ASCAP was not explicitly arguing for better compensation for musicians).[35]

Vaudeville did not particularly profit from the growth of sound recording and radio performance. Although early recordings made extensive use of vaudeville acts for their content, and although early phonographs toured the vaudeville circuit for a few years as a novelty, ultimately, the new sound media fed off vaudeville's success without becoming a source of additional promotion or revenue for vaudeville as an institution (still individual acts did occasionally benefit). David Nasaw attributes vaudeville's decline and disappearance to the compound effect of increasing audiences for recorded music and radio and the exploding motion picture business, which provided a direct alternative when audiences wanted to "go out" for their entertainment. Vaudeville did not collapse fully until the Great Depression, when its economic basis was completely wiped out.[36]

The sound media thus emerged from a whole network of industrial practices and concerns. Economically, they were tied into a whole range of business and entertainment concerns. Financially, they were backed by the same people. Technologically, they were rooted in many of the same processes and concerns. Considered as an object in its own right, the social field from which the sound media emerged offers a partial explanation for their plasticity. Viewed institutionally, technologies to reproduce sound were driven by two related imperatives: we might say that, to some extent, inventors like Edison and Bell believed in the cult of progress and "virtuosity values" of invention for its own sake, to borrow a phrase from Arnold Pacey.[37] But the funding that allowed for their apparently transcendental and intellectual interest in invention was very much driven by the motivation to create new media markets in an otherwise crowded industrial field. Thus, the socially and culturally salient aspects of a new medium—the social relations and practices that it actually embodied and promoted—mattered less to its early technicians and promoters than did its economic and technical function. As a result, the marketing of these technologies could take on an experimental character. Theoretically at least, social configurations of sound-reproduction technology could take on a "mix-and-match" flavor. This early plasticity did not mean, however, that new sound technologies emerged weightless into a space where they could float freely.

Marketing Sound Reproduction: Experimental Media Systems

The aspirations, conditions, and realities attached to early sound technologies depended on the social world that they inhabited, where and how they were disseminated. The early users of sound-reproduction technologies in the United States were overwhelmingly white and middle and upper class. Whether for work or leisure, early telephony, phonography, and radio were most often found in the emerging spaces of the new middle-class culture.

Their geographic development was also very uneven. Between about 1880 and 1910—precisely the first wave of sound reproduction's commercialization—the American professional-managerial class underwent tremendous transformations. What began as a genteel middle class largely made up of entrepreneurs, artisans turned businesspeople, shopkeepers, and other small capitalists was transformed in the course of two decades into a growing class of people who sold their labor for money, performing several kinds of "mental labor" for business firms, large and small alike (although the division between physical and mental labor was becoming less of a reliable marker of class distinction).[38] The different developmental paths of sound-reproduction technologies from 1880 through the 1920s come into relief when considered in terms of the contours of these broad transformations. What worked for broadcasting in the 1920s did not even exist for telephony and phonography in the 1880s. Victorian attitudes still held some sway in middle-class life at the turn of the century, while the consumer culture was more fully developed by the 1920s. Telephony and phonography both evolved through this broad cultural transformation, while radio was more of an artifact of its accomplishment.

This cultural context is essential to understanding the articulations of machines to forms of social organization, to their spatial and temporal modalities of use. Different social contexts produced different possibilities. In addition to point-to-point communication, early users of the telephone experimented with various forms of what we would call *broadcasting* and *information exchange*. Before the phonograph became a means for reproducing music, it was an office tool, a form of long-distance communication, and a home recording device. In other words, the functions of these new technologies shifted as they moved across cultural contexts and as they were embedded in different kinds of cultural practices, including the use of other sound-reproduction technologies. It is, thus, possible to find configurations of repeatable functions and relations—media forms—that sharply contrast with the ones to which we have become accustomed.

One of the most striking examples in this respect was the use of the telephone as a broadcast medium in various parts of Europe and the United States. Tivadar Puskás, who had worked with Edison in the United States and was one of the first to conceptualize a central telephone exchange, traveled first to Paris and then back to Budapest to experiment with telephone broadcasting systems. Puskás aided Clément Ader in setting up the Parisian system, which came to be known as the Théâtrophon. It broadcasted performances of the Parisian Opéra, the Opéra Comique, and the Théâtre

Française. These early concerts were probably the first stereophonic transmissions. Four transmitters would be set up—two at the edges of the stage and two toward the middle. The transmitters at the edge of the stage would capture the large orchestral instruments, while those toward the middle would capture performers' voices. Listeners would be given two earpieces, one from the far side of the stage and one from the near, thereby allowing them to perceive movement and also to hear clearly the orchestra and the performers: "The delighted audience professed to hear the words of the performers even more clearly than if they actually had been present at the Opéra." The audience was less delighted, however, with the fact that early concerts restricted listeners to five or even two minutes at an earset, so as to move as many people through as possible. These concerts were held in five rooms, with twenty pairs of earphones in each room. The telephone concerts were initially held at the Paris International Exhibition of Electricity and later moved to the Elysée Palace, where they became the basis for a telephone broadcasting system ten years later. The Paris Théâtrophon existed until 1925, when its circuitry was replaced with vacuum-tube amplifiers.[39]

As Puskás was helping organize the exhibition in Paris, he was also directing his brother to set up telephone concerts in Budapest. The brothers gave telephone concerts throughout the 1880s, and, in 1892, they founded the Telephon Hirmondó—perhaps the best-known and most successful telephone broadcasting system. The Telephon Hirmondó (literally, Telephone Herald) provided daily scheduled transmissions of stock prices, news, sports, and cultural programming to the Budapest elite. Although the service was relatively cheap, it broadcast entirely in Magyar (the language of the Hungarian ruling class until after World War I), subscription was limited to the political and cultural elite, and programming was geared toward their interests. The system had over 1,000 subscribers by the end of 1893 and over 6,000 by the end of 1896. The Hirmondó adopted a regular schedule of programming that remained relatively unchanged from 1896 through 1914. The service peaked in 1930 with 9,107 subscribers.[40]

The American trade press followed the Telephon Hirmondó with great interest, and a brief attempt to duplicate the Hirmondó's success surfaced in 1911 as the New Jersey Telephone Herald Company. The Telephone Herald failed, not because of lack of interest, but because of too much interest from subscribers. Investors were scared off because of legal problems that the company had been having, and the company was unable to install new equipment to keep up with customer demand—although it managed to get over twenty-five hundred contracts, it had only a thousand operating installations. When the Telephone Herald could no longer pay its musicians and (a month later) its office staff, they quit, essentially terminating the service.

Although the Telephone Herald represents the only major entertainment-based telephone network in the United States, telephone companies and subscribers experimented with other forms of telephonic entertainment—some of it approaching broadcasting—prior to 1911: concerts, sermons, speeches, and even impromptu jam sessions among late-night operators could occasionally be heard over telephone lines by the mid-1880s. Beginning in 1886, Milwaukee's Wisconsin Telephone Company offered its listeners orchestral music every evening and Sunday afternoons; other companies followed its example. Promoters experimented with relaying sporting events live to remote audiences by telephone as early as 1884, and, by 1889, telephone operators in Cleveland and elsewhere were kept abreast of changing baseball scores inning by inning, as a service to subscribers. As Carolyn Marvin argues, these anecdotes demonstrate the cultural and commercial feasibility of broadcast media long before radio broadcasting was even a possibility.[41]

The postal phonograph—and similar ideas, like "phonographic calling cards"—were also results of conceptualizing the use of sound-reproduction technologies outside the consumerist, domestic middle-class context within which they were popularized. A postal phonograph system would have had recording and reproducing machines at each post office so that those who could not write could speak into the machines and send the recordings in the mail to friends, family, and associates. This idea was in circulation before sound-recording devices were commercially available. Chichester Alexander Bell wrote in his Volta Laboratory notes of techniques "for making records of which copies are not required, as for example, messages on strips of paper to be sent by post."[42] Emile Berliner's idea of a gramophone office in every town implied that, once recordings were made, they could be mailed. An industry publication reported another plan for a postal phonograph in 1891: "Phonographs are to be used in Mexican post offices for the benefit of the illiterate. The sender will go to the office, talk his message into the receiver of the phonograph, and when the cylinder reaches its destination the person addressed will be sent for, and the message will be repeated to him from another machine."[43] Although I have found no records of fully functioning official postal phonograph systems, the mailing of cylinders was considered a significant potential use of the business phonograph. For instance, Bell's original plan for improving Edison's phonograph included experimenting with paper cylinders as paper would facilitate ease in mailing.[44] In fact, Bell mailed a graphophone cylinder to his colleagues at Volta in order to demonstrate the machine's readiness for commercial development.[45]

Even when the phonograph was understood as a machine that played music, its exact status remained unclear. Gramophones were difficult to classify for trade purposes. Canadian railroads offered added care and protection for the shipping of musical instruments, but, when Berliner's gramophone interests sought this special status for their merchandise on cross-continental trips, they were told in no uncertain terms that gramophones were not musical instruments.[46]

While early users of telephones and phonographs experimented with alternatives to their eventually dominant modes of use, radio too spent its first twenty years in a very different form than that which it would finally take: point-to-point radio accounted for most of its commercial, military, and amateur use into the 1920s. Even then, the nature of broadcasting itself had to be developed. Early broadcasts used "certified" amateur operators at the receiving end. Only later would broadcasters come to understand reception as an unsupervised activity.[47] Susan Douglas, Robert McChesney, and Susan Smulyan all have well documented the emergence of broadcast radio, and I will not rehearse their arguments here;[48] a single example should emphasize the number of variables surrounding the potential organization of broadcasting as a commercial enterprise, even for an industry giant like RCA.

On 2 July 1921, at the suggestion of Julius Hopp, RCA experimented with broadcasting a world heavyweight championship boxing match between the American Jack Dempsey and the Frenchman Georges Carpentier to remote audiences. The goal of the broadcast was twofold: first, to figure out the logistics of bringing a sporting event to a radio audience and, second, to determine whether such services would be profitable for the company. At the time, such undertakings were not widely considered to be commercially viable by radio professionals, and, after failing to find a commercial partner in the venture, RCA appealed to members of the Amateur Wireless Association for assistance. Members of the association found local theaters or other public spaces where they could set up a radio receiver and loudspeaker and in many cases charged an admission fee. The final report prepared by the company documented the extensive organization and planning that went into preparing to broadcast the fight: the certification of operators, the testing of the transmitter to discover the radius within which the broadcast could be heard, the selection of a narrator for the fight, and the planning of the event itself.[49]

The event was considered a tremendous success: over 300,000 people were estimated to have heard the broadcast, and many of the letters to RCA included donations of cash so that future events might be similarly broadcast. The report concluded that "the radiophone as an amusement device proved its practicability beyond any expectation. Commercially, it can be made very remunerative, and as a publicity device it should be rated far ahead of any other form of good-will building that could be used to reach the radio fraternity."[50]

Although this event was much less widely reported than KDKA's experimental broadcasting of the 1920 election returns, it shows the suspicion with which RCA and other corporations viewed broadcasting as late as 1922. While radio executives mulled over the possibilities of broadcasting, AT&T and other telecommunications companies had trouble conceiving of different roles for wired and wireless communication. Essentially, executives saw broadcasting as an event-specific practice, tied to important events only, not as a regular, everyday occurrence. There was no clear sense that one form of technology—wireless or wired—would be better than another for this purpose.[51] Moreover, this form of broadcasting still was based on an economic relation where the event itself is sold to an audience for profit. In fact, the radio industry, and the entertainment industries with which it had to collaborate, learned the hard way that this approach would be most difficult. In one famous case, a large number of ticket holders for a concert canceled their reservations when they learned that the music would be broadcast on a local radio station.[52] In contrast, the very idea of advertiser-supported broadcasting was both experimental and widely unpopular among listeners throughout the 1920s. The gradual inclusion of advertisements in radio programming initially drew strident outcries from even the most industry-friendly periodicals.[53]

Distinctions, Conveniences, and Changing Modes of Middle-Class Sociability

Although telephone and phonograph companies similarly sought monopoly status and worked to edge out competition, corporate plans did not necessarily dictate eventual uses. This is to say, even where there was a corporate structure in place trying to assert a particular set of relations, a particular form for a medium, actual use exhibited a great deal of plasticity as well. The environment of controlled experimentation for radio was in part an outgrowth of the move to regulation (and, in general, more top-down organization of radio communication) during and following World War I. While users' practical applications did help shape radio as a medium, they did not deviate from corporate expectations to the degree that people found unplanned uses for telephones and phonographs.

In the cases of both the telephone and the phonograph, corporations did not immediately make sense of the new sound media in the same ways in which users did. Although they later caught on to users' applications and adapted them to their own purposes, the early history of telephony and phonography is at least as much about the changing home and working lives of the middle class as it is about corporate planning and experimentation.

Initially, the boundaries between sound-recording and live-transmission media were not so clearly drawn. From the outset, the telephone and the phonograph were not clearly separated in the imaginations of their users and the schemes of their purveyors. While Carolyn Marvin shows that similar cultural concerns and anxieties surrounded the use of early sound media,[54] the question of use itself and its framing remained an open issue well into the twentieth century. Experiments in the deployment of phonography and telephony can be grouped into the two broad areas of business and leisure. Although business was an obsession for early telephone and phonograph executives, early users quickly discovered less goal-oriented uses for the machines. Because the telephone executives were largely educated in the telegraph industry, Western Union was the model for American Bell, and early telephone marketing was tightly focused on business uses. Although businessmen were the primary initial market, druggists and physicians were also common early users. Even residential phones were largely touted for their "convenience" and as laborsaving devices—for instance, in having the wife make social arrangements or do her shopping over the phone—but, until the 1920s, Bell Telephone openly discouraged social conversation on the telephone, considering it "trivial" and "frivolous." The telephone industry was officially opposed to social conversation until after the end of World War I. Yet the corporate opposition did little to curb socialization by telephone. Industry literature repeatedly urged users to keep their social calls brief, and, in 1909, a local Seattle manager went so far as to listen in on a sample of conversations going through a residential exchange to determine the extent of "purely idle gossip" on the residential telephone.⁵⁵ In addition to the concern with the telephone being a legitimate business tool rather than a toy, there was certainly a gender angle as well—often, these "idle" conversations were women's talk, outside the masculine world of offices and businesses.⁵⁶

In fact, the strongest evidence for this central use of the domestic telephone is in the protests against it. Although early telephone advertisements were largely geared toward men, women were, from a very early point, the primary users of telephones, especially residential telephones.⁵⁷ Newspaper articles, telephone company circulars and memos, magazines, novels, and other written material from the early years of telephony repeatedly complained of women's "frivolous" uses of the telephone lines, although, as Michèle Martin suggests, social conversation helped produce revenue for telephone companies by increasing traffic outside business hours.⁵⁸ Lana Rakow's interviews with women in a small rural Midwestern community similarly suggest that, although official sources derided women's uses of the telephone, their talk was in fact central to maintaining community life and was, therefore, intimately connected with other kinds of social activities.⁵⁹

If gender codes helped shape telephone conduct, conceptions of class helped shape accessibility to the telephone. Universal service was not an operating ideal for Bell Telephone in the nineteenth century, and, in several cases, the company raised its rates for the express purpose of maintaining the telephone as a primarily elite medium. With the proliferation of telephones in hotels, restaurants, and drugstores, telephone companies found more and more nonsubscribers using formerly "exclusive" telephone lines. Although some companies went so far as to try banning subscribers from lending their phones to nonsubscribers, the notion of the telephone as a public convenience and the potential for increased profit through pay phones and new subscribers generally outweighed these outcries against accessibility by the turn of the twentieth century.⁶⁰ Although telephones would not reach the majority of American homes until after World War II,⁶¹ the notion of *public convenience*—a notion rooted in an increasingly mobile and atomized middle-class public—coupled with an increasing sensibility of consumer entitlement no doubt contributed to a sense of the telephone's increasingly ubiquitous presence in the first decades of the twentieth century.

The convenience of the telephone, then, began as something of an affront to the exclusivity of the lines and wound up being a major selling point of the medium. In a way, this perhaps indexes changing constructs of middle-class self-identity—from Victorian notions of exclusivity and respectability to the more universal self-sense of the consumerist middle class. After all, the notion of convenience extended even further when we consider that the mandate of the Federal Radio Commission (FRC)—and, later, the Federal Communications Commission—was to license radio stations according to the "public interest, convenience, or necessity." This phrase was adapted from utility law, and some lawmakers thought that it was necessary for the FRC to have the constitutional power to grant broadcasting licenses. Lawmakers kept the language purposely vague, to allow the FRC maximum latitude in its policy decisions. As a result, the phrase has been hotly debated whenever licensing has become an issue.[62] This language of public convenience is significant because it signals a shift in both telephony and radio from a private concern to (potentially) a concern for everyone—hence the invocation of the public. Although the policy issues were in some cases vastly different in the telephone and radio industries, *public convenience* points to an understanding of telecommunications as connected with public life in the early twentieth century.[63]

As with the telephone, early phonograph promoters targeted a relatively exclusive elite. They cast it as a "serious" business machine, for use by a professional middle class. Business use implied a business class, a well-educated, well-funded, properly trained elite who would make "proper" use of the new sound technologies. Interestingly, this construct combines Victorian middle-class decorum (a "serious" attitude) with the new bureaucratic environments that helped shape the new middle class. Early marketing plans cast sound recording as a means of streamlining office work and, in many cases, replacing the task of stenographers in offices. Although Read and Welch suggest that Edison first saw the machine for its musical potential and was only later convinced by an assistant that its greatest commercial potential lay in business, the latter use almost exclusively dominated the industry literature in the early 1890s. The *Phonogram,* which ran from 1891 through 1893, was almost entirely devoted to business applications. *Phonogram* editors derided entertainment uses of the phonograph as trivial and possibly harmful to the machine's adaptation for "serious" uses.[64]

An 1891 phonograph directory for Washington, D.C., and surrounding areas provides a few clues to the early topology of phonograph use. Modeled after a telephone directory, the publication lists names of Columbia phonograph users and their addresses. The approximately three hundred entries are divided evenly between residences and business or government addresses, with just two listings for restaurants and a single musician.[65] The directory's very form demonstrates the extent to which the early phonograph industry remained undifferentiated from the early telephone industry, at least in the minds of those seeking to make money from it. Phonograph distribution was, in fact, modeled after American Bell's plan of granting exclusive licenses in each locality, with local companies leasing machines to subscribers (and American Bell, of course, developed and improvised on the basis of the success of Western Union's telegraph monopoly).[66] Largely, this was a result of the corporate structure of the phonograph industry at the time. A single investor, Jesse Lippincott, had founded the National Phonograph Company by purchasing the Edison patents and distribution rights and had acquired rights as the "exclusive sales agent" of the graphophone in the United States, with the exception of the District of Columbia and surrounding areas. In that region, the Columbia Phonograph Company agreed to operate as a local licensee in exchange for the rights to the Edison phonograph. Lippincott had effectively created a Bell-like monopoly by 1889, with the plan to push the phonograph as a business machine; by 1890, the first national conference of local phonograph companies had been organized.[67] As with the telephone, Lippincott imagined a national industry and a national communication network.

As with the telephone, many people found uses for the machines that were not officially promoted by the phonograph companies. Even in 1891, we can see that, despite the push for business use and the conviction among industry leaders that the phonograph could succeed only as a "serious machine" (as evidenced in the industry literature), close to half of all phonograph leases in one major market were residential. By way of comparison, in an 1891 New York City telephone directory, commercial subscriptions outnumbered residential listings by more than five to one—and many of the residential listings were for people like doctors or lawyers who were likely to have telephones at work as well.[68] Moreover, the District of Columbia was somewhat skewed toward business uses in that the federal government was the largest single customer for business phonographs at the time; phonograph directories for other cities would likely have yielded a much stronger bias toward residential use.[69] The massive and growing bureaucratic apparatus of the federal government made an excellent test market for the phonograph, and phonographs were used in a variety of stenographic fashions throughout the 1890s. Most widely reported was the use of phonographs for the congressional record. Records would be made on the floor of Congress and brought back to an office where they could be transcribed to typescript.

Promoters of both the telephone and the phonograph encountered problems with business uses. In addition to meeting resistance from stenographers rightly fearing for their jobs, the early phonographs were not particularly well designed for office use. They did not have an efficient start/stop mechanism (essential for transcribing) and were difficult to listen to. Early telephones often had unbearable levels of interference, especially after the introduction of electric systems in cities (since telephone lines were sometimes put up before electric lines, issues of shielding simply were not considered). More efficient techniques for grounding allowed the telephone system to continue its expansion as a business machine in the 1890s, while, by the turn of the century, phonographs were largely being marketed for entertainment uses—although a small minority were still being used for business purposes.

Plasticity, Domesticity, and Publicity

Already in 1890, frustrated phonograph merchants were turning away from business uses and toward the growing coin-in-the-slot business. By the mid-1890s, this was one of the main areas in which money could be made. David Nasaw locates the boom in the coin-in-the-slot business as part of a larger, emergent, middle-class culture of public and semipublic entertainments. Coin-in-the-slot machines, where a user could hear a song for a fee, were located in hotel lobbies, train stations, and arcades. As cities grew more spread out, a well-placed arcade could entertain commuters with a few minutes to kill and a few cents in their pockets. The boom period for this business lasted only a few years. Between the erosion of phonography's novelty to coin-in-the-slot users and a bottleneck in the manufacture and distribution of new recordings, the potential of arcade-style listening to support the industry died off in the first decade of the twentieth century.[70] Coin-in-the-slot machines persisted into the 1910s and 1920s, when new developments allowed the invention of the first machine that would be called a *jukebox* in 1927.[71]

The industry's changing attitude toward marketing the phonograph could perhaps be best illustrated by the shift in content among three major publications, the *Phonogram* (1891–93), the *Phonoscope* (1896–1900), and a second *Phonogram* (1900–1902). While the first *Phonogram* focused almost exclusively on business use, the *Phonoscope* focused on entertainment uses in public places, and the second *Phonogram* treated the phonograph largely as a means of domestic entertainment. Concurrent changes in middle-class domestic life during this period help set in relief the changes in the shape of phonography.

Since a medium is a configuration of a variety of social forces, we would expect that, as the social field changes, the possibilities for the medium change as well. The phonograph's history illustrates this quite well; the varying uses highlighted in the industry literature correspond to changes in middle-class sociability. Any discussion of the phonograph's possibilities would be incomplete without the list of potential applications offered by Edison in an early publication on the potential of the phonograph. Edison's list is a central facet of almost every history of sound recording, although there is no clear consensus on what conclusion to draw from it. Read on its own terms, it appears as nothing more than the product of brainstorming; potential uses appear in no particular order and with no relation to one another. Edison's list:

1. Letter writing and dictation without the aid of a stenographer.
2. Phonographic books for the blind.
3. The teaching of elocution.
4. Reproduction of music.
5. The "family record"—a registry of sayings, reminiscences, etc., by members of a family in their own voices, and the last words of dying persons.
6. Music boxes and toys.
7. Clocks that should announce in an articulate voice the time for going home, going to meals, etc.
8. The preservation of languages by exact reproduction of the manner of pronouncing.
9. Educational purposes such as preserving the explanations made by a teacher, so that the pupil can refer to them at any moment, and spelling or other lessons placed upon the phonograph for convenience in committing to memory.
10. Connection with the telephone, so as to make that instrument an auxiliary in the transmission of permanent and invaluable records, instead of being the recipient of momentary and fleeting communications.[72]

This list is usually cited by phonograph historians to suggest one of two things: that Edison was brilliant (or at least prophetic) because all of the uses on the list eventually came to pass; or, that nobody had any idea what to do with the technology when it was invented and, therefore, needed to be told. Neither reading is terribly compelling when set against the actual history of the machine—most of these uses came to pass, but the specific form that they ultimately took was determined by the changing world of their users. It was a matter not of fulfillment of prophesy, but of the changing ground on which the possibilities for phonography could be shaped.

Consider the uses numbered 4 and 5, the reproduction of music and the family record: a common and greatly oversimplified narrative of the phonograph's development has the early cylinder machines at a great disadvantage to the later disk machines because of how they worked. But this narrative works only insofar as historians privilege the mass reproduction of music—the "eventual" use—over a possible and immediately plausible use when the machine was first marketed: the production of the aural family album. Although the latter function is still present today in photographic practice, it had a much greater significance for the Victorian middle-class parlor culture than it did for the emergent consumer class.[73] Technological change is shaped by cultural change. If we consider early sound-recording devices in their contemporary milieu, the telos toward mass production of prepackaged recordings appears as only one of many possible futures.

Phonographs and graphophones commercially available in the late 1880s and the 1890s used wax cylinders as their medium. Prerecorded cylinders could not be easily mass produced for commercial sale: since each machine could record onto only a single cylinder at a time, performers would have to repeat a performance several times, even when several machines were employed during a recording session. In retrospect, we can say that Emile Berliner's gramophone, made public in 1888 and first marketed in 1895, changed all this. The gramophone is the direct ancestor of the phonographs most commonly used in the twentieth century: it uses a rotating flat disk on a horizontal plane. Berliner's machine was considerably louder than its immediate predecessors, but one of its most important differences was that its disks were reproduced through a "stamping" process and, therefore, easily mass produced.[74] The making of a master disk for stamping, however, was somewhat complicated and labor-intensive, involved etching and acid baths for the first copy and the matrix that would be used to stamp subsequent copies. As a result, gramophone records were easier to mass produce but much harder for people to make in their own homes.

The common narrative derived from these basic facts argues that the disk machines caught on *because* of the possibility for mass-producing content: essentially, that it was a better way to make money off phonography, from the family record to the musical record. Yet this is precisely where the changing status of the middle class comes into play. The domestic and social life of the emerging professional-managerial class was moving away from parlor culture by the 1890s. Whereas the parlor was a room in Victorian middle-class homes for formal presentation and the maintenance of family identity, where family albums and artwork would be combined with various styles of furniture and art to convey a certain identity to visitors and to family members themselves, the emergent consumerist middle class began in the 1890s to look on these practices as old-fashioned and sterile. Parlors largely populated with hand-crafted goods and family-specific cultural productions gave way in the early twentieth century to living rooms, which were considerably more informal in decor and arrangement and admitted more and more mass-produced goods.[75] The marketing of prerecorded music should be understood in this context. As phonographs became more widely available to a middle-class market, that market itself was changing. The middle-class consumer culture that would provide the cultural, economic, and affective basis for building collections of recordings and extensive listening to prerecorded music was only just emerging as these machines became available. As a result, both inventors and marketers hedged their bets, promoting phonographs as both machines with which a family could produce its own culture and mass-produced commodities that would put their users in touch with a larger public.

If the triumphalist narratives are to be believed, we would expect to find a sort of "Aha, now we can finally do it!" attitude toward the mass production of recordings once this was possible. But Emile Berliner's remarks in his first public presentation of the gramophone show precisely the opposite. He remains unsure as to how to think about the production of recordings: Who would record, under what conditions, and for what purpose? In his address to the Franklin Institute announcing the gramophone in 1888, Berliner moved freely among different notions of content—from the reproduction of mass-produced music, to an institutionalized variation on home recording, to an unrealized form of broadcasting, and back again:

Those having one [a gramophone], may then buy an assortment of phonautograms, to be increased occasionally, comprising recitations, songs, and instrumental solos or orchestral pieces of every variety.

In each city there will be at least one office having a gramophone recorder with all the necessary outfits. There will be an acoustic cabinet, or acousticon, containing a very large funnel, or other sound concentrator, the narrow end of which ends in a tube leading to the recording diaphragm. At the wide opening of the funnel will be placed a piano, and back of it a semicircular wall for reflecting the sound into the funnel. Persons desirous of having their voices "taken" will step before the funnel, and, upon a given signal, sing or speak, or they may perform upon an instrument. While they are waiting the plate will be developed, and, when it is satisfactory, it is turned over to the electrotyper, or to the glass moulder in charge, who will make as many copies as desired.

. . . There is another process which may be employed. Supposing his Holiness, the Pope, should desire to send broadcast a pontifical blessing to his millions of believers, he may speak into the recorder, and the plate then, after his words are etched, is turned over to a plate-printer, who may, within a few hours, print thousands of phonautograms on translucent tracing paper. The printed phonautograms are then sent to the principal cities in the world, and upon arrival they are photo-engraved by simply using them as photograph positives. The resultant engraved plate is then copied, *ad infinitum,* by electrotyping, or glass moulding, and sold to those having standard reproducers.

Prominent singers, speakers, or performers, may derive an income from royalties on the sale of their phonautograms, and valuable plates may be printed and registered to protect against unauthorized publication.[76]

Berliner's uncertain futurology offered a rich brew of potential media systems for the gramophone. While mass production was certainly an idea that seemed—and, indeed, proved to be—promising, it appeared alongside the idea of the local gramophone office, where people could go to make their own recordings. The gramophone office, essentially envisioned as a local, for-rent recording studio, suggests a system where home listening would mix original creations with mass-produced entertainment. Berliner's gramophone office nicely hybridizes Victorian domesticity with the new culture of "going out," to use David Nasaw's phrase.[77]

The appearance of the term *broadcast* as an adverb is also interesting here since we have since come to think of the mass production of recordings and broadcasting as two different things. Berliner's use of the term was probably closer to the agricultural sense of the word than to the sense that we would now associate with radio or television. Yet it suggests an interesting connection among the possibilities of dissemination that our current conventions of use do not emphasize. Berliner's *broadcast* indicated the dispersal of sound events over time *and* space. When we refer to radio or telephone broadcasting, we think only of dispersal over space. We can read into Berliner's usage, then, a sense of the plasticity of the sound event over time *and* space so central to modern sound culture. This potential for dissemination was perhaps the most salient quality of new sound technologies as they were being shaped into media.[78] This is one possible explanation for the relatively fluid boundaries among the point-to-point, broadcast, and archival functions in the minds of late-nineteenth-century inventors, promoters, and users.

The question of dissemination goes beyond simply dispersing sound events or messages in space and time. There was a sense of potential for cultural interconnection as well. It was no accident that royalty and the clergy figured heavily in early demonstrations of and discourses on sound reproduction—Bell, Edison, Marconi, and other inventors were also fond of presenting their inventions to royalty. As in Berliner's examples (in which lie an interesting tale about Jewish assimilation and passing), the new sound media presented an opportunity for cultural integration and the consolidation of authority. The pope can more easily get messages out to his followers; the word of kings can now be heard at the edge of the state. In fact, this model seems to cut across mass media as one possible understanding of the ways in which a medium can integrate a nation or some other large collectivity. For instance, Michael Warner writes that the contemporary media invoke a kind of mass public that turns and faces the image of a leader in order to help imagine itself. Through this move, Warner explores the importance of images of rulers in the mass media. His example was Ronald Reagan on television, but Berliner's pope suggests that the power lies, not in the *image* of the ruler, but in placing a trace of the ruler—image, voice, writing—within a larger network of communication in order to bring people together. Pace James Carey, an authoritarian streak may cut across models of communication that are based on community, ritual, and communing.[79]

Clearly, even when Berliner conceives of the mass production and dissemination of recordings, he does not necessarily envision a mass market for prepackaged music. That construct of recording would develop later. Lest Berliner's predictions appear as the fanciful meanderings of an inventor at a public speech, consider that, seven years later, a manual for the seven-inch American Hand Gramophone assured users that, "as quickly as expedient, gramophone recording offices will be established, which will enable you to have your own voice, and the voices of your friends and relatives, taken. Copies of such personal records can be furnished *ad libitum.*" New mass-produced sound recordings, on the other hand, would be made "available from time to time."[80] Similarly, the second *Phonogram* contained many references to making one's own recordings at home, even as it carried lists of new Edison cylinders in every issue. After the magazine discontinued publication (in 1902), the *Edison Phonograph Monthly* (which partially replaced it) reported that one of the *Phonogram*'s most popular features was the list of forthcoming records.[81]

Even when considered as primarily a playback medium, sound-recording devices were just as easily conceived as public entertainments as domestic entertainment appliances. The *Sears Roebuck Catalogue*, perhaps the emblematic publication of growing consumer aspirations in this period, marketed graphophones exclusively—and exclusively for entertainment uses. While mentions were made of recording diaphragms, there is a clear bias toward using the machine for listening to existing recordings, rather than making one's own. The 1897 catalog has a brief listing for "graphophones or talking machines," congratulating itself for making the instrument available and affordable to a broader public than ever before. Aside from the assurance that "the recording diaphragm can be used to make any records desired, which can be made on blanks furnished for that purpose," the rest of the copy is dedicated to the use of the graphophone for "home entertainment or exhibition outfit" and concludes with a list of available recordings.[82] Five years later, Sears Roebuck included a seven-page graphophone listing alongside those for stereopticons and moving pictures in their "Department of Public Entertainment Outfits and Supplies." The longer listing offers more expensive machines and many accessories in various "exhibition" outfits, including a combined graphophone/stereopticon setup that allowed for presentations of "illustrated songs." Although home use is mentioned with some of the cheaper models and all models are said to have recording as well as reproducing diaphragms, the focus is clearly and resolutely on public entertainment, rather than home entertainment or business use.[83]

The telephone in the middle-class home occupied a similarly ambiguous and changing status over the period. Anxieties over domestic relations were also framed against the telephone's presence in business and public places: a 1902 kinetoscope film, entitled *Appointment by Telephone,* depicts a man answering a phone call in his office, meeting a woman at a restaurant for lunch, and being discovered by his wife, who proceeds to beat him with an umbrella.[84] Although clearly meant as a joke, this film points to the ambiguity of "public" and "private" introduced by the organization of sound technologies into media. In a fashion similar to telegraphy (discussed in the last chapter), the telephone, organized as a medium, facilitated intimately personal connection *because* it was a massive network of connections. It was simultaneously intensely public and intensely private.

Apart from this bit of marital humor, it was well established that women were far ahead of men in using the telephone as a social device. The changes in attitudes toward residential and social telephone use over the first decades of the twentieth century seem largely to come from users rather than providers. With the growth of long distance, the continuing growth of the telephone system prior to the depression, and the further growth of the consumer culture, AT&T's move in the 1920s toward advertising the phone for social purposes seems more belated than anything else. Simply put, the telephone did not integrate well with the Victorian middle-class household. It provided a new level of accessibility: both of the outside world to the household and of the household to the outside world. The interior life of Victorian domesticity, with its formal and carefully circumscribed practices of self-presentation, was suddenly opened up to a ringing machine wired to the rest of the neighborhood and the rest of the city. As Michèle Martin writes, "The fact was that late-Victorian women were caught off-guard. The barriers that their society had built in order to preserve privacy did not work with the telephone, and there was no time to construct new ones. Yet, in spite of this inconvenience, women continued to use the telephone, and the system developed rapidly, especially from the early 1900s onward."[85] Although the telephone's rise corresponded with changes in practices of socializing, for instance, more planned and orchestrated visits and less "dropping in" and designated visiting hours, the telephone was largely integrated into the flow of domestic middle-class social life by the 1920s.[86]

The growth of radio in the middle-class home begins at the end of this period, and it must be understood through the same set of changes. The home radio set (first providing only Morse code, then later allowing for the broadcast of speech and music) has more in common with the present-day home computer and modem than it does with the phonograph in the parlor. Amateur radio was from the start based on a kind of cosmopolitanism: it allowed its middle-class users something over which they could gain a measure of technical mastery in a world of mass-produced consumer goods, and it facilitated communication with distant strangers. While a family making its own records would likely have played them for guests as well as for one another, the homemade phonograph record offered a more inwardly oriented set of identification processes: like the parlor itself, the home recording was a document of domesticity and togetherness. In contrast, amateur radio was—at least initially—oriented toward transcending the domestic space, opening up the family by making it a nodal point within the larger middle-class culture. (When, with the institutionalization of broadcast programming in the 1920s, radio reoriented itself toward family togetherness, its focus would be the family listening together as participants in a mass culture.) That sensibility would be manifested more widely through bringing prerecorded music into the home. The fetish of distance listening was both cosmopolitan (how *many* different places a listener could pick up) and exploratory (how *far* a listener could hear).[87] In a very real way, then, although amateur radio operators appeared as cultural producers in that they would create content on the airwaves, they had more in common with those people who used the phonograph to listen to "the music of the world," or with immigrants who listened to the music of the distant home culture that they had left, than they did with the hobbyists who made their own phonograph records. The latter activity was located in a disappearing Victorian domesticity; the others were all ways of belonging to an emergent consumer culture. The image of the family of the late 1920s, listening to a broadcast program while gathered around the radio, was an extension of this new form of middle-class belonging.[88]

Bureaucracy, Agency, Nationality

Issues of location and identity extended into the emergent work spaces of middle-class life in the early twentieth century, and the development of sound-reproduction technologies into media was very much conditioned by problems associated with these new work spaces. The business telephone and the business phonograph were both presented and understood as devices for managing and navigating the massive communication and information needs of growing, turn-of-the-century bureaucracies. Their purveyors hoped that these machines would appeal to the department manager, the clerk, the sales agents, organization men (and a few women) of all stripes working in the new, big corporations or in smaller firms whose reach was continually expanding. But marketers had to adjust as their target market itself adjusted to the new configurations of work and life in corporate, urban America. By the early twentieth century, both machines were presented as a new kind of fixed capital, a means for eliminating labor—both in the sense of reducing workloads and in the sense of eliminating workers. At the same time, these technologies were sold as a kind of new bureaucratic agency. Telephone advertising touted its capacity for "Multiplying Man-Power":[89]

The progressive manager has more than a TELEPHONE—he has a TELEPHONE SYSTEM and a definite TELEPHONE POLICY.

He realizes that the salary of an office boy or a clerk will pay for a private branch exchange and that the salary and expenses of one traveling salesman will more than equal the cost of the most liberal use of local and long distance service.[90]

The analogy is made explicit here: telephone service allows a manager to replace employees and to do their own work better. But this Bell ad also points to the difference between technologies and media—a technology is simply a machine that performs a function; a medium is a network of repeatable relations. The telephone was, thus, not simply a technology, but a shorthand name for a whole assemblage of connections, functions, institutions, and people. In fact, this campaign for Bell is pretty explicit—the agency offered by the telephone comes from the corporation conceived as a network. Here, we see a version of reification in reverse—to coin a phrase, a *strategic demystification*—where an advertisement highlights the relations embedded in a medium in order to promote the power that it presumably conveys to its users.

Command and control awaited the business telephone user: "In the simple act of lifting the telephone receiver from its hook every subscriber becomes the marshal of an army. At his service, as he needs them, a quarter of a million men and women are organized in the Bell system."[91] In this reversal of the fixed capital metaphor, we are offered a startling image of supplementation. One man (and the ad certainly was aimed at men) could now act as if he controlled an army; one can harness the power of the many. Where the office manager could replace his own workers, he could at the same time call on an entire workforce, organized into a system. This formulation works well as advertisement and as media theory: this fictional office manager could instrumentalize the more or less recurrent social relations embedded in the medium of telephony.

Beyond control lay the possibility of mobility: in its campaign for a national monopoly, AT&T reinvented and refined the metaphor of transportation for communication. The Bell System became "Highways of Speech" and "The Clear Track";[92] advertisements used the language of space and mobility to convey a sense of the telephone user's agency:

Every Bell telephone is the center of the system.

It is the point which can be reached with "the minimum aggregate travel," by all the people living within the range of telephone transmission and having access to Bell telephones. Wherever it may be on the map, each Bell telephone is a center for purposes of intercommunication.[93]

As an "Implement of the Nation," each telephone lay at the center of a vast network, itself sewing together the United States into a coherent field where the young bureaucrat could communicate with, travel to, effect changes in, and affect the activities of any other point in the network through the simple use of the telephone.[94] AT&T framed telephony as a form of agency and thus played on turn-of-the-century middle-class self-consciousness; it appealed to the culture and the power of bureaucracy. These ads also implied that business travel could be severely reduced or eliminated through the use of telephones (although this was not, in fact, the case). In AT&T's language, communication annihilated distance; a person could become a whole operation through the use of the network.

The pervasiveness of this model of telephony as agency, even before the turn of the century, is in evidence from its mockery in two short kinetoscope films from 1898 both entitled *The Telephone*. In the first, a telephone hangs on an office wall with the sign "Don't travel, use the telephone, you can get anything you want." A man rings up on the telephone and pulls a cup of coffee out of it. After enjoying the coffee, he rings on the phone again, this time getting a bunch of white powder in the face. The second film is set outdoors, with the same sign and scenario, except this time involving beer (first in a glass, then squirted in the actor's face) rather than coffee and white powder.[95] Both films clearly show that the promise of telephony to the middle class was understood but that the machinery was still seen as potentially unruly—far from AT&T's "thrifty habit," the telephone offered new possibilities, but the utility and cost of that potential were open to question.[96]

The business phonograph was presented in this period in much the same manner as the business telephone was. The business phonograph was "a faithful servant that will conduct business like a setting hen, and will never strike for higher wages."[97] The implication here is clear—replacing a skilled position, that of the stenographer, with a machine will save time and money and offer the user considerably more control. The attack on stenographers appeared frequently in phonograph literature. An advertisement entitled "Facts about the Phonograph When You Want a Stenographer" drew the analogy explicitly, claiming that the phonograph could meet and exceed the performance of a real stenographer in "reliability, accuracy and economy" for business purposes.[98] Stenographers rightly saw the business phonograph as an encroachment on their professional turf and fought its spread into offices and bureaucracies.[99]

The phonograph industry refined its advertising in response. In 1904, Edison produced a ten-minute film that, from today's perspective, can only be considered an infomercial for the business phonograph. Entitled *The Stenographer's Friend; or, What Was Accomplished by an Edison Business Phonograph*, the film offers a narrative sequence of scenes punctuated by intertitles. "Shorthand troubles" shows an office scene with a woman surrounded by a pile of papers and attempting to take dictation from two men at once. The woman breaks into tears as the men try to get her to stay even later (the clock on the wall shows 5:30). They let her go, and a salesman mysteriously appears. After giving the men a pamphlet, he proceeds to demonstrate at great length the workings of the phonograph: "The ease of Voicewriting. Dictation at any speed: instant repetition: a practical correction system." The film continues in this vein showing transcription and the shaving of wax cylinders for reuse. The final scene—"not yet five. Correspondence 'cleaned up' and everybody happy"—shows the same office, papers neatly filed, one man speaking into the phonograph while the woman types up the contents of another cylinder on another phonograph. The salesman stops by as the workers are about to leave, everyone shakes hands, and the film ends.[100]

The film is, of course, remarkable for its length and detail—few film advertisements of this duration or detail could be found from the same period. But the detail suggests that the operation of an Edison phonograph was not such an easy thing. That over half the film is devoted to the proper functioning of the business machine suggests that people were not quickly learning the machine and seeing its utility. The film is also interesting for its narrative content. While similar advertisements from the 1890s were directed entirely at men, the stenographer/secretary in this film is a working woman, and a large part of the contrived narrative is focused on her happiness and stress level. Whether the film was conceived as an attempt to speak directly to working women or simply to men's images of working women, the machine is now presented as a means of ameliorating unhappiness caused by the gendered inequities of office work.

Like telephony, phonography was also presented by its promoters as a means of overcoming other differences as well: according to the *Phonogram*, it could unite the nation and in fact supersede national boundaries through a universal language. This notion of sound recording would not really take hold until immigrants nostalgic for their home countries appropriated it in a very different form by buying phonographs in order to hear music "from home." Lizabeth Cohen writes that, along with refrigerators, phonographs were the only commodities that otherwise frugal immigrant workers were willing to buy on credit in the 1920s.[101] Like the construct of telephony in N. W. Ayer's fantasies of middle-class agency, this notion of phonography revolves around not a technology but a medium—a whole set of relations, interconnections, practices, institutions, and people. The possibility that a medium could connect a nation or even connect people across national boundaries requires a notion of social configuration and regularity at the heart of sound communication. Sound technologies became sound media as these imagined, planned, and real modalities of interconnection and articulation emerged.

Conclusion: Media First, Technologies Later

Sound media shared a common origin and common conditions of emergence, although of course they took different paths of development. This common origin offers a useful way of reframing the problem of a history of sound. Put simply, if, at certain points in history, sound-reproduction technologies were not well differentiated from one another, and if they share a common origin, then it should be possible to think of them together, to consider them as differentiated parts of a larger problematic.

The development of sound reproduction into recognizable media occupies a place among a whole range of social transformations in turn-of-the-century America. It was not, as inventors and their idolaters would have it, a matter of birth. The very possibility of sound media was structured by the changing economics and social organization of invention, the growth of corporate-managerial capitalism, and the concurrent move from

Victorian to consumerist forms of middle-class everyday life. Moreover, if the new sound technologies were not the products of some convoluted male birth scenario, neither did they arrive in the world fully formed. The shape of sound-reproduction technologies, the social arrangement of their functions, their development and differentiation into separate media, was itself a historical problem. This plasticity, in addition to the relatively stable forms later taken by sound media, is a defining aspect of modern sound culture: the social configuration of sound became both a problem and a field of possibilities, personal, cultural, economic, and technical.

Yet the story offered in this chapter, where technologies are organized into media, remains incomplete. In order even to imagine sound technologies functioning at all, one had to imagine them as part of a whole media system. According to AT&T, the office manager did not merely have a telephone; he had a telephone system: this is because a telephone was meaningless without the larger telephone system. The same could be said for sound recording and radio. For early users to imagine sound-reproduction technologies as working at all, there had to be a medium; there had to be multiple people, places, times, and machines. This was true even for inventors. Without a distant receiver awaiting the message, what good is a telephone or radio transmitter? Without the possibility of playback, what good is the recording of sound? In this way, we might say that, insofar as sound technologies are *ever* organized into sound media, the medium—or, at least, an imagined medium—precedes even the technology itself.

5 The Social Genesis of Sound Fidelity

I have noticed that the senses are sometimes deceptive; and it is a mark of prudence never to place our complete trust in those who have deceived us even once.
—**RENÉ DESCARTES**, *Meditations on First Philosophy*

Horace, this telephone thing, can you really hear the fellow on the other end?
—**WALT WHITMAN** to Horace Traubel

A 1908 advertisement for Victor Records pictures the opera singer Geraldine Farrar next to a Victor record player, with the caption "Which is which?" (figure 29). The ad almost taunts its readers: "You think you can tell the difference between hearing grand-opera artists sing and hearing their beautiful voices on the *Victor*. But can you?" Of course, the implied answer is a resounding "no": "In the opera-house corridor scene in 'The Pit' at Ye Liberty Theatre, Oakland, Cal., the famous quartet from Rigoletto was sung by Caruso, Abbot, Homer and Scotti on the *Victor,* and the delighted audience thought they were listening to the singers themselves."[1] If opera audiences in Oakland could not tell the difference, perhaps home audiences would be equally mystified. At least, they were dared to listen to the reproduction itself and to ask their record players to do a little philosophical work on their behalf.

Another version of this ad—captioned "Both are Caruso"—has the Italian tenor Enrico Caruso standing tall next to a Victor disk record. "The Victor record of Caruso's voice is just as truly Caruso as Caruso himself. It actually *is* Caruso—his own magnificent voice, with all the wonderful power and beauty of tone that make him the greatest of all tenors," declares the ad.²

In both cases, a simple picture and some straightforward ad copy offer the reader a deceptively complex set of philosophical claims about sound reproduction. The medium in question is recording, but the ads' claims index some fundamental issues surrounding sound's reproduction. The ads present a general equivalence of singer and recording—the two stand side by side; they are written of in parallel phrases. But consider the actual claims about Farrar and Caruso for a moment. In both ads, the person is placed side by side with the technology—either the record player or the record. In both cases, the person and the medium are the same size. The ads establish their equivalence by manipulating images and by providing text that is taunting and tautological (tauntological?): "Which is which?"; "Both are Caruso." The very claim that the Victor record delivers *all* the tone and power of Caruso anticipates the possibility that the recording might not in fact have *all* the tone and power of the singer. To assert that both are Caruso implies that one or both might *not* in fact be Caruso and that, in any event, the point requires some demonstration. Farrar, Caruso, records, and talking machines: these ads protest too much.

Although the philosophy of reproducibility in these ads and the questions that they raise are likely familiar by now, early discourses of and around sound fidelity reveal fights over the ground rules of reproducibility. Before sounds could be captured by electric devices for measuring signals, *fidelity* was an amazingly fluid term, signaling the plasticity of practices of and around sound reproduction. Even basic technical discussions—whether a given sound technology "worked"—were loaded down with implied theories of reproducibility. Functional, aesthetic, social, and philosophical issues were bound together from the very beginnings of sound reproduction. During the early history of commercially available sound-reproduction technologies, from roughly 1878 through the 1920s, these issues were both practical and philosophical. People had to learn how to understand the relations between sounds made by people and sounds made by machines. Over time, certain practical understandings would come to sediment around the process of sound reproduction and its attendant relation of original and copy. This chapter tells the story of how these understandings came to cohere and develop over time.

The discourse of fidelity is most common and most developed in discussions of sound recording, as opposed to other forms of sound reproduction. This is likely because of the increased ease of careful listening and

comparison made possible by recordings as well as by their mobility. Yet the centrality of recording to the discourse of fidelity should not be mistaken for a theoretical privileging of recording as such in my own account: my argument is that the problems described under the rubric *sound fidelity*—some of the key questions that orbit around the concept *reproducibility* itself—apply equally, although in slightly varying ways, to the other kinds of sound reproduction.

Although I will argue that the logic of "original" and "copy" does not adequately describe the process of sound reproduction, one cannot deny that questions of the relation between originals and copies have formed a central preoccupation of twentieth-century theories of communication and

Figure 29. "Which Is Which?"—Victor Talking Machine advertisement, 1908 (courtesy Archives Center, National Museum of American History)

culture.[3] Conventional accounts of sound fidelity often invite us to think of reproduced sound as a mediation of "live" sounds, such as face-to-face speech or musical performance, either extending or debasing them in the process.[4] Within a philosophy of mediation, sound fidelity offers a kind of gold standard: it is the measure of sound-reproduction technologies' product against a fictitious external reality. From this perspective, the technology enabling the reproduction of sound thus mediates because it conditions the possibility of reproduction, but, ideally, it is supposed to be a "vanishing" mediator—rendering the relation as transparent, as if it were not there.[5] Inasmuch as its mediation can be detected, there is a loss of fidelity or a *loss of being* between original and copy.[6] In this philosophy of mediation, copies are debasements of the originals.

Everywhere we turn in the search for true fidelity, the desire to capture the world and reproduce it "as it really is" yields a theory of correspondence between representation and that which is represented. While the locution *perfect fidelity* suggests that there is no loss of being between an original sound and its copy—as do the Victor records ads discussed at the beginning of this chapter—today this sensibility has few philosophical adherents.[7] The problem is commonly conceived in this fashion: reproduction, the technological transformation of an original into a copy, introduces a potential or real loss of being in the original sound. Eric Rothenbuhler and John Peters offer a meditation on recording as mediation in a fascinating essay entitled "Defining Phonography." In comparing analog and digital recording technologies, they write that "the phonograph record and magnetic tape do contain traces of the music": "There is an unbroken chain from the sound in the living room to the original sound as recorded." In other words, analog recording technologies have an authentic relation with the "original" behind the recording because—in their estimation—sound bears a causal relation to the analog recording. Digital recording, meanwhile, converts sound into a series of zeros and ones to be reconstructed as sound at the moment of reproduction. For Rothenbuhler and Peters, digital recording is, therefore, more ontologically distant from live performance than analog recording. While their thesis that phonography is ontologically different than digital sound recording is certainly a fascinating proposition, their definition of *phonography* assumes that recording captures sounds as they exist out in the world. In essence, they argue that mediation is an ontological problem brought about by the technology of sound reproduction itself.[8] In contrast, this chapter argues that mediation is a cultural problem and only one possible way of describing sound reproduction.

Rothenbuhler and Peters anticipate and answer an Altmanesque objection that asserts the material heterogeneity of recorded sound, but my point here is slightly different. Rather than asserting that, by virtue of their physical location, all sounds are *different* sounds,[9] my argument is historical in scope: the "original" sound embedded in the recording—regardless of whether the process is "continuous"—certainly bears a causal relation with the reproduction, but *only* because the original is itself an artifact of the process of reproduction. Without the technology of reproduction, the copies do not exist, but, then, neither would the originals. A philosophy of mediation ontologizes sound reproduction too quickly. Therefore, a notion of sound fidelity based on a fundamental distinction between original and copy will most likely bracket the question of what constitutes the originality itself. In emphasizing the products of reproduction, it effaces the process.

I argued in the last chapter that, insofar as sound technologies are *ever* organized into sound media, the medium—or, at least, an imagined medium—can be said to precede even the technology itself. But by *medium* I do *not* necessarily mean to imply a philosophy of "mediation." To consider the products of reproduction—original and copy—separate from the process, even in a philosophical exercise, is to confuse a commercially useful representation of reproduction with the ontological character of reproduced sound itself. "Original" sounds are as much a product of the medium as are copies—reproduced sounds are not simply mediated versions of unmediated original sounds. Sound reproduction is a social process. The possibility of reproduction precedes the fact.

Sound fidelity is much more about faith in the social function and organization of machines than it is about the relation of a sound to its "source." "We have to break from the common procedure of isolating an object and then discovering its components," wrote Raymond Williams. "On the contrary, we have to discover the nature of a practice and then its conditions."[10] From the very beginning, sound reproduction was a studio art, and, therefore, the source was as bound up in the social relations of reproducibility as any copy was. Sound fidelity is a story that we tell ourselves to staple separate pieces of sonic reality together. The efficacy of sound reproduction as a technology or as a cultural practice is not in its keeping faith with a world wholly external to itself. On the contrary, sound reproduction—from its very beginnings—always implied social relations among people, machines, practices, and sounds. The very concept of sound fidelity is a result of this conceptual and practical labor.

Debates about the authenticity of copies with respect to a truly authentic original miss a more fundamental issue: the very nature of originality and authenticity is transformed in the context of reproducibility. Walter Benjamin's famous essay "The Work of Art in the Age of Mechanical Reproduction" makes this argument with respect to film. At first blush, Benjamin appears to advance the "loss of being" hypothesis since he coins the term *aura* as "that which withers in the age of mechanical reproduction." Aura is the unique presence in time and space of a particular representation, its location in a particular context and tradition. In freeing that which is reproduced from a particular time, space, and tradition, mechanical reproduction destroys aura.[11] The copy is that which is similar to the original but has failed to be the same, "the pretender who possesses in a secondary way."[12] But to stop here would be to miss the point of the entire essay.

Benjamin immediately qualifies his definition of *aura* in a note: "Precisely because authenticity is not reproducible, the intensive penetration of certain (mechanical) processes of reproduction was instrumental in differentiating and grading authenticity." In *this* formulation, the very construct of aura is, by and large, retroactive, something that is an artifact of reproducibility, rather than a side effect or an inherent quality of self-presence. Aura is the object of a nostalgia that accompanies reproduction. In fact, reproduction does not really separate copies from originals but instead results in the creation of a distinctive form of originality: the possibility of reproduction transforms the practice of production. This is the expressed purpose of Benjamin's analysis of film in the second half of the essay. His fascination with film lies in its composition, its artificiality. He writes of shooting and editing practices "in the studio": "The mechanical equipment has penetrated so deeply into reality that its pure aspect freed from the foreign substance of equipment is the result of a special procedure, namely, the shooting by the specially adjusted camera and the mounting of the shot together with other similar ones. The equipment-free aspect of reality here has become the height of artifice; the sight of immediate reality has become an orchid in the land of technology." Now, clearly, Benjamin connects artifice with artificiality. Nature, as "immediate reality" disappears for him. But this is because reproduction highlights the possibility of reality having an immediate self-presence in the first place: authenticity and presence become issues only when there is something to which we can compare them. For Benjamin, this process is best embodied in the cinema; it was "unimaginable anywhere at any time before this."[13] It is also an apt description of the condition of originality in the age of reproduced sound—whether we are considering recording, telephony, or radio. As a studio art, sound reproduction developed shortly before and then alongside film.[14] The possibility of sound reproduction reorients the practices of sound production; insofar as it is a possibility at all, reproduction precedes originality.

Nowhere is this more clear than in our anachronistic use of the word *live* to describe performances that are not reproduced. As Sarah Thornton has written, the term *live* as we apply it to music (and potentially to all face-to-face communication) entered "the lexicon of music appreciation" only in the 1950s, as a part of a public relations campaign by musicians' unions in Britain and the United States. Although our accepted idea of *live* music emerged from a serious labor struggle over musicians' abilities to make a living at their trade, the term has taken on a life of its own in aesthetic discourse, completely abstracted from its original context. At the time of the unions' public relations campaigns, the word *live* was short for *living,* as in *living musicians:* "Later, it referred to music itself and quickly accumulated connotations which took it beyond the denotative meaning of performance. . . . Through a series of condensations . . . the expression 'live music' gave positive valuation to and became generic for performed music. It soaked up the aesthetic and ethical connotations of life-versus-death, human-versus-mechanical, creative-versus-imitative."[15]

As they are anguished over in the discourses of fidelity, speculations on the relation between original and copy operate as placeholders for concerns about the social process of sound reproduction itself. By restoring this sense of process to sound theory, we restore sociality and contingency to our theories of reproduced sound. In short, we treat reproduction as an artifact of human life instead of as an ontological condition. To do otherwise is to take one "socially constructed practice of sound production and reception" as the ground for discussing and evaluating all others.[16] We should consider sound events in terms of their own social and cultural location, rather than beginning our analysis of reproduced sounds by treating each as a contender for the right to reign over the domain of *all* sounds.[17]

This question of location is urgent in early discourses of fidelity. The early history of sound fidelity, as an operative concept, a technical principle, and an aesthetic, is a history of beliefs in and about sound reproduction as well as a history of the apparatuses themselves. Fidelity, after all, is the quality of faithfulness to some kind of pact or agreement. The very choice of the term *fidelity* (first applied to sound in 1878) indicates both a faith in media and a belief in media that can hold faith, a belief that media and sounds themselves could hold *faithfully* to the agreement that two sounds are the same sound. We need this faith to have a sense of equivalence among originals and copies since, in addition to the philosophical quagmire introduced by the locution *perfect fidelity,* identity between original and copy is impossible from a purely technical standpoint—leaving aside the larger metaphysical question.[18]

Nevertheless, the term *sound fidelity* has become a kind of technicistic shorthand for addressing the problems of sound's reproducibility—a gold standard for originals and copies, an imagined basis for the currency in sounds. Today, perhaps, *fidelity* connotes a measurable correspondence between two different sounds—implying finely graded electronic or digital measurements of frequency response and amplitude—but the tools used to make these measurements were not even available until the 1910s and 1920s, and the terms themselves were not formalized among electricians until the 1930s.[19] Thus, the concept as we are likely to understand it today is far removed from the concept as it was understood at the turn of the twentieth century. At that moment, a precise technical definition of *sound fidelity* was simply unavailable. Instead, an entirely different set of concerns shaped the development of the concept during the period that I consider here. At stake was the relation between original and copy and the fundamental conditions of sound's reproducibility. Far from standing outside sound reproduction in order to describe it, the discourse of fidelity is a key part of the history of sound reproduction.

Histories of sound fidelity usually begin at the end, with the achievement of perfect fidelity and flawless sound reproduction. Narratives of technological change and the transformation of technical specifications are folded back into an aesthetic and technological telos: the latest technological innovation equals the "best-sounding" or "perfect" sound reproduction. The progress narrative is ultimately untenable: the transformation of practices and technologies stands in for a narrative of vanishing mediation, where sources and copies move ever closer together until they are identical. The nature of what is heard and the very conditions of reproducibility are thereby presented as if they spring forth from the technology. This is a convenient narrative for advertisers with new hardware to sell, but it is not an especially compelling historiographic frame.

Even if it were, we would be confronted with an intractable descriptive problem: after 1878, every age has its own perfect fidelity. The Victor ad campaign that began in the first decade of the twentieth century extended through the 1920s—making identical claims for vastly different phonographs. A 1924–27 ad campaign for Victor's Orthophonic Victrola would assure its readers that "the human voice *is* human on the New Orthophonic Victrola" (figure 30).[20] It is a short step from Caruso's voice in 1913 to the human voice in 1927. The later ad raises the same philosophical question raised around Caruso in 1913: in which times and places might the human voice *not* be fully human? As I argue below, using the case of the Orthophonic Victrola as my example, the idea of "better" sound reproduction was itself a changing standard over time.

Rather than following a narrative history, the chapter proceeds conceptually, starting with a discussion of the apparatus of sound reproduction—technologies, institutions, and people organized into networks. On the basis of this history, I argue that sound reproduction is "always already" a kind of studio art. Even the most basic functional questions about sound technologies presume this relation. If we consider media as recurring relations among people, practices, institutions, and machines (rather than simply machines in and of themselves), we can then say that the media of sound reproduction are the material conditions for contemplation of the original/copy distinction. Both the discourse of true fidelity and an alternative discourse of artifice in sound reproduction develop out of the social configurations embodied in the new sound media.

If, as I have argued, the physical, practical, and social formations of sound reproduction were themselves historical artifacts, then the very proposition that sound technologies actually worked must also be historicized. In the realm of mechanical reproduction, even the most basic level of "function" implies an imagined or realized set of social relations. One could also add that people *wanted* the machines to work—so much so that, in many early cases, sound-reproduction technologies worked only with a little human help.

The second half of this chapter considers the development of audile technique in relation to sound fidelity. Any notion of sound fidelity or construct of sound reproduction as a form of mediation requires some kind of audile technique: a particular kind of listening for detail and a particular relation between listener and instrument. As I argued in chapters 2 and 3, this technique developed over the course of the nineteenth century in a variety of different contexts and was later articulated to the media of sound reproduction. Audile technique is prior to the possibility of any "faithful" relation between sounds. But an established set of audile techniques does not necessarily lead to the apprehension of sound fidelity as a natural

Figure 30. "The Human Voice *Is* Human"—Victor Victrola advertisement, 1927 (courtesy Archives Center, National Museum of American History)

outcome. Unlike the sounds of the body heard through the stethoscope or the staccato Morse code, the reproduced sounds apprehended through audile technique were now supposed to be transparent, that is, *without a code,* and therefore immediately apparent to any listener who knew the technique. Ultimately, this would lead to a conflicted aesthetic of reproduced sound, where the ideal state for the technology as vanishing mediator would continually be set in conflict with the reality that sound-reproduction technologies had their own sonic character. The aesthetic notion that the best medium was one that was the "least there" thus served as an inverted image of sound reproduction's social existence: the more "there" it was, the more effective it could be.

Studios and Networks as Diagrams of Sound Reproduction

Any medium of sound reproduction is an apparatus, a network—a whole set of relations, practices, people, and technologies. The very possibility of sound reproduction emerges from the character and connectedness of the medium. Early graphic representations of sound reproduction frequently gesture to this "networked" aspect of the seemingly mechanical process of reproduction; the orders depicted in early graphic representations of sound reproduction show sound-reproduction technology to be embedded in networks that are simultaneously social and technical.[21] Consider the many social connections in even a simple representation of radio broadcasting. RCA's depiction of how radio works (figure 31) shows how even a "merely" technical network implies a larger field of social relations. The singer and the hearer are at two opposite ends (although there are many more than two people) of a vast technical network. The sound becomes electricity, is manipulated as electricity, and is remade as sound. Although the imagery is heavy on wattage and vacuum tubes, it also implies a larger social sphere: the theater, RCA's control room and broadcast transmitter rooms, the broadcast area, home reception. Implied just beyond the picture are the theater audience, the technicians servicing the microphones and transmitters, the radio network people setting up the concert as a radio performance, the musicians' unions, the store where our domestic listener purchased her radio set, and the rest of the radio listeners within the hundred-mile radius of the station. In short, this picture uses the technology of radio as a placeholder for a recurring relation among people, practices, institutions, and

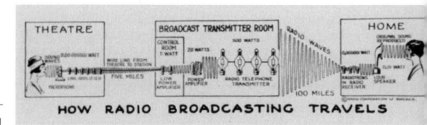

HOW RADIO BROADCASTING TRAVELS

machines. The medium does not mediate the relation between singer and listener, original and copy. It *is* the nature of their connection. Without the medium, there would be no connection, no copy, but also no original, or at least no original in the same form. The performance is for the medium itself. The singer sings to the microphone, *to the network,* not to the woman listening at the other end.

The network depicted in the center of this diagram is very much the centerpoint of sound-reproduction practices. Sounds themselves come to exist in the first place in order to be reproduced through the network. They are not plucked from the world for deposit and transmission. This is a crucial distinction. The medium is the shape of a network of social and technological relations, and the sounds produced within the medium cannot be assumed to exist in the world apart from the network. The "medium" does not necessarily mediate, authenticate, dilute, or extend a preexisting social relation. This "network" sensibility is widespread in the iconography of reproducibility—there are many images of performances for the network, sounding to a machine so that the machine might then reproduce the sound. Nineteenth-century depictions of telephone communication use a similar "network" iconography (figures 32–33).

It is significant—if I may be allowed a brief excursis—that women appear on the listening end of the three network images discussed thus far. At the end of chapter 3, I argued that eroticized images of headphone listening were about eroticizing physical distance and suggesting the depth of interconnection made possible by bodily absence. These networked images of telephony and radio carry some of that audile-erotic sense, but the

Figure 31. "How Radio Broadcasting Travels"—RCA diagram, 1920s (courtesy Archives Center, National Museum of American History)

Figure 32a–c. Advertising cards that depict telephone networks, ca. 1881 (courtesy Archives Center, National Museum of American History)

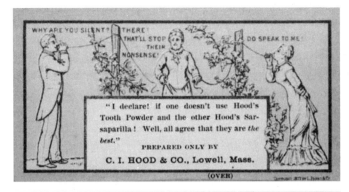

32a–c

participation of the listening white woman goes beyond eroticizing the connection itself. Andreas Huyssen's famous argument that mass culture is represented as female seems apropos here as well, but with a twist. These images of women participating in the sound-reproduction network are not metaphoric—as are Huyssen's examples. Rather, they are literal. The emergent media of the late nineteenth and early twentieth centuries grew alongside a whole class of women who were full participants in mass culture—and that participation was on an unprecedented scale. The listening white woman thus supplanted the image of the Victorian woman expressing herself and entertaining the family at the piano. This change was as much a result of real participation of women in emerging networks of sociability—including the networks of sound reproduction—as it was a result of the "image" of mass culture and new media as somehow feminized.[22]

Although popular and commercial images featured women, lab drawings and patent caveats tended more to feature men. Since research labs were almost entirely men's spaces at the time, this should come as no surprise. Still, these images also amply illustrate the "network" sensibility articulated around sound-reproduction technologies. In a series of lab draw-

Figure 33. Artist's illustration of Elisha Gray's telephone, 1890

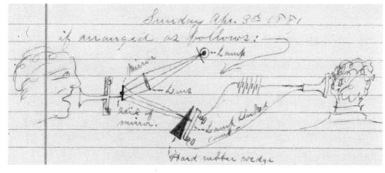

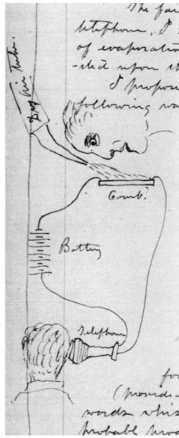

Figure 34. Drawing from the home notebook of Charles Sumner Tainter, 3 April 1881 (courtesy Archives Center, National Museum of American History)

Figure 35. Drawing from the home notebook of Charles Sumner Tainter, 19 November 1882 (courtesy Archives Center, National Museum of American History)

ings of various sound-reproduction scenarios, Charles Sumner Tainter affixes disembodied mouths, heads, and ears to the machines pictured at the center of the images. Published representations of the graphophone and the photophone follow this style (figures 34–40).[23] Like the ear phonautograph—where a human ear literally becomes part of a machine[24]—these depictions of the telephone, the photophone, and many graphophones all

Figure 36. Dictating and listening to the graphophone, 1888 (courtesy Division of Mechanisms, National Museum of American History)

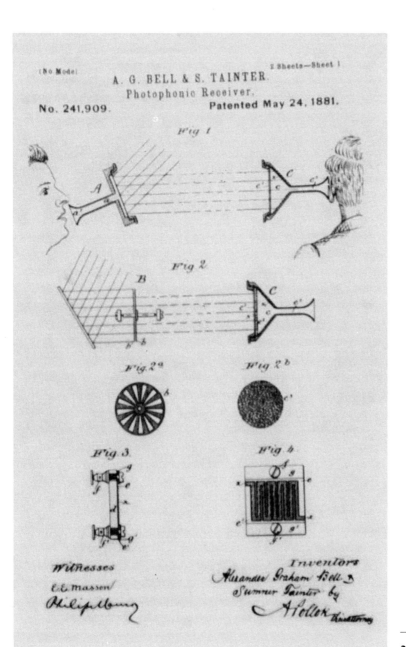

Figure 37. Patent drawing for photophonic receiver, 1881

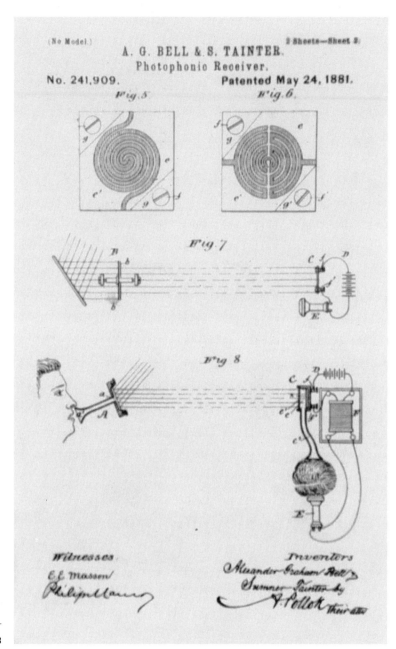

Figure 38. Another patent drawing for photophonic receiver, 1881

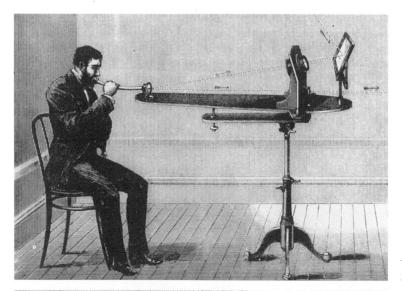

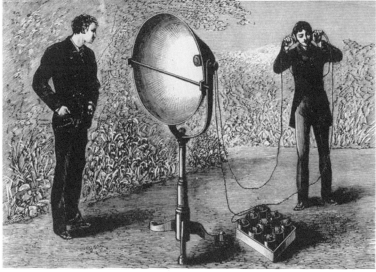

involve ears and mouths in close proximity. Although they are now attached to living people—and modeled in the machine itself—they retain a clear functional importance in the images. They are strictly interchangeable from the perspective of the machine's functioning; they are parts of the process. The tympanic function is now doubled—once immediately inside

Figure 39. Speaking to the photophone, 1884
Figure 40. Listening to the photophone, 1884

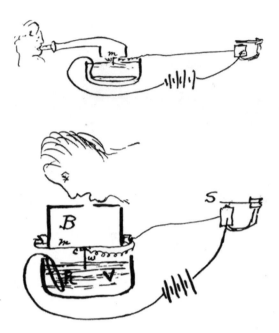

41

the machine and once immediately outside the machine. Arranged in a network, the tympanic machines supplement the faculty of hearing. Bell's "machine to hear for them" now included the hearing of those who could hear as well. If the network can hear you, others can too.

The standard practice of sound reproduction did not differ much from the manner in which Alexander Graham Bell had first portrayed the telephone (figure 41): the sounds are made for and through the network. The network is at the center of the picture. Speakers and auditors take their voices and ears to the network. This is clear from the most common model of telephone conversation or any other point-to-point communication like early radio, where a person speaks into the apparatus specifically in order to have his or her speech reproduced elsewhere in the network. As a 1923 AT&T ad put it, "An effective telephone personality is to-day a business and social asset. . . . The Bell System maintains for telephone users the best facilities that science, modern equipment, skilled operation and careful management can bring to telephone speech. But these facilities can be fully effective only when they are properly used."[25] In this quote, we find the studio and the network logics clearly connected—because the Bell System

Figure 41. Telephone drawings from Alexander Graham Bell's notebooks, 1876

is a specialized network of relations, users must acclimate themselves to it. Sonic events produced for mass redistribution through recording or broadcast similarly operated as studio undertakings from the very beginning. The iconography of reproducibility suggests a relation of sound reproduction where networks connect people, rather than machines mediating or extending existing auditory relations.

Another possible interpretation of these images immediately arises: that, in these pictures, the people (or parts of people) pictured at either end of the network are the important elements; the network is, thus, merely a means to an end. Certainly, if one were to read these images in terms of authorial intention, this is very often the case. RCA put wattage numbers instead of the names of technicians and departments in their pictures because they wanted to highlight the technical and perhaps even "automatic" aspects of the process. Tainter and Bell put the machines in the middle of their drawings because they are primarily working on the machines, not because they thought that they held a place for a larger set of social relations. This much should seem clear and plausible. But these images are also maps; they make some connections and imply still others. This is not to move from a "manifest" to a "latent" content of the images; it is simply to draw out a set of connections of present interest.

Alongside images of sound reproduction, we have accounts of sound reproduction that follow the same implied logic of "the network" as that found in the images reproduced above. People performed for the machines; machines did not simply "capture" sounds that already existed in the world. While the modern recording studio is largely an invention of the mid-twentieth century, recording has always been a studio art. Making sounds for the machines was always different than performing for a live audience. Live or on-site recording of music or reproduction of sound via radio or telephone was extremely rare until the 1920s. Even in so-called live situations, the machine required a certain amount of attention, care, and technique. Sometimes, actual spontaneity would interfere with the recorded appearance of spontaneity: for instance, a particularly well-selling Victor record of a London street scene came out of a civic group's efforts to reduce traffic noise. Their intention in making the recording was to provide Parliament with a sense of what the traffic sounded like. The recording took over twenty takes because the police constable standing near the recording gramophone kept interfering by making comments into the recording gramophone's horn like, "That's *so* unnecessary," and, "By God!" The people who commissioned the recording clearly felt that the constable's spontaneity interfered with the "spontaneous" capture of the street scene. Recording did not simply capture reality as it was; it aimed to capture reality suitable for reproduction. Sponteneity was spontaneous only through artifice.

Considered as a social process, sound reproduction has irreducible social and spatial components. Without studios, and without other social placements of microphones in performative frames that were always real spaces, there was no independent reproducibility of sound.[26] The studio in particular implies a configuration of bodies and sounds in space, a particular ordering of practices and attitudes. Its significance is at once technical, social, and spatial. The studio becomes a way of doing things and a social frame for reproducibility. As with any cultural practice, sound reproduction has an acutely spatial dimension, and the space of the studio was radically different from other performance spaces. Performances for reproduction focused on bodily disposition and affective states. This is contrary to the often-made claim that reproduction decontextualizes performance and deterritorializes sound. Considered as a product, reproduced sound might appear mobile, decontextualized, disembodied. Considered as a technology, sound reproduction might appear mobile, dehumanized, and mechanical. But, considered as a process, sound reproduction has an irreducible humanity, sociality, and spatiality.

From the very beginning, recorded sound was a studio art. From before the technology was commercially available, users were aware of the special conditions of sound production accompanying reproduction. In the midst of experimentation, Chichester Alexander Bell wrote of the physical contortions necessary to get one's mouth close enough to the mouthpiece to get a good recording: "With the mouth in such a position, not only is it very difficult to talk in a natural manner, but it is obvious that sound waves within the mouth-piece must interfere with each other."[27] Even a cramped loft studio was better than the best spontaneous conditions. Eldridge Johnson, commenting on his work with the gramophone, remembered: "We had no place for the singer to record except in a loft that you got to with a ladder. I would scurry around and get some poor devil to come and sing for a dollar in real money and then I'd push him up the ladder and try to get a record."[28] The studio was a necessary framing device for the performance of both performer and apparatus: the room isolated the performer from the outside world, while crude soundproofing and physical separation optimized the room to the needs of the tympanic machine and ensured the unity and distinctness of the sound event being produced for reproduction. As Steve Jones points out, sound engineers quickly learned to prefer studio recording to on-location recording because the studio allowed them to control the acoustic environment much better—and thereby to control the actual sound of the recording.[29]

Like those created for sound recording, the sound events broadcast by radio were primarily not existing ones but manufactured ones. An early account of the broadcasting of opera emphasizes the qualities of studio work: the smallness of the room, the abstraction of the music and the singing from the rest of the operatic performance, and the special training of the singers. After getting the singers to abandon all visual aspects of their performance—facial expressions, movements, costume—the issue of "maximum tonal effect" became paramount: "This was accomplished by introducing a shifting process, each singer having a fixed position from which he moved forward, backward, and sidewise according to a prearranged scheme, precisely like a football line that opens and shuts and moves by a code of signals."[30]

The title of the article from which this quotation was taken eliminates any doubts about the author's view of the difference between live performance and performance for reproduction: "How Opera Is Broadcasted: Difficulties That Must Be Overcome in Order to Obtain the Best Results; How Singers Must Be Especially Drilled and Grouped, and How the Opera Must Be Revised, Interpreted, and Visualized to Make Up for the Lack of Action, Costumes, and Scenery; Artists Are Put in a Musical Straitjacket; Moving, Whispering, Even Deep Breathing a Crime." Clearly, the author had the standard disdain for recording shared by some performing artists of the time. But analysis can disentangle the description of the event from its aesthetic evaluation. Although you or I might like recorded music much more than this author does, his description of the recorded operatic performance is essentially correct. The physical placement of performers during the recording process is different from that during live performance, as is the entire presentation of the opera. This is the salient point for all reproduction: it is not just eavesdropping on live performance; it is a studio art.[31]

Location was everything. Studio work was widely understood as a practice entirely different from live performance. Early accounts of singers' performances for reproduction frequently focused on bodily disposition and affective state. Thomas Watson wrote that, when a manager replaced him with a hired singer for a telephone demonstration, the singer was "handicapped for the telephone business by being musical, and he didn't like the sound of his voice jammed up in that way."[32] Singers making recordings for the phonograph had to contort themselves as well:

> Now he [the singer] throws back his head, now thrusts it forward, now poises it this way and now that. All this would look ridiculous to an audience, but is necessary before the Phonograph. The force of the note must be accommodated to the machine. If the composition calls for unusual force in propulsion, the singer must hold his head back so that his voice may not strike the diaphragm of the Phonograph too violently; if, on the contrary, the music is soft and gentle, the head must be brought nearer the receiving horn, so as to make the due impression on the wax.[33]

The physical affect could invoke different states of mind in performers as well:

> It is often difficult to get the proper attitude on the part of the singer. Curiously enough, some of those who seem to lose themselves when on a stage, confronting an audience, appear to be terribly self conscious when they face the machine.
>
> There is such a thing as "stage" fright in performing for the phonograph. I do not know how to explain it, whether it comes from the thought that the record will be reproduced far away from the singer's presence and perhaps long after he is dead or from some other reason. But I have observed it many times and in some noted persons.
>
> There are some singers from whom it seems impossible to get a perfect record. You know it is only recently that the voices of women singers have been taken to any extent. Their high and fine tones are apt to shrill and shatter when transferred to the rolls.[34]

Getting a good recording was, thus, a matter of tonal response *as well as* conditioning the performer to an entirely different kind of performance. Early phonograph and radio performers reported unusually intense stage fright before the apparatus; even the editor of a magazine dedicated to promoting sound recording publicly confessed that fear overtook him the first time he attempted to make a record.[35] These reactions were largely responses to the physical environment of the studio and both the imagined and the real social relation of reproduction. Perhaps the fear was a reaction to the unfamiliar surroundings; perhaps it lay in the idea of a massive, remote audience instead of an immediate and close one. Regardless, seasoned performers' fears signaled the distinctiveness of the studio space.

The singer Leon Alfred Duthernoy described his first experience performing on radio as evoking horror, which gave way to tremendous gratification. In this way, it is emblematic of the agonistic relations that live performers had to studio performance. In his narrative, Duthernoy emphasizes his successive affective states. He reports that he had expected the performance to be more or less like a live performance but found the difference between performance for a physically present audience and performance for reproduction terrifying. On entering the studio, he was immediately frightened by the thought of performing for a huge anonymous mass. The description of his distress is particularly apt: "In my mind I visualized a life-size map of the United States, and in every town, every hamlet, every cross-roads, there was *nothing but ears*. And all of these countless thousands of ears were cocked and pointed in my direction. I could see ears sticking out from behind library tables, book-cases and sideboards: the handles were ears, the glass knobs were ears, *and they were waiting for me.*"[36]

Duthernoy's account, along with the accompanying pictures, provides a narrative version of the various diagrams of sound reproduction. Mouths and ears fit loosely at either end of the network, with the machine in the middle. Duthernoy and his thousands of distant ears were joined by the medium of radio. His imaginative vision of his performance draws from the same sources as Tainter's representations of sound reproduction in his lab diagrams. This was not a permutation of live performer-audience interactions but something else entirely. Radio performance offered a peculiar—and, at that moment, terrifying—configuration of bodies and spaces. As Duthernoy stepped to the microphone, he made his own map of the network for which he was about to sing. Moments before his studio performance began, his very active imagination offered a striking representation of sound reproduction as an eminently social relation.

The studio struck Duthernoy as an incredibly uncomfortable environment—a room entirely unadorned except for the potato sacks hanging from the ceiling to dampen the sound and the "tin can" microphone in the center of the room. The announcer spoke in a perfectly modulated voice and instructed him to move closer to the microphone when singing quietly and farther away when singing loudly. That was all, and, after the announcer's short introduction,

I sang the aria to the tiny tin can. When I had finished, the room seemed dead. The piano had stopped reverberating and there was not the slightest sound. . . .

The attendant then went over to the transmitter and announced that I would sing two songs. . . . This I then proceeded to do. At the end, there was the same dull, empty silence. I would have given anything for even a pathetic pattering of applause. It was my meat and drink—my board bill. But no—not a sound, not a flutter of a programme. I felt like a bell tinkling in a vacuum—you know the example we used to have in high school in physics. I swore to myself that of all the stupid experiences, singing through a tin can was the most stupid.[37]

Duthernoy's experience had a happy ending—a few phone calls to the station requesting encores assured him that the audience was listening and afforded a degree of interactivity, even if it was delayed. Performer-audience interaction was replaced with listener response and delayed gratification: "When unseen and unknown people clamor to hear you sing, it is far more to be desired than the roaring applause in the concert hall."[38] Here was a Protestant ethic for performers! Perhaps that potato-sack-filled room resembled for a few moments the "iron cage" of which Weber had written seventeen years previously.[39] Perhaps Duthernoy had a glimpse of a future, hollow existence. But he put his faith in the machinery, the medium, and his link to listeners, and he felt that his faith was rewarded. Thomas Watson writes similarly of his brief experience performing on the telephone for lecture audiences. Once he mastered performing for the machine, he "always felt the artist's joy when I heard in [the telephone] the long applause that followed each of my efforts."[40]

This truly aural universe of reproducibility, where the ears hang off radios and household furniture alike, points to the strangely human artifice of sound reproduction. Duthernoy's experience remains fairly typical of live performers who first come to a studio situation. He immediately apprehended the difference between performance for reproduction and live performance. All elements of the sonic event are isolated from one another and recombined in an entirely different form of experience. This experience, in turn, is rooted in the sociality of the event. Duthernoy went in assuming that live performance and radio performances are simply two instances of the one social practice. His fear was a result of the realization that they are *not* in fact the same thing. His gratification returned when he managed to convince himself that reproduced is in fact better than live—"far more to be desired." The "live" and the "reproduced" performance practices exist in relation to one another, but they are not the same thing. This was what our novice radio singer learned.

The Artifice of Authenticity

If its reproduction exists even as a possibility, sound production is oriented toward reproduction from the very moment sound is created at a "source." Sound reproduction always involves a distinct practice of sound production. As in Duthernoy's case, the sound event is created for the explicit purpose of its reproduction. Therefore, we can no longer argue that copies are debased versions of a more authentic original that exists either outside or prior to the process of reproduction. Both copy and original are products of the process of reproducibility. The original requires as much artifice as the copy. Philosophies of sound reproduction that reference a prior authenticity that is neither reproduced nor reproducible are untenable since their point of reference—an authentic original untainted by reproduction—is at best a false idol.

Even in this kind of account, authenticity does not disappear altogether, although it does change. Sound reproductions that are acknowledged as wholly artificial by performers or the audience (or both) can still come to have a sense of authenticity. But *this* notion of authenticity refers more to an intensity or consistency of the listening experience. It is a claim about affect and effect, rather than a claim about degrees of truth or presence in a reproduced sound. Certainly, listeners desired reproduced sounds that bore a purely mimetic relation to the events that they purportedly captured. Certainly, performers strove for what they would come to call *realism* in their effects. But, as many critics of film and photography have shown us, reality is as much about aesthetic creation as it is about any other effect when we are talking about media.

The case of a recording that purported to contain the last words of Harry Hayward illustrates the simultaneous desire for early sound recordings to capture events as they happened, the impossibility of that happening, and the resulting artifice of authenticity.[41] Hayward, a member of a prominent Minneapolis family, was hung in 1895 for the murder of Catherine Ging, a young dressmaker who had come to Minneapolis to establish a shop. As collateral on a loan from Hayward to start the shop, Ging had taken out a life insurance policy naming him as the beneficiary. The trial attracted considerable press coverage, as did Hayward's fortune after being convicted. His final days were reported in great detail, and several publications appeared detailing his "confession and criminal life." Following his execution, the *Minneapolis Journal* reported that, on the date of the execution, two men had entered the jail with a large package. Supposedly, this package contained a phonograph that would record Hayward's gallows speech immediately before his execution. Yet the logistics of recording the speech would likely have prevented any intelligible imprint on the record. The execution room of the Hennepin County Jail was two stories tall, with an immense gallows in the middle, and, at the time of the execution, it was crowded with people. The collector Tim Brooks estimates that the phonograph could not have been closer than twenty feet from the gallows, in which case it likely would not have been able to capture Hayward's speech in any detail. Moreover, Hayward's full thirty-thousand-word confession—given to a stenographer prior to his hanging—would not have fit on a single phonograph cylinder. On the basis of his research into the case and comparison with the actual recording, an examination of the cylinder that purportedly captured Hayward's last words, Brooks concluded that it was a composite of statements made at entirely different times by Hayward and widely reported in the press. Far from being a reproduction of the actual event, the recording was a "re-creation."

Many early cylinders contained this kind of sensational material. In addition to re-creations of the confessions of murderers (like the Hayward cylinder), re-creations of famous speeches (such as William McKinley's final speech, of which there is no extant recording) were popular, as were re-creations of Civil War battles, public festivals, and other well-known events. Re-creations were quite common in early recordings, and they served at least three useful purposes. To some extent, they advertised to listeners the affective and aesthetic potential of the medium—re-creations suggest that it would offer listeners kinds of experiences not previously available to them. This, in turn, fueled a kind of media tourism fostered in many late-nineteenth- and early-twentieth-century media: stereoscopic pictures or photographs of distant places and exotic events; films of important news events from around the world.

In all these cases, the goal was not necessarily mimetic art; it was about crafting a particular kind of listening experience. This is the old argument that realism is, at its core, a set of arbitrary artistic conventions designed to have a particular aesthetic effect. These recordings were all about the compartmentalization and commoditization of experience, rendering experience mobile and available for repeated consumption as pleasurable, shocking, or merely diversionary. Jacques Attali wrote that recordings offered a way of stockpiling "other people's use-time." By this he meant that listeners could both eavesdrop on others' experience and stockpile a set of possible experiences for themselves by taking advantage of others' sonic labor. Playing on the classic Marxist distinction between use value and exchange value, Attali framed recording as multiplying the *possibility* of use value without necessarily actualizing it. But, when applied to these re-creations, his example of stockpiled recordings suggests once again that the distinction between use value and exchange value is difficult to make in practice. It goes even beyond that—insofar as they are re-creations, these early recordings are not simply other people's use time; they are use time made for stockpiling. In the middle-class world at the turn of the twentieth century, the difference between production and consumption was becoming more apparent and felt by the day, and the sorts of portable experiences offered by early record re-creations offered listeners an easily compartmentalized form of experience, even as they highlighted the possibilities of the medium itself. As Attali puts it, *"People must devote their time to producing the means to buy recordings of other people's time,* losing in the process not only the use of their time, but also the time required to use other people's time. . . . Use-time and exchange-time destroy one another."[42]

Attali's observation should be tempered, however, because people actually did make use of the particular modalities of experience afforded by recordings for all sorts of purposes. William Kenney writes that recorded music allowed people to experience concert music that they might otherwise not have encountered: "If they couldn't soothe their mates with their piano virtuosity, wives could always slip an appropriately calming and/or uplifting record on the parlor phonograph." Although Ruth Cowan and other feminist writers have criticized the notion of laborsaving devices, it is clear from this example that sound recording offered a particular kind of use time to listeners and that the uses of recordings would be shaped by gendered and classed aspects of social life. It is almost irrelevant whether listeners often thought that they were hearing the real thing. Early recordings offered a kind of "sample" of experience in three-minute doses. In a way, this question of the uses of realism foreshadows current discussions of "the virtual"; and, as Greg Wise has argued, virtuality is above all a kind of intensity or modality of affect. It is a form of experience. The same can be said for audio realism from its beginnings.[43]

The art of reenacting events for the machine was the foundation of a now-forgotten recording genre called the *descriptive specialty*. Somewhere between a contrived re-creation of an actual event and a vaudeville sketch, descriptive specialties offered their listeners "tone pictures" of different places and events. The following sampling of descriptive specialties from the 1904 Columbia catalog gives some idea of the breadth of the genre:

> Anvil Chorus from "Il Trovatore" (with anvil effect)—Verdi
>
> Arkansaw Husking Bee, An—Pryor
>
> Capture of the Forts at Port Arthur (a scene from one of the Russian forts, with cannonading, and shriek of shells. The Russian Band is heard playing the National Anthem. The Japanese approach, headed by their band playing their National Air, and take possession of the forts, amid loud cries of "Banzai")
>
> Charge of the Light Brigade March
>
> Chariot Race March (with whistling solo)—Paul
>
> . . .
>
> Evening Chimes in the Mountains (with bell solo)
>
> Forge in the Forest, The (with bells, cock and crow and anvil effects)—Michaelis
>
> Indian Chase, An (gallop)—A. E. Loetz
>
> . . .
>
> McKinley Memorial (introducing President McKinley's last speech, and "Lead Kindly Light" by Brass Quartet)
>
> Mr. Thomas Cat (March comique, trombone imitations)—Zimmerman
>
> . . .
>
> Tone Pictures of the 71st Regiment leaving for Cuba—F. W. Hager[44]

Clearly aimed at a middle-class market, the range of pieces echoes the range of subject matter in other middle-class entertainments: as in vaudeville, there were imitations and comedies; as in stereoscopes and films, there were representations of distant events available for domestic consumption. Depending on the recording, sound reproduction was treated as a form of mediation and representation or an extension of the senses. In each case, the point was not to get as close to reality as possible but rather to establish a kind of auditory realism and, through that realism, present a distinct aesthetic experience. No matter how real the descriptive specialty may have seemed to its listeners, it was a sophisticated artifice. It was use time and exchange time rolled together in a cylinder.

The recordings were very much limited by the parameters of the available technology: narratives were short and to the point, effects rudimentary, and dialogue brief. The experience of hearing these recordings is difficult to retell in print. Some recordings were essentially medleys of music interspersed with brief dialogue and sound effects. Others consisted mostly of dialogue, interspersing the fabricated noises of a horse race or yells of victorious soldiers. Still others re-created actual events such as Theodore Roosevelt's inauguration (which concludes with a spectator saying that he has seen "every inauguration since Andrew Jackson's and this one beats them all)" or fictional scenes such as night in a clock store.[45]

Descriptive specialties were the predecessor of more enduring audio arts, such as Foley effects in film and the use of sound effects in radio drama. Rudolf Arnheim wrote that radio art allowed for "natural sounds" to be raised to a "super-realistic level."[46] While Arnheim had in mind the radio drama, the principle of "imitative art" in sound creation that he distilled was present from the very beginning. Many of the techniques of "imitative" art later standardized for sound film and radio drama were first developed for descriptive specialties. Moreover, descriptive specialties emphasized for listeners the "realism" of the medium, even if audiences were aware of the fabrication of the actual performance on record. Like "primitive" cinema, where the camera's ability to document motion is highlighted, descriptive specialties emphasized the possibility for sound reproduction to present realistic and fanciful accounts alike of events over time.[47]

Of course, it did eventually become possible to reproduce events as they happened for a listening audience. But, even here, the auditory reproduction of the actual event is highly contrived; the audience hears not so much the event itself as a performance concurrent with the event. RCA's experimental broadcast of the Dempsey-Carpentier fight is an excellent example of this: from possibly the first sporting event broadcast to a mass audience, play-by-play reporting—filtering and shaping at the point of production for reproduction—was essential to the success of the endeavor. RCA received hundreds of enthusiastic letters from those who heard the broadcast. One listener wrote, "The broadcasting of the fight was simply wonderful. Even the gong sounded plainly as could be.... Never expected to hear a 'world crier' by radiophone. You must have been heard over thousands of miles. Some 'Town Crier,' I'll say. Almost thot [sic] I was in the front row at the ringside when you counted Carpentier out. It was realistic and impressive to the highest degree." The criterion here is realism, not reality itself. An internal RCA report was clear on this matter: RCA broadcast a "voice description" of the fight, not the fight itself.[48] Once again, to borrow Roland Barthes's language, the point of the artifice is to connote denotation, to construct a realism that holds the place of reality without being it.[49] Like the descriptive specialty, the live radio broadcast did not so much capture the event as it became an event in itself. Even today, voice descriptions of sporting events are a crucial part of their audio and audiovisual broadcast.

The goal in reproducing live events was not reproducing reality but producing a particular kind of listening experience. Early sound reproduction—whether live or wholly contrived—sliced up reality in order to fashion a new aesthetic realism. The point was never to capture the event in its positivity but rather to create a new form of sonic realism appropriate to the events being represented and to the listeners auditing them. The desire for sound-reproduction technologies to capture reality and faithfully reproduce it thus quickly gave way to the use of those technologies to fashion an aesthetic realism worthy of listeners' faith. But, even with this level of stability, a set of conventions was required for an original to be regarded as real or at least realistic. If this was true for casual listeners enjoying a version of aesthetic realism, it was even more the case for expert listeners trying to determine whether sound-reproduction technologies could effectively reproduce sound at all.

Functional Aesthetics and the Very Fact of Reproduction

It is perhaps obvious to state that early users and experimenters began by asking whether the machine "worked." But embedded in such an apparently innocent functional question were social and aesthetic issues: could the various sound-reproduction technologies function as media? There is no telephone, phonograph, or radio without telephone, phonograph, and radio "systems." Even in a highly atomized and exceptionalized state (where there were only a few in the world), telephones, phonographs, and radios could be understood as (at least potentially) part of networks, however small. This was clear in the technology as well as in its use: telephones and radios required transmission and receiving apparatus; phonographs required production and reproduction styli. This is to say that, at a very basic, functional level, sound-reproduction technologies need a great deal of human assistance if they are to work, that is, to "reproduce" sound. This is tautologically the case when we consider technologies as media since recurrent human relations are a core component of any medium. But it goes deeper still: even when testing these technologies to see whether they worked, their users provided all sorts of assistance to the machines. This is to say, when sound-reproduction technologies barely worked, they needed human assistance to stitch together the apparent gaps in their abilities to make recognizable sounds. This is something of an inversion of

Bruno Latour's "delegation" thesis. Latour argues that we delegate certain social functions to technologies. They do things for us.[50] But, in the case of early experiments in and demonstrations of sound reproduction, we can also see the converse, where people delegate their skills to technology in order to help it work.

At the most basic level, sound-reproduction technologies presumed some kind of social network, a coordination of people and actions over time and space; they were *partial* machines that, from the outset, depended on the presence or possibility of other machines. In these earliest moments, listeners could not assume even a basic level of functioning. It is something of a wonder that they extrapolated from early sound-reproduction technologies—machines that could barely reproduce sound!—great possibilities for their future as media. If we consider those moments when the threshold of reproducibility is under scrutiny, when auditors were considering whether the machines worked at all, we begin to understand how social these apparently technical beasts really were. Although inventors rarely highlighted the fact in their own writing, early versions of sound-reproduction technologies constantly required human assistance in order to reproduce recognizable sounds. From the outset, they had a little help from their human friends.

Of course, early accounts of sound reproduction focused on the bare fact of reproducibility: could a listener hear a copy of a sound or not? But, in the process of creating and testing a machine designed to reproduce sound as such, certain types of sound were privileged as ideal testing material—specifically, easily recognizable forms of human speech. This kind of speech was limited and particularly conducive to reproduction, that is, easily understood by a listener with relatively few explicit cues to go on: rhymes, popular quotations, newspaper headlines, queries as to the effectiveness of the transmission (such as, "Can you hear me?"), and instructions for action were among the most commonly used "tests" for reproducibility. In other words, conventionalized language helped the machine along in doing its job of reproducing. It enacted the *possibility* of reproduction before that function could be fully delegated to the machine. From its very beginnings, sound reproduction required a certain level of faith in the apparatus and a certain familiarity with what was to be reproduced.

Early functional accounts of the telephone illustrate this point, and not just because Alexander Graham Bell was a noted elocutionist. The point is, however, worth making: what speech could be *more* fit for reproduction than that of an expert in elocution? Bell's first success with the telephone is well-known: American children still learn in school that Bell said, "Mr. Watson—Come here—I want to see you," although this interaction did not become public knowledge until 1882, six years after the telephone was introduced to the public. This famous remark brought Watson downstairs to Bell to repeat what he had heard, verbatim. The men then changed places, and Bell listened to Watson read a few passages from a book. Bell wrote in his notebook two days later: "The effect was loud but indistinct and muffled. If I had read beforehand the passage given by Mr. Watson I should have recognized every word. As it was I could not make out the sense—but an occasional word here and there was quite distinct. I made out 'to' and 'out' and 'further'; and finally the sentence 'Mr. Bell do you understand what I say? Do—you—un—der—stand—what—I—say' came quite clearly and intelligibly."[51] Intelligibility was clearly linked to conventionality at this early stage. Speech that could be easily interpreted on the basis of little actual audio information—a call, a query, a cliché—was more likely to be understood over the telephone's lines. Bell's "if I had read beforehand" qualification shows the degree to which early technological reproduction relied on the human capacity for linguistic reproduction: had he known in advance what was to be said, he might have heard what was said! Listeners were lending their memories to machines.

The telephone's public debut at the Philadelphia International Exposition on 25 June 1876 offers an even more elaborate example of this connection between functional demonstration and aesthetic choice. In the southeast corner of the main building, Alexander Graham Bell set up a small table, where he laid out his apparatus for the exposition: a harmonic telegraph, a modified König manometric flame, and the receiving apparatus for a rudimentary telephone (the transmitter was about a hundred yards away at the north end of the wing). Bell spent the day touring the technical exhibitions with a group of over fifty people, including the exhibition judges, noted scientists, and the emperor of Brazil. Bell's turn came at the end of the day and in one of the hottest parts of the hall. Without the emperor's interest, there might have been no demonstration of the telephone that day. Of course, one can only speculate as to why the emperor was so particularly interested in the telephone. But, as I noted in the previous chapter, the very modern-seeming idea of new media bringing people together has an analogue in a very old idea of the figure, voice, and presence of a ruler bringing people together.[52]

Whatever the emperor's motivation, the result is now a canonical story in telephone history. After a demonstration of harmonic telegraphy and the electric modulation of a vibrating diaphragm through the manometric flame, Bell moved to the demonstration of his telephonic apparatus. Sir William Thomson, a well-known Scottish scientist, was the first to listen to the apparatus, as the noted elocutionist Bell retreated to the north end of the gallery for the first ever "mediated" public vocal performance. Thomson first heard singing and then made out the words, "Do you understand what I say?" which he shouted in repetition. Running to Bell to confirm his audition, he quickly returned to the receiver for more. Emperor Dom Pedro II of Brazil then took his turn: Bell was now reciting *Hamlet*—"to be or not to be." Bell spoke, the emperor heard, the emperor repeated. Even Bell's later competitor Elisha Gray heard: "I listened intently for some moments, hearing a very faint, ghostly, ringing sort of a sound; but, finally, I thought I caught the words, 'Aye, there's the rub.' I turned to the audience, repeating these words, and they cheered."[53] The marvel of the machine was not that it reproduced sound well (of course it didn't) but that it reproduced sound at all; this was cause for applause in and of itself: here was an aesthetic of function.

Perhaps it should be no surprise that early aesthetic choices aimed precisely at proving functionality. At the same time, we must not forget that these choices are aesthetic, not simply instrumental. That fall, *Scientific American* authenticated early telephone experiments and demonstrations by printing transcripts maintained by Bell and Watson at the Boston and Cambridge ends of the connection, respectively. The *Scientific American* writer reported that "articulate conversation then . . . took place. . . . The sounds, at first faint and indistinct, became suddenly quite loud and intelligible." Apart from a few minor gaps—Bell's "I think we were both speaking at the same time" in Boston became Watson's "I think . . . at the same time" in Cambridge—the transcripts were more or less the same.[54] Bell and Watson *could* have spoken about anything. That they chose to have a conversation on the phone about having a conversation on the phone highlighted the process, the medium, the relation, and the possibilities of interconnection. Moreover, the simplicity of the conversation topic allowed Bell and Watson to lend a little assistance to the telephone itself. At the level of semantics and reference, there was little meaning to the conversation. At the level of practice, the meaning of the conversation was that the medium worked.

Stories of Edison's invention of the phonograph follow a similar narrative logic. As Andre Millard recounts, Edison's staff put together a rudimentary phonograph—a telephone speaker, an indenting stylus attached to the diaphragm, and a strip of paper coated with paraffin wax that was run under the stylus as Edison spoke into the machine. Visual examination of the strip showed that the stylus had clearly made an irregular indentation in the wax. When the strip was pulled back under the stylus, the group of men in the workshop could faintly hear Edison's voice. Although the sounds were inarticulate, the staff knew that they had discerned a principle by which speech could be reproduced. Edison wrote in his notes that "the spkg vibrations are indented nicely & theres no doubt that I shall be able to store & reproduce automatically at any future time the human voice perfectly [sic]."[55] Nearly six months later, when Edison's lab was able to construct a fully functional phonograph (one that allowed individual words to be made out distinctly), Edison's famous test quote was again language easily remembered and easily understood:

Mary had a little lamb
Its fleece was white as snow
And everywhere that Mary went
The lamb was sure to go.

These laboratory demonstrations were the model for public performances later on.

Demonstrations of the telephone and the phonograph in public lectures from the late 1870s and through the 1880s were oriented around the machines' limited and in some sense borrowed capacities: to reproduce language that the audience either already knew or could otherwise readily understand or did not need to understand (such as reproductions of music that the audience had not heard). During the spring and summer of 1877, Bell and Watson went on the lecture circuit, with Bell demonstrating the virtues of the phone and Watson performing (and managing other performers) on the other end of the line. Telephone concerts consisted of pitched circuit breakers, coronet solos and sometimes a small brass band, and an electrical organist (figure 42a–b). The star performance, however, was Watson's; he would prove that the telephone could "speak and sing" through his own vocal performances: "I would shout such sentences as 'How do you do?' 'Good evening,' and 'What do you think of the telephone?' which they could all hear, although the words issued from the mouthpieces rather badly marred by the defective talking powers of the telephones of that date. Then I would sing. . . . [My] repertoire always brought down the house."[56]

Watson's self-deprecating humor aside, his was a major accomplishment. Proving that the telephone could speak was no easy feat: telephone company representatives would have to demonstrate the possibility of telephony to skeptical audiences personally and assure non-English speakers that the telephone "spoke" their language as well.[57] A newspaper report of a May 1877 performance acknowledged the difficulty of hearing the telephone concert and mused that an announcement of the program beforehand meant that "there was no imagination to help out" the audience.[58]

One could easily dismiss the conventionality of the language and the use of imagination in these performances. To play on McLuhan, the medium was the message: the point was to demonstrate that the technology actually *could* reproduce sound. But that would be to miss the point—any kind of banal speech would (theoretically) have served this function in public performance. By the use of clichéd and conventionalized language, early "performers" of sound reproduction helped listeners help the machine reproduce speech. Considered in retrospect, the fascinating aspect of automated reproduction is not in the machine's automatic function, as is often noted. Instead, what is truly fascinating is the automatic response of the speakers and listeners: to *help the machine.* This speaks to a matter of desire—not desire in any deep psychoanalytic sense, but simply the desire for the machine to work. Of course, these early demonstrations would suggest that they were primarily aimed at marketing and promoting the new technology, making the machine and the process as desirable to audiences as possible. But the performances were also about the technologies' possibilities as media: their *potential* to be linked together in technical and social networks. In writing of communication between people, John Peters calls the longing for connection with distant others the *eros* of communication—for him, it explains why people want to communicate and their fascination with media.[59] But, in these early moments of sound reproduction, not only did the eros of communication radiate from person to person, and not only was it a register of distant interpersonal longing: it also radiated from people to machines. If there is a story of love and longing in these early performances of sound reproduction, it is love and longing for the machine and the process that it enacted. Performers and audiences collaborated with machines.

GRAND CONCERT,

Vocal and Instrumental,

COMPLIMENTARY TO

Miss Clara O. Willard,

AT THE

PRESBYTERIAN CHURCH, HIGHLAND PARK,

Tuesday Even'g, Dec. 29, 1874.

THE BLANEY LODGE QUARTETTE,

FROM CHICAGO.

W. C. COFFIN,	First Tenor
W. M. GOODRIDGE,	Second Tenor
D. A. KIMBARK,	First Bass
C. C. LEFLER,	Second Bass

And other Favorite Vocalists will assist, and a UNIQUE AND EXTRAORDINARY FEATURE will be the first public exhibition of ELISHA GRAY'S

ELECTRIC TELEPHONE,

By means of which, a number of familiar melodies, transmitted from a distance, through telegraphic wire, will be received upon Violins and other instruments, within the room.

CONCERT WILL COMMENCE AT 8 O'CLOCK, PRECISELY.

Tickets of Admission, adults, 50 Cents; Children, 25 Cents; Reserved Seats 25 Cents extra. May be secured at Cummings' Drug Store St John's Avenue, or at Dr. J. C. Dean's Dental Office, 67 Washington St., Chicago.

Chicago Evening Journal Print, 159 and 161 Dearborn St.

THE

STEINWAY HALL PROGRAMME.

VOL. I. NEW YORK, MONDAY, APRIL 2D, 1877. NO. 94.

TELEPHONE CONCERTS.

TRANSMISSION OF MUSIC BY TELEGRAPH.

TRIUMPH OF AMERICAN SCIENCE.

Mr. MAURICE STRAKOSCH has the honor of announcing that the first public performance on Professor ELISHA GRAY'S marvellous

TELEPHONE

will take place on MONDAY EVENING, April 2d. Musical Melodies will be performed in Philadelphia and distinctly heard by the audience in New York.

The following melodies will be performed MONDAY EVENING, April 2, on the Telephone:

1. "HOME, SWEET HOME."
2. "COME GENTIL," (Don Pasquale).
3. "THEN YOU'LL REMEMBER ME," (Bohemian Girl).
4. "LAST ROSE OF SUMMER."
5. "M'APPARI," romance (Martha).
6. "THE CARNEVAL OF VENICE."

The following eminent Artists will appear:

MME. CARRENO SAURET, THE CELEBRATED PIANISTE.
MISS FANNY KELLOGG, THE CHARMING YOUNG SOPRANO.
MISS SALLIE C. REBER, THE HIGHLY GIFTED AND BEAUTIFUL YOUNG VOCALIST.
SIGNOR TAGLIAPIETRA, THE GREAT BARITONE.
THE CELEBRATED YOUNG APOLLO CLUB, of New York, (Mr. W. F. WILLIAMS, Director).
MR. F. BOSCOVITZ, THE RENOWNED PIANIST, who will perform on the "*Telephone*" in Phil'a.

Conductors: Mr. MAX LIEBLING and Mr. W. F. WILLIAMS.

Steinway & Sons' Pianos used.

Mr. STRAKOSCH is happy to announce that Professor ELISHA GRAY will be present and superintend personally the performance of the TELEPHONE.

ADMISSION, ... ONE DOLLAR.
RESERVED SEAT, . . ONE DOLLAR AND A HALF.

Reserved Seats can be secured at STEINWAY HALL, also at SCHIRMER'S, 701 Broadway, at RULLMAN'S, 111 Broadway, and at SCHUBERTH'S, 23 Union Square.

FOR EXTENDED PROGRAMME SEE NEXT PAGE.

42b

Figure 42a–b. Posters for telephone concerts (courtesy Archives Center, National Museum of American History). The "telephone" in the left poster was probably Reis's telephone, as neither Bell nor Gray had invented their "speaking" versions of the telephone in 1874.

We can extend this analysis back into the laboratory as well—because experimenters lent the very same types of assistance to experimental sound technologies. Certainly, the laboratory is not the same kind of promotional context as the Philadelphia Exposition or a public performance of the telephone in Boston. Again, we could begin with a vaguely economic (or even just psychobiographical) reading that would locate the desire for the machine to work properly in experimenters' hopes for personal advancement, enrichment, success, and fame. We could also tell a story about the departure from scientific method—an excess of enthusiasm in the experiment that leads to experimenters giving their machines unwarranted assistance. But a more interesting reading again considers the eros directed at the technology itself. Latour and Woolgar's analysis of laboratory conversation suggests as much: the difference between a statement of solid fact and a statement that proves to be a figment of the researcher's imagination is not to be found in the referents belonging to each. The difference between fact and imagination is itself manufactured through *reflection* on the events under consideration, and reflection is always shot through with human feelings, tensions, hopes, and prejudices.[60] Certainly, these machines made sounds of differing characteristics, but what mattered in subsequent human activity—and what matters for our current purposes—is the statements surrounding them.

So, in addition to public demonstrations and famous firsts, experimentation with sound reproduction largely had the machines reproducing easily remembered and imitated language. Test sounds in the Volta Laboratory consisted of trilled *r*'s, samples from familiar or easy-to-guess passages in newspapers and books, commands such as, "Professor Bell, if you understand what I say, come to the window and wave your hat," and easily recognizable speech by members of the lab and their guests:

The phonogram that we have been using in all the experiments this past week was made on Monday, July 4th and was as follows: "Several trilled r's—then—'Mary had a little lamb, whose fleece was white as snow, and everywhere that Mary went—the lamb was sure to go.'—several trilled r's—then 'How is that for high?'—trilled r's—and—one—two—three—four—five—six—seven—eight—nine." Every word upon this phonogram could be easily understood . . . if the ear was placed close to the jet, with a pressure of less than one atmosphere, and with the air pressure at 180 lbs. per square inch the sounds were audible all over the room.

And elsewhere:

There was a girl named O'Brian
Whose feet were like those of Orion
To the Circus she would go
To see the great show
And scratch the left ear of the lion.⁶¹

The use of heavily conventionalized language helped make reproduced sound intelligible to its earliest listeners. Additionally, conventionalized language probably struck lab staff and their guests as an obvious choice for experimental recordings since it was easily performed and came to mind with little thought. This was transmission without a message. Or, rather, the message was simply that sound was being reproduced; any propositional content was purely irrelevant. But, again, the very conventionality of the language helped the machines along in their task of reproducing speech.

The use of highly conventional and therefore easily imitated language helped lower the threshold at which reproduced sound became comprehensible and still proved the possibility of mechanical reproduction of all language. A particularly clear example of this differentiated process at work was Bell's notation of some photophone work the lab had been doing. Phrases like, "Hoy—hoy—hoy," "Do you hear me—do you understand what I say," and "No extra charge for reserved seats," were easy to make out, but longer sentences with propositional content were much more difficult. Charles Bell, who was on the other end of the line, could make out only the proper nouns from his cousin Alexander's "We must note our results and give them to Professor Baird at the Smithsonian in a sealed package." Songs and commands were heard clearly by all, but Alexander could not tell the difference between "good piece of bread" and "put me to bed." He would later note, however, that a grown man yelling "put me to bed" in the middle of the day did attract the interest of the neighbors. If we read this as the Bell cousins testing a new medium as well as a new technology, their inability to hear each other at least had the side effect of demonstrating to their neighbors the difference between everyday conversation and photophonic conversation.⁶²

Certainly, these early sound-reproduction devices were barely understandable in most cases, and any practical use would take considerable practice—and this is precisely the point. Practice met intelligibility halfway. The neighbors and the Bell cousins already knew that sound reproduction requires distinct practices of sound production *and* audition. As an early telephone ad put it, conversation over the phone can be easily accomplished and understood "with practice." Perhaps, then, this is a Puritanical account of sound reproduction because only with love came fidelity.

Fidelity and the Extension of Pure Audition

Both practiced performance and practiced audition were necessary for perfect reproduction. As a studio art (even when within earshot of the neighbors), sound reproduction entailed distinctive practices of sound production. As with mediate auscultation and telegraphy, sound reproduction also required the development of audile technique. Even the earliest experiments were a form of listening practice, and, while this listening practice extended the constructs of audile technique developed earlier in the nineteenth century, it also developed them in new and interesting ways. Although there are few available reflections on *how* one should listen to an experimental phonograph, we can learn a great deal about technique by considering how early users thought about the *sound* of sound reproduction. The members of the Volta Laboratory were some of the few people in this early period focusing on the sound of sound-reproduction technologies, practicing their audile technique while shaping sound technology. We can see this in the relation between Charles Sumner Tainter's list of stated goals for improving the phonograph and his analyses of different experimental apparatus. Tainter had sought a kind of acoustic transparency in sound reproduction: ideally, the medium would disappear, and original and copy would be identical for listeners. In practice, however, this would require listeners to separate foreground and background sounds, to treat the apparatus of sound reproduction as merely incidental to the sounds thereby perceived. In other words, listeners were helping the machines reproduce sound "perfectly."

Tainter's goals for improving sound recording were straightforward enough: durability, accuracy, increased recording time, reproducibility, density (more sound in a small surface area), and ease of use. Here, he echoed Edison's own early concerns about the future development of the phonograph: Edison's list is almost identical to Tainter's, although Edison separates the practical issue of recordings' durability for multiple playings from the conceptual issue of recordings' permanence for the purposes of preservation.[63] Tainter's practical listening work as discussed in his notes was largely directed at functional issues. Evaluations of sound were written in a fairly cursory manner and without much detail: articulation of speech was clear, or it wasn't; the volume of the recording was loud, or it wasn't. Tainter wondered whether he could discern the distinctive qualities of the speaker's voice on the recording, whether he would understand the recording if he did not know what had been recorded beforehand. He sought the "best" surfaces and materials, the best kinds and temperatures of wax, the proper materials for styli, and ideal levels of pressure of stylus on the recording surface.[64] Notes from this period by Tainter's boss, Alexander Graham Bell, are remarkably similar: "It gives a loud sound"; "got a better reproduction by keeping efficient part of the record at a constant distance from the point."[65]

Tainter's aesthetic of transparency committed him to a project of erasing the medium. His ideal was a machine that produced an exact correspondence between original and copy, leaving no mark of its own process. But, as with performers, so it was with listeners: in order to accomplish this transparency, listeners had to lend a little help to the machines. Audile technique required a certain amount of faith—a belief that the machine does or at least will work. In remarkable similarity to Laennec's hopes for the stethoscope, Tainter hoped that elaborate audile technique combined with an effective technology and an appropriate practice of sound production would lead to "perfect" reproduction. But there were some important differences. Laennec posited the human ear as in need of supplementation for the purposes of listening to the body and, therefore, could cast the stethoscope's modifications of the perceived sound as "beneficial." Tainter and his colleagues wanted a supplement that would erase itself: the problem was how to supplement the recording silently—how to bring it in and out of the audible world.

One failed attempt to solve this problem led Tainter to a bizarre physics. Tainter's criteria for a "perfect" record were that changes in the physical character of the recording should correspond exactly to changes in the sonic character (e.g., loudness) of the recording. Tainter assumed that the correspondence between changes in loudness and changes in the physical character of the recording surface would translate into a correspondence between original and copy—and that this correspondence, in the guise of fidelity, would become the gold standard of reproducibility. He concluded that, since the physical contact of the stylus with the recording surface itself has an effect on the tone of the reproduction, "the conditions necessary in order to produce a perfect-record, and an exact reproduction of the sounds seem to have to be the following: the body acted upon by the voice should be suspended in the air without being supported in any way."[66] Interpreted literally, Tainter's goal was impossible at the time, but this curious physics of a recorded surface suspended in air without any physical support corresponds exactly to an aesthetic of transparent recorded sound. Tainter's aspirations toward disembodying sound led him to a theory of the apparatus itself that would disallow touch since, even in a purely mechanical form, touch would somehow distort the perfect fidelity between original and copy (both suspended in space like Tainter's untainted recording surface).

Through this logic, Tainter places the *device* as somehow outside the universe of sound reproduction. Noises made by the machine are "exterior" or "outside" sound. And these sounds must be either eliminated or tempered enough to ignore: at one point, Tainter writes, "I believe the sounds of the voice would have been audible had the outside sound been eliminated."[67] We can find a similar line of thinking in Graham Bell's notes: "In point of loudness this was all that could be desired, but it was accompanied by a loud rushing noise and was nearly as clear and sharp as the [other] sounds from the apparatus."[68] Elsewhere, Tainter complains when the recordings sound too much like recordings: *hollow, ringing,* and *musical* are just a few of the terms he uses to describe the grain of the apparatus as it is applied to the reproduced voice.[69] *Musical* is particularly interesting as a criticism here since it suggests a manifest difference between a musical instrument and a recording device: the former is supposed to shape sound, the latter to reflect it. Again, the ghost of Laennec haunts later sound reproduction: the machine must inaudibly supplement the sound, yielding only an increased effect of realism. Tainter's vision of the recording apparatus was conditioned by a desire to experience its effects while ignoring its presence.

Given that the sound quality of even a "perfect" recording in this period would be limited in volume, compressed in tonal range, and very scratchy, Tainter's description is at least as much a description of an approach to listening to the apparatus as it is a description of how the machine actually sounded. In fact, Tainter's approach to listening was so well formed and so clearly framed through a hope for pure fidelity that it does not differ greatly from the sensibility that would come to dominate practices of listening to reproduced sound. As John Corbett argues, even current listening practices perpetuate this distinction between interior and exterior sound in the recording so central to Tainter's approach:

Imagine several partitioned cubicles, each of which contains a headphoned student who faces an amplifier and a turntable; on each platter spins a record of Beethoven's Ninth Symphony. One student lifts his needle to run to the bathroom; another listens twenty times to a difficult passage; a third is frustrated by a skip in the record and proceeds directly to the next movement of the symphony; at the same time another finds it difficult to concentrate due to the volume of her neighbor's headphones. Even as they do these things that are made possible only by the technology of recording, these students are required to develop a historico-theoretical interpretation as if the technical means through which the music is accessed—right there, staring them in the face—are of no significance whatsoever.[70]

Before the apparatus even worked, Tainter was concerned with how to listen to it. Although the sounds of the machine could index the very possibility of the experience itself, they were to be treated as if they didn't exist. Film theorists have long commented that cinema's particular psychological and ideological effects were predicated on the erasure of the medium.[71] We can make a similar argument in the case of sound reproduction. The sounds of the medium in effect indexed its social and material existence—the machine could stand in metonymically for the medium. Wishing away the noise of the machine then suggests wishing away the noise of society. The relations and functions that made possible the moment of sound reproduction were labeled *exterior,* outside the act itself. (There were also moments when the machines themselves were highlighted, as in the early public performances of the telephone and the phonograph. But, even there, as I discuss both above and below, the goal was to encourage a fascination with the technology as having an agency all its own.)

When transparency was the goal for listening, the fact of reproduction was instrumentalized and ignored (or, alternatively, fetishized) in order to assert the primacy and independence of the original-copy relation that it was said to engender. While one could argue that this is how it worked in an experiential sense, there is no such thing as innocent experience—experience is always already intensely social.

Audile technique—and especially the separation of foreground and background sound into *interior* and *exterior*—was, thus, presupposed by the most basic functional criteria for sound reproduction. The point was not to produce a perfectly silent apparatus (which would not happen for decades). Rather, it was to produce an apparatus that listeners could *pretend* was silent, a machine that could hear anything but with no voice of its own—a surrogate appropriate to the audile technique to be employed. Of course, this imagined silence was itself already contingent and framed—listeners have spoken of skips on records and "telephone voices" from the very beginning—but even these "failures" indexed the possibility of perfect fidelity in reproduction. Before it could be a measurable entity, sound fidelity relied on the construction of a social correspondence among different sounds through audile technique, elevated to an almost metaphysical attitude. Sound fidelity was sound's own unique "dismal science"—it was ultimately about deciding the values of competing and contending sounds.

If "perfect" reproduction was initially reproduction that could be barely understood, eventually this pragmatic view would move from the minute details of experimental procedure to the shape of the experimental enterprise as a whole. After five years of experimentation on the graphophone, function was sufficient for marketing—further improvements of tone could come later. When Alexander Graham Bell wrote to his colleagues at the Volta Laboratory that it was time to move from experimentation to development for commercial use, he dictated his letter to a graphophone cylinder and mailed it as *both* a typed page and a graphophone cylinder to the lab. Although the articulation of the graphophone was imperfect, Bell argued that its commercial use was a practical possibility:

> I have no doubt that the articulation of the telephone can be greatly improved and yet it [*sic*] in spite of the imperfections of the instrument—hundreds of thousands of telephones are in daily use.
>
> I am quite sure also that the Phonograph in its present form may be made of great use—and I would, therefore, urge upon you both the importance of devoting attention to the mechanical details of the apparatus rather than spend all your time in attempting to improve the *character* of the articulation. . . .
>
> I hope you understand all that I have said. Spoken to Mr. Tainter's Paper Cylinder phonograph this 14th day of June 1885.
>
> In proof thereof witness my voice!
>   Alex. G. Bell[72]

Bell's voice arrived as an afterthought to the letter, and the recording was less durable than the paper and typewriter with which the written portion of the message was conceived: the written version is the only surviving part of the letter.[73] The letter thus carried with it a double signature as a fail-safe redundancy ensuring that its point would come through: the letter ensured the intelligibility and authenticity of the recording; the recording "proved" the claims in the letter.

As the insurance embodied in this example suggests, the commercial development of sound-reproduction technologies often *preceded* practical or mechanical reliability. The business phonograph was a colossal failure in the early 1890s largely because it was not a reliable machine or even very good at what it did.[74] Conversely, early telephone conversation was a learned skill, and developing a good ear for telephony was essential for making use of it. Early sound-reproduction technologies were oriented around an aesthetic of transparency, but it was enough to produce an apparatus that people imagined could work perfectly.

From Phone Tests to Tone Tests: Machines to Believe In

It was one thing for early developers to exercise their imaginations and come up with a construct of "perfect fidelity" and another thing for everyday listeners to do so. Listeners did not necessarily consider sound-reproduction technologies simply to mediate between a sound and its representation. *Mediation, schizophonia, the separation of sound and source*—these were neither foregone nor necessary descriptions of the process of sound reproduction. But they were commercially useful ways of thinking about reproducibility, and they had to be elaborately demonstrated for listeners. Early public performances of telephones and phonographs emphasized their novelty—the magic was in their working at all. Later performances, however, had a different task. As with Tainter's graphophone cylinder suspended in midair and his identification of "exterior noise," later performances sought to erase the medium (ironically, by highlighting the technology), to render it transparent, and to turn the question of reproduction into an issue of equivalence between original and copy.

The Edison Phonograph Company's "tone tests" offer an excellent opportunity to consider this matter because they contrast so clearly with earlier public performances of sound-reproduction technologies. While the earlier demonstrations simply had to convince audiences that the machines worked *at all,* the tone tests expressly sought to establish for their audiences an equivalency between live performance and a sound recording. Moreover, the tone tests presumed and made use of the series of prior conditions discussed thus far in this chapter: (1) the studio and the network as foundations for the efficacy of any sound reproduction; (2) the existence of an imitative art, a distinct form of "originality" or performance suitable for reproduction; (3) a widespread desire for the machines to work; and (4) a basic level of audile technique, especially listeners' abilities to separate foreground and background sound in the reproduction itself. Only on meeting these prior conditions could the tone tests even pretend to convince listeners that "live" and "reproduced" were so similar as to be indistinguishable.

From 1915 through 1925, the Edison Company conducted over four thousand tone tests in front of millions of listeners throughout the United States. The flow of a tone test is well characterized in an advertisement for the Edison phonograph from the *Ladies' Home Journal*. The advertisement introduces a new performer—Signor Friscoe—and then walks the reader through his performance in a tone test. He begins playing solo; then the phonograph starts to play with him; he stops playing, and the phonograph continues. Finally, a curtain is raised that reveals the phonograph to the audience. The ad suggests that audiences would not be able to tell the difference between the performer and the recording. Most tone tests were conducted by relatively unknown performers, like Signor Friscoe.[75] The Edison Company provided exacting standards for the performers, and, although it is quite unlikely that these standards were met in every case, they did tend to result in the company employing performers who were able to "play like the recording": Signor Friscoe probably played with a limited dynamic range, his timing was no doubt metronomic, and he certainly did not improvise in his performance. He was probably very skilled at repeating the *exact same* nuances and flourishes in performance after performance—the same nuances and flourishes that were on the record with which he played. That he was relatively unknown to the audience was an advantage for Edison, as auditors would not have had a prior memory of his performances or recordings to judge against the tone tests. The company, meanwhile, did its best to control the ways in which its tone tests would be covered in the press. Edison furnished press releases and advertisements to run both before and after the event, to ensure the right kind of publicity.[76]

We need not look much further than the blocking for the performance to understand the significance and message of the tone tests. Here, for the first time, was a live performer and a sound-reproduction device presented onstage—*together as equals.* The metonymic logic was clear enough—if these great performers can share a stage with the Edison phonograph, then live musical performance and recording can be understood as two species of the same practice. The staging and history of the tone tests thus shows the elaborate work necessary to convince listeners of a correspondence between two different sounds—of the fact that the machine was merely a mediation of the "authentic" event.

Although the stated goal of the tone tests was to demonstrate that the Edison phonograph had such high fidelity that audiences could not tell the difference between live and recorded performance, the most important result was in convincing audiences that one was comparable to the other. It was a logical extension of the high-fidelity advertising campaign that characterized Edison phonographs. In this way, the tone tests extended a controversial strain of thought and practice that had been around since the advent of commercial sound reproduction.

Lore recounting individuals conducting their own tone tests had been around since the advent of sound recording. For instance, a speaker wrote in a phonograph magazine that he bewildered his audience at a toastmasters' club by having a phonograph perform his prerecorded speech from behind a curtain while he and his audience sat quietly in the next room. After much confusion during the delivery, the curtain was drawn back to show that a phonograph had in fact been doing the talking.[77] The Edison tone test campaign added to this already extant impulse by providing a greater degree of organization and ideological coherence to tone testing, which in turn helped define the way in which sound fidelity has been thought down to the present day. Even as the Edison Company was playing a new sonic game, it was working to convince audiences that the same old rules applied: that a good reproduction is the same thing as a live performance.

In her history of the tone tests, Emily Thompson notes that, in many cases, audiences and reviewers were initially much more interested in seeing the live performer than the machine. Some newspaper reviewers even rejected the premise of the comparisons altogether: one reviewer remained unconvinced that the Edison was different from other phonographs; another offered Benjaminian speculations on whether it was possible to recreate in a recording the "element of personality" that was so central to audience-performer interactions. These reviews were the exception: most written accounts of the tone tests either took the Edison Company at its word or at least considered the tone tests on the terms on which they were intended to be considered—live versus reproduced.[78] While recording had been the mere shadow of live performance in most people's minds during the late nineteenth century, as phonograph ownership spread people were able to hear music that they otherwise would not have. Once an audience became familiar with music through a recording rather than through live performance, it was possible to conceive of different relations between performance and recording. As a Victor ad would put it, there was "no need to wait for hours in the rain" to hear an opera or a concert.[79] The tone tests, and, later, ads that privileged recording as an alternative to live performance, could be read as one of the first series of performances guided by recordings—especially since, as noted above, the Edison Company would hire only performers who could "play like the recording." By 1915, the recording industry had made celebrities out of otherwise unknown performers and helped the careers of others, like Enrico Caruso. But the tone tests embodied the relation between live performance and recording that the company hoped to demonstrate to audiences: tone tests were organized in much the same way as record labels organize rock and pop groups' concert tours today—to promote new recordings—while existing recordings helped promote live performances (or helped promote the sale of phonographs).

The tone tests were not the first time that the Edison Company advertised recording as the equivalent of live performance. A particularly well-titled Edison ad, "The Acme of Realism"—depicting a child destroying a phonograph—carries the subtitle "looking for the band."[80] The ad clearly emphasizes that there was no experiential difference between hearing a live performance and hearing a recording. In a particularly effective ideological inversion, it suggests that, rather than learning that recorded and live music were "the same thing," children had to be taught the *difference* between live and recorded sound. Some phonograph owners clearly identified with this rhetoric and sought to impress its "obviousness" on others: "He said it was pretty good for an imitation. 'Imitation!' said I, 'why, Mr. T., that was no imitation; that was the genuine article—the Phonograph *never* imitates, it reproduces the actual music as played by the performer,' but I could see by his manner that he did not accept the statement; even though it was backed up by his wife. We had a good laugh about my friend's mistake afterward."[81] The writer finds his friend laughable because of the faith that he has already invested in the machine. The writer's emphatic assertion of obviousness is in this case itself ideological: *of course* the machine is capable of authentic reproduction because the writer has already framed it in the tropics of authenticity.[82] Obviousness is central to the whole artifice: to admit that it is *not* obvious to any listener that sound reproduction "reproduces the actual" calls the authenticity of the whole enterprise into

Figure 43. "Dr. Jekyll and Mr. Hyde at the Telephone"—AT&T advertisement (courtesy Archives Center, National Museum of American History)

question. It would suggest that the apprehension of mediation is not immediately understood—it is learned.

In a fashion similar to the Edison Company's campaign for the fidelity of its machines, and around the same time, Bell Telephone also sought to convince users that telephone conversation was the same thing as face-to-face conversation. Its ads suggest that people had to be convinced that telephony was simply a mediation of a face-to-face interaction. In response to operators' complaints and the company's own concerns about misbehavior on the telephone, Bell Telephone hired an agency to produce advertisements urging telephone users to ignore the difference between telephone and face-to-face interaction. Entitled "Dr. Jekyll and Mr. Hyde at the Telephone" (figure 43), one ad emphasized the need for courtesy because the "human element" was essential to telephone service: "Discourtesy on the part of telephone users is only possible when they fail to realize the efficiency of the service. It will cease when they talk over the telephone as they would talk face to face."[83] Another ad urged that "one who is courteous face to face should also be courteous when he bridges the distance by means of the telephone wire." Telephone ads exhorted users to "only talk as in ordinary conversation" since "undoubtedly there would be a far higher degree of telephone courtesy, particularly in the way of reasonable consideration for the operators, if the 'face-to-face' idea were more generally held in mind. The fact that a line of wire and two shining instruments separate you from the person with whom you are talking, takes none of the sting out of unkind words."[84]

Like the Edison tone tests and the Victor ads discussed at the beginning of this chapter, these ads protest too much. They suggest that readers might not have automatically assumed that face-to-face conversation and conversation on the telephone were comparable or two species of the same thing. Interestingly, the last quotation verbally represents the network (in the form of wires and "shining" telephones) as it simultaneously instrumentalizes it. Mediation was simply one possible description of the experience of sound's reproducibility. It was not a necessary or an automatic outcome of sound's reproducibility.

Variable Verities

For all the claims about "true fidelity," immediately on their entrance into the commercial market sound-reproduction devices were understood to produce a variable tone. Listeners had to be trained to use sound-reproduction technologies "correctly." The ads explaining telephone courtesy illustrate this well and connect rules for use with the aesthetics of transparency:

There is a most agreeable mode of beginning a telephone conversation which many people are now adopting, because it saves useless words and is, at the same time, courteous and direct. It runs thus:

The telephone bell rings, and person answering it says "Morton & Company, Mr. Baker speaking." The person calling then says: "Mr. Wood, of Curtis & Sons, wishes to talk with Mr. White."

When Mr. White picks up the receiver, he knows Mr. Wood is on the other end of the line, and without any unnecessary and undignified "Hello's," he at once greets him with the refreshingly courteous salutation: "Good morning, Mr. Wood." That savors of the genial handshake that Mr. Wood would have received had he called in person on Mr. White.[85]

This etiquette training suggests that transparency could be accomplished only after a set of ground rules had been established and a set of practices had become routine. Although transparency may have been an operative aesthetic for listeners and users, what counted as transparency was itself open to question. Telephones required skill to understand, and users had to develop proper modes of address for telephonic conversation. Early phonographs could be adjusted in order to vary speed, volume, and tonal character; different models of telephone transmitters each had different tonal characteristics; and kinds of radio loudspeakers varied widely from one to the next in tonal qualities.

All this meant that listeners were expected to learn to discern among the various types of sound-reproduction technologies. They were to hone their audile technique, to become connoisseurs of the various shades of perfection in tone, thereby learning to distinguish between truth and falsity, or at least to be able to construct their own auditory realities. Listeners knew very well that it was impossible to create a truly transparent sound-reproduction technology. It was obvious that different machines had sounds all their own. Thus grew a whole set of techniques for discerning the various qualities of sound alongside the discourse of fidelity. The hope for perfect equivalence between original and copy lay in tension with the knowledge of sound reproduction's situatedness. The motor force driving listeners to move between the poles of this duality was their desire to hear and thereby connect with the machines and with one another.

A constant playback speed was ostensibly required for "faithful" reproduction, although early exhibitors would routinely speed up and slow down records to impress their audiences. Early on, Bell and others noted that increased speed in reproduction would impart "a nasal metallic quality" to the human voice and that slowing down the record would achieve a "hollow, resonant effect" and a lower pitch.[86] In fact, maintaining pitch was something of an issue for phonograph and gramophone users. One of Tainter's major innovations in the graphophone was a device called a *governor* that took the irregular motion of a hand crank or foot pedal and converted it to regular motion for the cylinder's rotations. In fact, some early graphophones used treadles from sewing machines, which operated on a similar principle.[87] Machines without a governor, like the early gramophone, required other strategies if consistent results during operation were to be obtained. Sales tags from an 1894 gramophone advise the potential user not to "get discouraged if the machine doesn't give entire satisfaction *at once*. It will take a *little* practice to turn the machine according to the directions coming with each machine."[88] First performance took practice; then listening took practice—now the process of reproduction itself takes practice. The directions themselves were elaborate and designed for people who had never heard reproduced sound before. In their specificity, they make clear some of the central elements of reproducibility such as constant speed that we now take for granted:

The American Hand Gramophone reproducer is a talking machine which is both simple and effective, and will not easily get out of order, provided that the following directions are carefully kept in mind: 1. Place the machine before you, as shown in the picture [the cover pictures Berliner's daughter sitting in front of a gramophone], resting the arm fully upon the table, and turn the hand-wheel with a *wrist* movement at the rate of about 150 times a minute. To acquire this regularity of motion, practice it a number of times with the level and sound-box lifted off from the turn-table.

Hold the handle loosely, so that that it slides readily through the fingers. 2. The standard velocity of the center turn-table for 7-inch plates is about 70 *revolutions a minute.* The more rapid motion will raise the pitch and sharpen the sound; a slower motion will deepen the same. *First get the speed and then place the reproducer and needle into the outer groove or the next one.*[89]

Of course, users could not possibly count seventy revolutions a minute; their best guess as to the proper pitch for playback was as close as they would get. Later gramophones would add a governor.[90] The manual that provided these instructions took nothing for granted: every nuance of tone and volume was spelled out in great detail. An adjusting spring that could be tightened or loosened allowed for variation in volume and clarity (the tone became less and less clear as the volume grew louder). The frequent replacement of needles was recommended, on the grounds that a duller needle would produce a louder, less-articulate sound and would eventually wear down the record. Instructions for use included a quick lesson in acoustics: pointing the machine toward the wall would deepen the bass response; large wall hangings would dampen the sound. Finally, this particular manual emphasized that preferences in tone were individual—some people would prefer hearing tubes and others the horn.

In spelling it all out for the user who had never heard a reproduced sound before, this gramophone manual tells us a lot about the difference between the rhetoric of sound fidelity and the practice of reproducing sound. The manual painstakingly takes the novice user through all the possible variables in the machine's tonality: the needle; the rotation speed; the placement of the machine in the room; the type of horn or ear tube used; clarity; and even personal preference. While the advertising language of fidelity suggests that "perfect fidelity" occurs when the machine mysteriously effects the sound and no longer affects it, this manual for use offers us something closer to a sonic nominalism. Our gramophone novice was being taught that the sound of sound is entirely situated, depending on physical space, timing, personal preference, and the idiosyncrasies of any given sound event.

Inspectors' manuals for phonographs offer a sort of inverted account of the gramophone manual. Edison inspectors were advised as to the character of improper sonic performance—"scratchy reproduction; poor recording and reproducing; sounds too weak, or failure to articulate properly"—and each sound was indicative of a mechanical malfunction. The inspector's manual reads like Laennec's *Treatise on Mediate Auscultation,* this time abridged and for machines. Not only were the sounds of the apparatus cataloged; they were medicalized like the sounds of the body. The goal was to eliminate the tones of the machine except for those deemed preferable and, therefore, labeled *transparent* (or, more likely, *easier to ignore*).[91] A working phonograph for the Edison Company was like Hegel's "vanishing mediator"; it organized sonic relations and faded away into nothingness.

Despite this desire for transparency, the ideas of "preferred tone" and consumer choice were useful selling points for phonographs and, later, for radio parts. In a 1913 advertisement, Victor claimed that its "system of changeable needles gives you complete musical control." The copy masterfully blended this idea of consumer choice with the transparent aesthetic of pure fidelity: "A changeable needle is the only system that positively guarantees a perfect point for playing every record; a changeable needle adapts the different selections to the requirements of different rooms, and to meet the tastes of different people; a changeable needle enables *you* to hear every record just as *you* want to hear it.... Always use Victor Machines with Victor Records and Victor Needles—*the combination.* There is no other way to get the unequaled Victor tone."[92] "A perfect point for playing every record": if fidelity had been a gold standard for sound reproduction, Victor's ad agency had just floated the dollar. As with post-1973 American currency, the ad enjoins us simply to have faith in the process. The unequaled Victor tone is, in this case, presumably no tone at all except for the sound of the recording, yet this ad offers its readers *four different* "no" tones—presumably so that listeners could match both the variations in their musical tastes and the variations in their states of mind. Perfection becomes situational. Instrumental reason and technical control congealed together in the hand that changed the needle and the ear that perceived the difference between versions of "unequaled" tones: the practiced listener was to become the connoisseur of true fidelity; the "best" tones became the "truest" in the ear of the beholder. Other manufacturers followed Victor's lead in turning tone over to the listener.[93]

As with sound recording, radio systems actually advertised tonal differences even as they claimed to have achieved true fidelity. Radio loudspeakers claimed that the speaker was the most important part of the sound, a good speaker providing "roundness" and "cello-like" tone. Amplifying transformers were advertised as ending the howls of a radio set that would otherwise make "the squalls of a two-year-old sound like music" in comparison; Formica insulation would prevent "buzzing and sizzling."[94] Telephony, largely a monopoly business in the United States, did not focus so much on tonal differences between telephones; instead, it presented a progress narrative from Bell through Edison, Dolbear, and Berliner, each improving the sound-transmitting and -receiving capabilities of the telephone. Each new sound technology thus was presented as an even greater refinement in tone than the previous one, even as listeners had to be taught how to hear the differences. But not all listeners took the tone advertising seriously.

Sound quality *was* an issue for early telephone users, but it was not so much a matter of differences between different brands of telephone as it was a matter of the quality of the telephone system itself. An 1877 cartoon (figure 44a–b) lambastes telephonists' message-taking errors in comparison with telegraphers'. The telephonists' big ears stand in for both the physical difficulty of hearing and the need for auditory skill to master the new instrument.[95] A friend wrote to the inventor Elisha Gray and complained that a telephone concert in Westminster, England, was "the *worst* music I ever listened to."[96] Telephone paraphernalia (including some ad-

Figure 44a–b. "Ye Telephonists of 1877"—cartoon (courtesy University of Illinois Libraries)

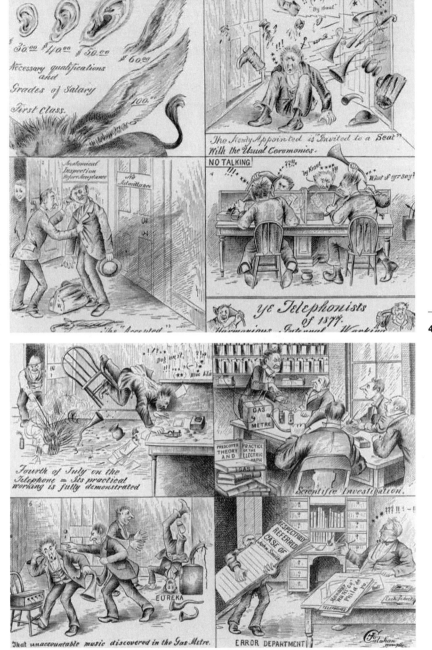

44a-b

vertising) would make fun of poor connections and the difficulty of hearing over the wires.[97]

Sound-recording technologies received their share of printed criticism as well. A 1906 story written from a child's point of view lambasted phonograph music as "an awful mechanical orgy. . . . I screamed and howled with rage, shame and terror. I tried to wrap the bear-skin rug around me as protection."[98] Victor's advertising campaign for the Orthophonic Victrola was parodied:

You hear the deep boiler-factory cacophony of the bass, the shrill shrieking of the trebles. Every instrument sounds like a skeleton's Charleston on a tin roof. Orthopunic reproduction will make you a firm believer in birth control.

The new Orthophunic Sictrola brings the elite of the nation's boiler factories to your home. It makes even a classical symphony sound like jazz—if you don't like jazz.[99]

The parody works because all the essential components are there in reverse: instead of treble and bass being good, they are harsh; the machine sounds terrible; and it adds its own sonic character to everything it touches. In essence, the joke tells the truth that nobody wanted to hear but everybody knew—sound reproduction shaped the sounds that went through the network. It could not be otherwise.

Radio listening suffered similar criticism in cartoons and articles. The noise that advertisers claimed to be able to prevent was a common characteristic of amateur radio listening. One cartoon carried the caption "A piercing shriek rang through the mansion as Madeline discovered to her horror that her husband, with whom she had just been united in wedlock has an appetite for the radio," the frame juxtaposing the unhappy bride, in the background, watching from behind a drawn curtain, and a groom, happily listening to a cacaphonic radio loudspeaker.[100] Writers coined the term *listeneritis* or *radioitis* for the phenomenon of the radio operator hearing something in the static where there was nothing at all: "Broadcasting of the radioitis static fantasie should be given a place among the horrors."[101] Cartoons made fun of listeners tolerating the radio's noise in search of that tiny snippet of programming or the popular myth that radio transmitters were so sensitive that they could pick up "a fly's footsteps." These are all images of audile technique wildly out of control—too fine a sense of foreground and background led listeners to discern sonic indices of nonexistent distant events (figure 45).

Listening for fidelity was structured by the desire to hear something through the network and the fear that the network would not work. Another kind of pitfall of radio listening was mistaking local sounds for distant ones. One writer confessed to thinking that he had heard a distant wireless conversation when in fact he had grounded his radio to the same wire as that used for the household telephone—which was a party line. Similarly, 1923 cartoon made fun of a woman who heard people singing hymns through her headphones: "Oh Ted, I bet I got HEAVEN!" "Baby you're a nut—it's that church revival meeting next door!" replied her male companion as he opened the window.[102] The cartoon's lesson is that one must refine her audile technique lest she find a false salvation in the technology.

If this last anecdote suggests that the technology was shot through with gender relations, the hopes and dreams for sound recording and radio show their connection to other kinds of gendered and sexualized longings. These technologies were understood as having the potential to break through social boundaries and at the same time to help enforce social norms. The idea of audiosurveillance was captivating to early users. One dealer wrote of a woman customer who wanted to "catch her husband at it"; he replied that the phonograph would not catch sounds through a keyhole.[103] Yet the industry press seemed at least as interested in the possibilities as in the facts. Carolyn Marvin has collected a whole range of anecdotes: operators listening in on conversations; men having affairs with women operators; cases of mistaken identity; and marriage by telephone and phonograph.[104]

Figure 45. "Winnie Winkle, Breadwinner"—1924 cartoon (courtesy Archives Center, National Museum of American History)

The desire for sound-reproduction technologies to capture the true world as it was and transmit it perfectly was, thus, tempered by the knowledge that the apparatus had its own grain; the supposedly mute machines had many voices of their own. The dream of the perfect machine as a vanishing mediator came in conflict with the practical knowledge of sound reproduction as a specialized social network that shaped the sounds going through it. The desire for a slice of reality coming through the diaphragm of the machine gave way to a somewhat variable aesthetics of auditory realism.

True Fidelity at Last

That the idea of sound fidelity could never really be just about the sound points us back toward the network, toward fidelity as marking a kind of faith in reproducibility. Sound fidelity was, thus, more about enacting, solidifying, and erasing the relations of sound reproduction than about reflecting on any particular characteristics of a reproduced sound. If perfect fidelity simply meant a set of social and sonic relations in which participants could have faith, it would be no wonder that we find repeated declarations of perfect reproduction from the 1870s on down to the present. Alexander Graham Bell wrote in 1877, "Mr. Williams has finished his line to East Comerville. I went into his office this afternoon and found him *talking to his wife by telephone.* He seemed as delighted as could be. The articulation was *perfect*—and they had no difficulty in understanding one another."[105] *Perfect* here is synonymous with *worthy of my faith* since the early telephone by no means produced a perfect reproduction of sound. Add to this faith the ideology of progress and the widely known variability among machines, and you find a shifting standard of sonic perfection. Advertisements hailed each new improvement in sound-recording and -reproducing technology as yielding perfect or near perfect reproduction. Bettini diaphragms were hailed in 1900 as a "True Mirror of Sound."[106] The Columbia Grafonola was, in the first decade of the twentieth century, declared "nearly perfect in its sound reproducing qualities," the last step in the march of progress: "The grafonola marks the culmination of human inventive genius in the science of sound reproduction."[107] This rhetoric was not limited to phonographs—a 1922 advertisement for the Richtone Loud Speaker asserted that, with the product, "every tone is reproduced with the utmost fidelity. . . . Voices from miles away sound as though the singer is in your room."[108] The constancy of advertisers' descriptions of sound should alone call into question the accuracy of *true fidelity* as an empirical description. The very use of the term shows that it is clearly bound up with the institutional and cultural practices of sound reproduction.

As listeners became more familiar with sound reproduction, they were encouraged to distinguish among the sonic signatures of different machines and technologies. In other words, the variability found among machines invoked the opposite—an impossible gold standard. *The best available* or *the preferable* became a stand-in for *the true*. This is well illustrated by the Victor Talking Machine Company's advertisements for the Orthophonic Victrola—a new kind of phonograph introduced in 1924 and designed to sound more like radio. Although the machine represented in these ads was something of a departure from previous phonographs, the copy describing it was completely typical. As a product of forty years of fidelity rhetoric, the ad series brings together several key elements in the discourse of sound fidelity: the narrative of technical progress; the belief in the transparency of reproduced sound; the desire for pure tone; and the equivalence of live and recorded music.

The Orthophonic ads, which ran from 1924 to 1929, featured musicians praising the new machine for its technological sophistication and tonal qualities. The Orthophonic was a "stupendous advance over former recording instruments," integrating radio technology into phonographic reproduction.[109] "To me the new reproduction seems uncanny in its faithfulness. It was as though the artist in person was in the room, giving life to the voice coming from the instrument."[110] Artists were quoted as praising its warmth and tonal richness, in ads claiming that it perfectly reproduced the human voice and the full orchestra: "I felt as though my voice was being reflected back to me."[111] "The full orchestra, the chorus, the single voice—each seems to surpass the other in richness and purity of tone."[112] These advertisements sought to convince their readers that artists found the tonal quality of the Orthophonic to be pleasing and faithful. They also emphasized that the experience of listening to recorded music was *the same* as listening to live music: "In listening to it, the artist is conscious that he is hearing again, as though in an echo, the ideas and emotions which he has sought to express in his rendition"; the Orthophonic record was to be "a musical experience equaled only by the personal performance of the musicians."[113]

It is quite possible that the Orthophonic Victrola did strike its listeners as a "new" and "better" sound when they first heard it, but this has more to do with aesthetic preferences and tonal distinctions *among* sound-reproduction technologies than with the fidelity of a copy to an original. The difference between Edison's tone test ads and the Orthophonic ads was that the latter could presume a comparability between live and recorded music—their goal was to show that the Orthophonic was closer to the gold standard of true fidelity.

As a machine, the Orthophonic was the result of considerable technical innovation, innovation started by Western Electric and completed by Victor, Columbia, and other phonograph manufacturers. During the mid-1920s, the sound-recording industry faced a significant challenge from radio. By all accounts, audiences preferred the sound of radio—which used vacuum tubes and electricity to receive, transmit, and reproduce sound across space. Electric recording grafted radio's electric reproduction technology onto sound recording's hitherto acoustic mechanism: it allowed for a wider frequency spectrum to be heard and also for a greater degree of amplification than in acoustic recording. The Orthophonic was one of the first machines to use electricity to play back recordings. It was also one of the first to play back recordings made with equipment modeled after radio's—using vacuum tubes and electricity in the recording process as well as the reproduction process. Electric recordings had more treble and bass and could be played at a considerably higher volume.

Listeners were starting to notice the difference, and the introduction of electric recording was in part an attempt to co-opt some of the enthusiasm for "the radio sound."[114] Orthophonics combined this electric recording process, first, with a folded horn speaker and, later, with an electric speaker. Both represented radical departures from previous horn construction. The folded horn allowed for a much "longer" horn in a smaller space, thereby improving frequency response and power—essentially emphasizing the tonal characteristics of electric recording. The electric-magnetic voice coil speakers that replaced folded horns allowed for even greater amplification and tonal range. These technical changes are usually cited as the reason for a preference for the Orthophonic and electric recording in general over mechanical recording—what historians have called *acoustic* recording.

These historical shifts in the definition of what constitutes high fidelity were not, however, unidirectional. It was not a matter of each new machine simply setting a higher standard. The standards themselves were contested. For all the many mechanical differences that could be listed, we cannot automatically assume that wider frequency response is a necessarily desirable characteristic of sound reproduction. In other words, listeners do not always or automatically understand improved technical specifications as resulting in "better" sound. Examples abound of people preferring older sound-reproduction technologies to newer ones (presumably more technically perfect than their predecessors). As I write, many self-appointed heirs to the McLuhanite throne offer that we are undergoing a digital revolution. Yet there is a booming market in used and vacuum-tube equipment among musicians, sound engineers, and studios; other companies are reintroducing old vacuum-tube designs for audio gear and developing "tube-emulation" software for digital audio. Nobody disputes the clarity of digital reproduction, but, to many ears, the old vacuum-tube equipment sounds better, and the result has been an explosion in the vintage and used market. Similarly, audio recording engineers routinely select microphones with an incomplete and uneven frequency response in order to record a more appealing sound. John Mowitt makes a similar point about noise reduction and the move toward "silent" digital recording: "The fetish of noise reduction has gone hand in hand with the aggressive marketing of distortion boosters and other less obvious instrumental sources of noise." Similarly, Peter Manuel argues that some Indian audiences have become accustomed to the sound of cheaply reproduced tapes and, therefore, have come to prefer that sound in their recorded music.[115]

The same can be said of the moment marked by the Orthophonic's introduction. Two noted phonograph historians, Oliver Read and Walter Welch, wax nostalgic (pardon the pun) for mechanical recording because they believe that mechanical recording was more representative of its source. In mechanical recordings, sounds vibrate a stylus, which in turn indents or engraves a surface, recording those vibrations, which can then be reproduced by reversing the process. Electric recording and reproduction add a stage by transforming sound into electricity. Read and Welch reason that mechanical recording is more properly "acoustic" (and faithful to the original) because it has one less stage of transformation. But both processes involve turning sound into something else in order to reproduce it. One may have a preference for the sound of one process or the other, but, in either case, the recording is not truly acoustic—since it transforms sound into something else—and it does not retain the original in its initial form. Contemplating whether one is "closer" to the original than the other after such a transformation has taken place is a purely semantic exercise.[116]

While Orthophonic advertisements promoted electric recording as progress, getting closer to "truth" in reproduction, Read and Welch read it the opposite way. Fidelity is, thus, confused with aesthetic preference.

The implication should be clear for an understanding of electric recording and the Orthophonic Victrola in relation to the history of sound fidelity: the preference for electric recording is *not* necessarily a natural outgrowth of a technically improved process. On the contrary, it may in this case simply reflect the popularity of radio—and the sound of radio—at the time and have nothing whatsoever to do with the inherent tonal qualities of the machine. The claim that the perceived "better" sound sprang forth from the increased frequency response offered by the machine and its increased loudness sounds reasonable from a commonsense standpoint, but it is modulated by the historical evidence. Technical improvements were not always welcomed aesthetic changes.

In some cases, opposition to specific technological innovations was simply a matter of an entrenched, oligopolistic industry fighting to preserve a solid market share. RCA's suppression of Edwin Armstrong's invention of frequency modulation (FM) is a particularly well-noted case. FM provided a much greater signal-to-noise ratio and was, therefore, considerably less noisy than AM radio—it was easier to tune in and easier to hear. By today's standards, it is widely accepted that FM sounds "better." But major players in the radio industry, concerned with competition from AM and also with preserving a large chunk of the electromagnetic spectrum for television, chose not to develop FM. FM sets built in accordance with Armstrong's patent became useless when the Federal Communications Commission moved FM to a higher frequency range. As a result, although FM could have been in wide use by the early 1940s, it was not really developed until after the popularization of television.[117]

In other cases, industry concerns collided with competing aesthetic sensibilities, as was the case with increases in volume for phonographs, graphophones, and gramophones before the Orthophonic. Although increased volume was a stated goal of inventors, their efforts met with varying responses. The graphophone was slightly louder than Edison's original tinfoil phonograph, and increased volume was among the goals of Volta's innovation of the phonograph. It was also a concern with radio and telephony.[118] Although early graphophones and phonographs were audible, they were not particularly loud. Berliner's gramophone, first widely marketed in the mid-1890s, was partly distinguished from its predecessors by its greater volume: "The striking character of the instrument is power."[119] An *Electrical World* article on the Gramophone also called the gramophone's articulation "remarkably clear" on the basis of its ability to amplify.[120] Berliner went even further in the pursuit of volume, designing the "multiphone": a series of gramophones hooked up together to play in unison. By adding more machines playing more records, the volume could be increased proportionally. Plans included a future model that would somehow use a giant single horn to amplify all the records together.[121]

Increased volume was not, however, considered an unmitigated good. Competing aesthetics were mixed freely with competing corporate interests behind competing technologies. Writers sympathetic to competitors lambasted the gramophone precisely for its loudness and timbre just as easily as Berliner could promote it on those bases. Consider this piece written by the editor of *Phonoscope,* a coin-in-the-slot industry periodical with Edisonian connections:

A wax record [as used by the phonograph or graphophone] is reproduced by a small ball tracing a groove in the velvety surface of a wax cylinder. A rubber record [as used by the gramophone] is reproduced by the scratching of a carpet tack or some similar device, in the granulated groove eaten by acid on the rubber disc. The one must be and is pleasing to the ear. The other sounds first like escaping steam. You listen more attentively hoping for better things and you are next reminded of the rumbling of a horse-less carriage. Finally, when the attempt to reproduce a voice is begun, you are forcibly compelled to liken the noise from the Gramophone to the braying of a wild ass.

. . . Its blasty, whang-doodle noises are not desired by citizens of culture. There is one Gramophone in use, however, in the coal mines at Carbondale. It is properly used by the miners there.

. . . Wax records are not intended to be so loud as to blow off the side of a man's face; but wax records are pleasing to the ear.[122]

Here, volume is clearly cast as a class and respectability issue: excessive volume becomes a form of noise and, therefore, a kind of social disruption, suitable for miners but not for men of culture. But, apart from the obvious hatchet job that the author has undertaken, we see that increased volume and distinctiveness of articulation are not automatically to be understood as an improvement in overall sound quality. The quotation also presumes an earlier listening practice of putting one's ear up to the horn, which was sometimes necessary for quieter machines—hence the comment about "the side of a man's face." For the louder gramophone, one could presumably sit back and listen with both ears. The *New York Times* ran a similar article a decade later, praising a St. Petersburg (Russia) prefect who banned the use of gramophones, first, near open doors or windows and, eventually, in the whole central city. In praising the prefect's efforts to save the population "from the torture of its metal voice," the anonymous *Times* writer concluded that "martial law has its advantages."[123] The *Times* neglected to report a similar event in the United States ten years earlier when a New Orleans judge ruled that phonograph and graphophone users (gramophones were not widely available, if at all, in 1899 New Orleans) could be fined for disturbing the peace if neighbors complained.

The *Phonoscope* and the *Times* did not represent unanimous views on the question of volume: they were neatly contradicted by another phonograph magazine equally concerned with middle-class decorum but considerably more willing to consider the potential of a machine that could "blow off the side of a man's face." The *Phonogram* ran what was for it a lengthy feature on a phonograph built in England that could be heard ten miles away: "You can whisper a sentence into the machine's little funnel-shaped mouthpiece and it will repeat it in tones that are more deafening than the shrieks of a liner's steam siren."[124] No doubt, the piece was more fantasy than reality, but the desire for amplification is clear. A month later, the same magazine ran a piece entitled "An Instrument of Satan," decrying the excessive volume of the gramophone, and a short letter from a Bishop, Illinois, man who was "very well pleased" with his machine, "especially on account of loudness."[125] Clearly, what constitutes "good," "better," and "faithful" reproduction in actual social practice, especially in the early history of sound reproduction, was based on highly variable judgments. The job of advertisers and manual writers was to train listeners.

One of the most common moves in twentieth-century American advertising is to "educate" readers in order to persuade them that the product being hawked is superior. This has been widely noted in ads for soap and other hygiene products.[126] The same kind of logic was at work in advertisements for reproduction equipment—teaching readers to *listen* for fidelity (or at least a particular dimension of sound called *fidelity*) was the expressed goal. This set of operations is rendered abundantly clear in a sixty-page pamphlet put out by Federal-Brandes in 1927. It combined a discussion of audile technique and sound fidelity with an extended advertisement for Kolster radios and a lengthy discussion of the aesthetics of concert music. In short, this "manual" explained how to listen and what to listen for in both the music and the medium. Perhaps it can be read as a kind of audile hygienics.

The pamphlet opens with the "novel sense" assertion so common to scholars as well as advertisers: "Modern life has educated the eye far more than the ear. We receive much of literature and the [sic] drama by way of the motion pictures [still silent at this point]. Magazines and newspaper have become largely pictorial. 'Visual Education' is common in our schools; but the human ear still remains comparatively neglected." The implication is that the ear *needs* aesthetic training—a need that Kolster intends to fulfill. With this justification in place, Sigmund Spaeth—a music educator who had previously published this essay elsewhere—highlights listening as an aesthetic skill to be cultivated through careful attention and practice: "Many people who think they are listening to music are really not listening at all." [127] Spaeth's reasoning moves carefully and quickly: beginning with a comparison of music to the everyday sounds of the city, suburb, or country, he argues that, if a person can distinguish the sounds of his or her environment, it is equally possible to distinguish the sounds of an orchestra and learn the differences among the instruments. Audile technique is now within the reach of everyone. On this basis, Spaeth introduces the concept of timbre ("tonal color") and spends the rest of the pamphlet discussing how to discern among the timbres of voices and instruments.

Throughout, as a part of the advertisement, the publishers have inserted footnotes in the form of prominently placed pull-quotes. Taken together, these footnotes point clearly to the object of Spaeth's lessons in audile technique—the sound of the radio itself. Consider this selection:

1. How many of the Announcers can you identify over your set before they tell you who they are? Kolster Radio owners get the complete identifying timbre of the voice with every vowel and consonant.
2. Make these listening experiments on your Kolster set and you will be surprised at the skill and power to differentiate which your sense of hearing develops.
6. Have your radio checked against a Kolster set for fidelity of reception. Most people are amazed at the difference. Any Kolster dealer will do this for you.
7. This personal element is the individuality whose rich presence prevents ear-fatigue and boredom. These last come from "untrue" sounds which are the reasons why an inferior radio tires you in a short time.

9. Every Kolster set is tested for the complete gamut of vibrational detail. Nothing is left undone to assist the assuagement of your ear's hunger for the complete pattern of beautiful sound in any composition.[128]

The aesthetic of the detail (pace Horkheimer and Adorno) reigns over Federal-Brandes's description: the cultivation of audile technique becomes a mode of personal distinction (and conscious consumerism) through the perception of detail and difference in a radio broadcast. Audile technique depends on the supplementation of listening through both technique and technology so that listeners can hear leading musicians in their own homes. The notion of sound fidelity is necessary for the medium to function as a vanishing mediator and thereby construct a relation of social correspondence among the sounds emanating from a musician's instrument and the sounds emanating from radios in listeners' homes.

Spaeth wrote in a post–Philadelphia Exposition, post–Volta Laboratory, and post–tone test world, where the networks enabling sound's reproducibility were well established, and where the language of mediation appeared to explain the medium (almost) naturalistically. In this world, listeners needed to worry less about the reproduced aspect of the music than about the aesthetics of reproduction. Sound fidelity ultimately became a shifting standard for judging reproduction—a way of judging the sound of the technology of reproduction—but it had little to do with correspondences between reproduced sounds and sounds that existed outside networks of reproduction. From its inception through its maturation, the concept of sound fidelity was about audio realism, audile technique, and the artifice of reproducibility. Sounds could neither hold faith nor be faithful—that task was left to listeners and performers.

Breaking the Faith

The very idea that a reproduced sound could be faithful to an original sound was an artifact of the culture and history of sound reproduction. Copies would not exist without reproduction, *but neither would their originals.* Sound fidelity was a story about sound reproduction that proved useful for selling machines and amenable to thinking of the medium through a philosophy of mediation. The shift accomplished in this chapter—from evaluation of products to the process of reproduction—recasts the question of what reproducibility means and how it works. The history of sound fidelity is really a story about reproducibility itself—specifically, the ways in which reproducibility was, from its very beginning, shot through with social relations. Sound reproduction was oriented around and through the twin diagrams of the network and the studio—the former emphasizing its irreducibly social character, the latter embodying the transformation of sound production in the events of reproduction. Sound fidelity was, ultimately, about faith and investment in these configurations of practices, people, and technologies. It posited the technology to reproduce sound as a vanishing mediator—a means that would obliterate itself in achieving its end. Throughout the early history of sound media, performers and listeners lent some of their own mimetic powers to the machines so that they might be dazzled. In developing their audile technique, listeners learned to differentiate between sounds "of" and sounds "by" the network, casting the former as "exterior" and the latter as "interior" to the process of reproduction. They had to be convinced of the general equivalence of the live and the reproduced. Even when the *sounds* of sound-reproduction technologies were explicitly discussed, it was with an eye toward finding new ways for the medium to erase itself. When listeners were familiar and comfortable with sound media, they were still hailed by advertisers and an industry that sought to educate and orient their hearing toward greater and greater refinement of audile technique.

Sound fidelity thus embodied something of a contradiction. The discourse of fidelity and the philosophy of mediation that it upholds were, in fact, central to the ways in which sound-reproduction technologies were developed, marketed, organized, and used. At the same time, this discourse was repeatedly presented as something *outside* the history of sound reproduction, as something that was a relatively accurate description of what was happening. In this sense, we can see the prescriptive and descriptive moments in language coexisting. We might then ask what it is that the discourse of sound fidelity actually prescribes.

The discourse of sound fidelity is as much a product of and a player in cultural history as are the machines that it purports to describe. The possibility that a reproduced sound could be faithful required that listeners and performers have faith in a network: a set of social relations, technologies, and techniques. This combination of technologies, practices, and social relations in sound reproduction opened up the relation between two sounds (original/copy, copy/copy) as a problem specifically for sound. This combination offered the possibility that sound reproduction *could* move sound itself over time and space. This was, after all, supposed to be what sound reproduction was supposed to do; it was the fantastic desire behind sound-reproduction technologies. But, like other fragile commodities, sounds had to be packaged for transport. The social process of reproduction transformed the practice of sound production.

In essence, the discourse of fidelity takes sound reproduction out of the social world and places it in the world of magic. This is very useful if one's goal is to make a new sound technology appealing to listeners. At a theoretical level, this is a useful move for privileging face-to-face speech and live musical performance because it explains sound reproduction through a theory of mediation: through reproduction, a hitherto "unmediated" phenomenon (the original, in all its presence and fullness) becomes mediated. The implicit goal of the mediation is to reproduce the fullness of presence found in the original. Not surprisingly, sound reproduction usually fails on this score.[129] Reasoning based on constructs of mediation or correspondence theories of representation—reasoning that takes for granted a certain kind of original/copy relation—not only results in positing an inevitable loss of being in moving from original to copy, but also, ironically, posits mediation only in the hope that it will later vanish, yielding a perfect or transparent copy. The failed hopes for mediation then stand in as the promise of interpersonal presence.

Sound fidelity has traditionally been conceptualized as the faithfulness of machines, their products, and their users to a reality that exists prior to and outside reproduction. Copies were understood as debased versions of originals, as sonic events that had experienced a loss of being. Yet, as I have shown, the reproduction of sound invariably involves a specialized process of sound production. The logical conclusion of this reasoning should be clear by now, as it was to audiences listening to early demonstrations of telephones and phonographs: "face-to-face" or "live" sound events are a social practice fundamentally different from technological sound reproduction and its attendant forms of sound production. It was not possible to sample the acoustic world, to audit an event, without participating in it. This is not to say that the two spheres of practice have nothing in common or even that people may experience them as two versions of the same thing—they are, of course, deeply connected today. Not only do various forms of face-to-face communication reference (and get referenced by) sound reproduction, but they also often coexist in the same room, in events as simple as a conversation between two people watching—and hearing—their television. (In fact, with the increasing use of sampling devices and computers in live musical performance, their connections are deeper than ever.) But the imagined correspondence between live and reproduced had to be invented along with the sound media; listeners had to be convinced of this equivalence. Since true fidelity could never be achieved (since a copy would under all circumstances suffer some loss of being from the original, however small), a set of procedures and aesthetics had to be developed to stand in for reality within the system of reproduced sounds. Through the conventions of realism and the rhetoric of fidelity, listeners could collapse the difference between live and reproduced into a single continuum of likeness and difference. Sound fidelity became an ever-shifting standard for the functioning of sound-reproduction technologies, a means by which to measure the distance between original and copy: it was an impossible vantage point from which to assess the fidelity of the machines to a fictitious external reality.

The discourse of fidelity is very much alive in commercial contexts as well as in some cultural theory (as I discussed at the beginning of the chapter). The idea that the mediation of sound inhered in the technology to reproduce it could be read as a "modernization" thesis: reproduction technology modernizes sound, rupturing a prior and ancient fullness and self-presence in the sonic world. Such a formulation demands a price from the present in order to pay the past: the loss of being, the disappearance of aura. The mediation thesis requires us to believe its opposite, not once, but twice. First, it posits the moment of unmediated sonic reality prior to sound's technological mediation. Then, it posits the ideal form of mediation as a vanishing mediator—where the medium produces a perfect symmetry between copy and original and, thereby, erases itself.[130] I have said throughout this chapter that the construct of reproduction as mediation takes sound reproduction out of its social milieu. But this sort of thinking does suggest a larger perspective on social reality: a privileging of the small, the interpersonal, and the face-to-face alongside a coincident hostility to large-scale forms of social organization. In this way, the discourse of sound fidelity is not significantly different from other philosophical accounts of communication that privilege some version of physical and/or metaphysical presence in binary contrast with absence.

The conceptual split between original and copy was itself an ideological project, a resolution to a cultural problem. Faith propelled the history of the machines, even as it denied their historicity. The promise of better fidelity has always been a Hegelian promise of synthesis and supersession—that *this* incarnation of reproducibility will finally capture the essence of some prior unreproduced reality. The perfect mediator would vanish in doing its work. But that moment of perfect correspondence never comes, and, because it never comes, theories of mediation posit sound reproduction as a failure, a sham, and a debasement of a more fundamental live presence. Accounts of reproduction that presuppose an ontological split between original and copy offer only a negative theory of sound's reproducibility, where reproduction can reference only that which is not reproduced. Like advertising, philosophy promised a synthesis that the thing itself could never deliver.

6 A Resonant Tomb

"That shows that the phonograph can be . . . for a very long time."—**THE VOICE OF JESSE WALTER FEWKES** on a test cylinder, ca. 1890, as heard ca. 1980

"I'm going to put in a fresh cylinder and do a little of it myself, and I want you to. When either of us is dead and gone, the survivor will be tickled to death to hear the other's tones again. This, you see, is a cylinder. It is composed of a waxy substance. Now it's in, and you've only to begin talking. Whatever you say this evening may be listened to by people 10,000 years hence."

"But I'm not going to talk."

"Why not?"

"Because I've got nothing to say to interest people 10,000 years hence."

—**C. B. LEWIS**, "Mr. Bowser's Tribulations" (ca. 1901)

If there was a defining figure in early accounts of sound recording, it was the possibility of preserving the voice beyond the death of the speaker. If there was a defining characteristic of those first recording devices and the uses to which they were put, it was the ephemerality of sound recordings. The epigraphs to this chapter illustrate the difference between the imagination and the practice of sound recording in its early days. In a manner both geologic and poetic, some of Fewkes's own ruminations on the preservative power of sound recording have eroded from the surface of his own recording.[1] The transcribing engineer could no longer hear what he had to say. Lewis's fictional dialogue between husband and wife illustrates another tension between imagination and practice: the husband promises his wife a chance to address future generations; she wonders what they will want to hear.²

From the moment of its public introduction, sound recording was understood to have great possibilities as an archival medium. Its potential to preserve sound indefinitely into the future was immediately grasped by users and publicists alike. Yet the early practice of sound recording was significantly different—the first recordings were essentially unplayable after they were removed from the machine. Later wax cylinder recordings and even metal or shellac disks were often treated by their makers and users as ephemera. As D. L. LeMahieu writes of the gramophone in Britain:

> The hope for immortality on shellac often became lost, however, in the continual and often extraordinarily rapid turnover of records. For commercial culture, the wonder of this new technology lay not in historic preservation but in mass production. . . . Popular records became almost as transitory in the market-place as the ephemeral sounds which they preserved. Moreover, high turnover in a perpetually changing market led to an indifference, even contempt for earlier, more primitive technologies and the often less sophisticated products they created. Within a few generations, records produced by the thousands and millions became rare items. Many were lost altogether.

Sound recording did as much to promote ephemerality as it did to promote permanence in auditory life. If we consider sound recording on the basis of its technical possibilities, repeatability is as much a central characteristic of the technology as preservation is. In fact, the former is a prior condition of the latter. Inasmuch as we can claim that it promoted permanence, sound recording also helped accelerate the pace of fashion and turnover in popular music. As LeMahieu notes, "Songs which a few generations before might have remained popular for decades now rose and fell within a year, or even months."³

Today, early wax cylinder recordings are considered *incunabula* (the earliest and rarest artifacts of a medium) by archivists and collectors alike. Many are notoriously fragile and difficult to hear. They require careful and attentive storage and ginger use. Listening to those early cylinders often brings confusion and clarity in equal doses to the auditor—hence the missing adjective or clause in the transcription from Fewkes's 1890 test cylinder. The two epigraphic ruminations on the possibilities of sound recording's preservative power thus illustrate the disjuncture between the imagination and the practice of early phonography. Writers imagined that the technology finally set free the voices of the dead, but this permanence in the technology or the medium was more imagined than real. If anything, permanence was less a description of the power of the medium than a program for its development. That this is so renders the early declarations of sound recording's preservative power all the more fascinating.

Although it is perhaps most pronounced in phonography, death is everywhere among the living in early discussions of sound's reproducibility. The spirit world was alive and well in telephony and radio. "The telephone has always been inhabited by the rhetoric of the departed," writes Avital Ronell.[4] One *Washington Post* writer speculated during an interview with gramophone and microphone inventor Emile Berliner that radio would eventually allow for communication with the dead since it picks up vibrations in the ether and the dead "simply vibrate at a slower rate" than the living.[5] The logic is impeccable—if sound reproduction simply stratifies vibration in new ways, if we learn to "hear" other areas of the vibrating world, then it would only make sense that we might pick up the voices of the dead. The writer simply failed to mention that the frequency of the dead's vibrations approaches zero, thereby rendering them difficult to hear. In this formulation, the medium is the metaphysics. The metaphorization of the human body, mind, and soul follows the medium currently in vogue. Current and fashionable comparisons of computers with the human brain are not very different in spirit from the *Post* reporter's speculations. In this line of thinking, media are forever setting free little parts of the human body, mind, and soul. If the voices of the dead were, indeed, free agents, perhaps they could then be enticed back into the world of the living.[6]

We now dwell without comment among these voices of the dead. The sounds of many dead musicians and singers have casually graced my ears in the time spent writing this book—they commingle with recorded music made by artists still living. If this experience is unremarkable today, it seems as though it *demanded* commentary one hundred years ago. Despite the ephemerality of the recordings themselves, death and the invocations of the "voices of the dead" were everywhere in writings about sound recording in the late nineteenth and early twentieth centuries. Without regard for genre or context, writers repeatedly produced tracts on the possibilities for hearing the voices of the deceased as some kind of guarantee or signature for the cultural and affective power of recorded sound. The chance to hear "the voices of the dead" as a figure of the possibilities of sound recording appears with morbid regularity in technical descriptions, advertisements, announcements, circulars, philosophical speculations, and practical descriptions.

Is this the ultimate and shocking power of sound reproduction—that it finally set the voice free from the living and self-aware body (if only for a few moments)? This is the tale often told about sound reproduction. In this formulation, death appears as a philosophical limit case for sound reproduction, and sound recording becomes a philosophical index for sound reproduction in general. The reasoning goes like this: when recorded, one's voice was abstracted from one's body, and, once so abstracted, the voice could be preserved indefinitely on record. The ultimate case of this scenario is, of course, the voice's persistence through recordings after the death of the speaker. *The voices of the dead* is a striking figure of exteriority. Because it comes from within the body and extends out into the world, speech is traditionally considered as both interior and exterior, both "inside" and "outside" the limits of subjectivity. In constrast, the voices of the dead no longer emanate from bodies that serve as containers for self-awareness. The recording is, therefore, a resonant tomb, offering the exteriority of the voice with none of its interior self-awareness.

We can date this emergent construct of sound as *exteriority* to the early nineteenth century and probably earlier. It most certainly predates the phonograph. As exteriority, sound was primarily understood as an effect or force in the world rather than as a manifestation of an internal and enveloping bodily force (such as the human voice). From Auenbrugger's drumming of the body, to Chladni's sand figures and Laennec's diagnostic signs, through Helmholtz's theory of sound as an effect in the ear, then on through emergent practices of sound telegraphy and, later, telephony, we can find a whole panopoly of resolutely exterior constructs of sound. The phonograph, after all, was a tympanic machine.[7] The telephone facilitated the hearing of a voice physically *absent* to the listener. The phonograph took this a step further by dramatically facilitating the audition of voices absent to themselves. This made it special in the minds of its first auditors and philosophers.

But arguing that sound recording transformed the experience of death and the voice is to tell the story in reverse, at least in part. It turns out that there is another cable between sound recording and death, one running in the opposite direction: *for its early users, death somehow explained and shaped the cultural power of sound recording.* This, in turn, raises the cultural status of death itself as a problem within the history of sound recording. Death is not the same everyplace, everywhere, and for everyone. Everybody dies, but not at all in the same way. To understand the cultural significance of "the voices of the dead," we must question the meaning of death itself. This chapter explores the ways in which sound recording, even as we know it today, bears the residual traces of late-Victorian death culture in the United States and the United Kingdom (and possibly elsewhere).

Any contemporary account of the death imagery surrounding sound recording must attend to the difference between late-nineteenth-century attitudes about death and the ways in which we talk about death today—or, rather, the fact that we try to avoid talking about death. John Peters writes that, "what sex was to the Victorians, death is to us: the ultimate but inescapable taboo. . . . We chuckle at Victorian primness, congratulating ourselves on our liberalism on topics sexual, but nothing is so veiled to us as death, so cloaked in euphemisms—or as pervasive in popular culture."[8] Death was everywhere in Victorian society because it had not yet been consigned to the "nowheres" that it currently inhabits in American society: basements of hospitals; the sterile, professional spaces of funeral homes. It had not yet been fully professionalized and cordoned off from everyday life. It was the age of great cemeteries and funeral processions: a good funeral was something to which every middle- and working-class person aspired. Spiritualism, that strange mix of religion and science, was a major cultural force among the middle classes and something in which even respectable intellectuals publically dabbled.[9] Perhaps, then, we should exhibit no surprise at the death imagery surrounding sound recording since it was already everywhere in Victorian culture. But, surprised or not, we should be *interested* in the death imagery because those early writers believed that there was something special about the relation between sound recording and death.

It is well-known that contemporaries of the early phonograph were not the first people to associate a medium with death: "The realm of the dead is as extensive as the storage and transmission capabilities of a given culture." Spirit photography had been around for decades. Images of work in print shops as a "dance of death" began appearing centuries before—shortly after the emergence of the printing press itself. Even some of the older arts of representation created materials that could potentially outlast the human body. Sculpture, architecture, painting, and writing all have long-standing associations with death in the imaginations of philosophers.[10] But it is a radically different death that explained the new mechanism of sound recording to those who contemplated it at the turn of the last century. When sound recording first appeared in 1878, it entered a vastly different cultural milieu of death than even the one surrounding early photography, which had preceded sound recording by only a few decades. Recording was the product of a culture that had learned to can and to embalm, to preserve the bodies of the dead so that they could continue to perform a social function after life. The nineteenth century's momentous battle against decay offered a way to explain sound recording. The ethos of preservation described *and prescribed* the cultural and technical possibilities of sound recording.

Better Listening through Chemistry

John Philip Sousa's famous remark that phonograph music was "canned music"[11] may have been meant as an aesthetic criticism, but, as a metaphor, it suggests the line of historical reasoning that I follow here: the practical and imagined possibilities of recording's permanence existed as part of a longer history and larger culture of preservation. In Sousa's statement, the possibility of recording sound is just one more form of preservation, and chemical preservation was one of the major innovations in nineteenth-century American culture. By the time the phonograph was patented in 1878, Americans were familiar with the idea of consuming food that had been physically transformed for the purposes of preservation, orderly handling on a large scale, and mass production. They were also familiar with and increasingly interested in all manners of preserving the dead—hopes that phonography would preserve the voices of the dead were only an extension of a larger, emergent culture of preservation.

Sousa's analogical connection between the engraved or later etched recording surface and chemically transformed food may not have been completely parallel, but, culturally, it made a world of sense. Modern canning, which began early in the nineteenth century, did not become widespread until shortly before and during the American Civil War. The mass production of tin cans, beginning in 1849, Borden's method for canning milk patented in 1856, and the inventions of screw-top mason jars and bell jars in 1858 helped stimulate the spread of artisanally and industrially produced canned goods. As, with sound technology, preservation technology did not have an autonomous cultural life or an inherent cultural impact; canning was an artifact of spreading industrialization and increasing migration. The former helped separate the spheres of production and consumption— canned food is, thus, an early artifact of an emerging consumer culture. Events that displaced large numbers of people also helped stimulate the spread of canned goods. By 1850, gold miners in northern California were eating large quantities of canned fish, shellfish, tomatoes, and peas. During the Civil War, canning businesses thrived through U.S. government patronage, as canning was the only efficient means of providing large quantities of food to large numbers of soldiers in the field. Afterward, the canning industry benefited from exploding industrialization in other areas. H. J. Heinz went into business selling jarred pickles, horseradish, and sauerkraut in the 1870s. Franco-American and other canned food companies went into business in the 1870s and 1880s. As the historian Ruth Cowan points out, although the national output of canned goods in the United States was only about 5 million in 1860, it grew to 30 million by 1870 and increased fourfold again by 1880.[12]

The canned music metaphor was, therefore, particularly timely. For the first two decades of the phonograph's existence, there was an explosion in the quantity of canned food available in the United States. Sousa thus chose one widely known and experienced kind of transformation for preservation to describe another. Clearly, he meant the comparison to be derogatory—canned food did not taste as good a fresh food, and canned music was not as good as "fresh" music.[13] As I discussed in the previous chapter, experiencing live music and experiencing recorded music were not even initially comparable practices for many listeners. But the metaphor is actually quite apt: in canning, the food is preserved through a chemical transformation; in recording, the sound performance is preserved through a practical transformation.

While the metaphor of canning was used to denigrate the "flavor" of recorded music, the voices-of-the-dead figure was used to promote the process of recording, and, as I will show, it was actually more of a program for developing the technology of recording than an actual description of the technology as it first emerged. This is why we need to move beyond canning to another mode of preservation that gained popularity during the course of the nineteenth century—embalming. One could protest that a closer cultural analogue to recording is photography: others like Amy Lawrence, James Lastra, or Theodor Adorno have pointed to photography as something of a cultural model for sound recording.[14] This is a profitable line of reasoning. Certainly, photography involved preserving images beyond death (especially with photographs of deceased babies and other sentimental Victorian keepsakes). My interest here, however, lies in changing attitudes about and practices of death, and, here, embalming comes to the forefront. Changing practices of preserving the bodies of the dead prior to the invention of the phonograph laid a foundation for the trope of the voices of the dead. The desire to hear these voices had to be learned: it was not a given. The death imagery surrounding early sound recording marked emerging changes in attitudes about death and especially the preservation of the dead body, of which the voice was in some sense a logistic extension. This is why the nineteenth-century boom in embalming is a key part of the prehistory of sound recording.

Both the introduction of embalming and the introduction of recording are significant moments in that larger set of transformations that Michel Foucault and others have provocatively termed *the history of the body.* Fantasies of speaking to the not yet born and hearing the dead cast phonography as a species of biopower, a modification of the relations between life and death.[15] The connection between phonography and embalming is interesting because, in many ways, attitudes about the voices of the dead are extensions of attitudes about the *bodies* of the dead. In defense of embalming, the historian Christine Quigley writes, "While it doesn't always promise permanence, embalming protects mourners from the sights, smells, and sounds of the decay of their loved ones."[16] As we will see, sound recording was also presented as a way of protecting its present and future auditors from the experience of decay. Both embalming and recording "protect" future audiences by transforming a substance in the present in anticipation of the future: the chemical transformation of the body was to have its analogue in the physical transformation of sound in the process of its recording. To understand the meaning of this "protection," then, it is worth considering for a moment the practice and history of embalming.

The practice of embalming is straightforward enough. Embalming is a chemical process, usually effected by a combination of evisceration and injection. Embalming is "the fixation of the tissues by chemical means. The action of changing the proteins of the tissues is compared to the change in consistency when a raw egg is fried."[17] The goal is to alter the corpse in such a way as to slow its physical decay and thereby render it presentable for viewing, at least for a time. Preservation for viewing is a distinctly Christian practice since many Christian funerals include an open casket. Modern embalming for Christian burial is also frequently accompanied by cosmetic treatment of the corpse—this has become a crucial step in making the embalmed corpse suitable for public viewing.

The chemical embalming of corpses was not a new practice in the 1870s: there are widespread accounts of embalming in Europe from the Middle Ages on. Many funeral customs at the time required that measures be taken to preserve the bodies of the dead, whether because the bodies of the elite usually lay "in state" for a week prior to burial or because it took the poor some time to raise money for a burial. American practices were quite varied as well, but some need for (at least temporary) preservation had always existed—either because it could take time for relatives to arrive for a funeral if they lived far away or to transport the body to a distant family gravesite.

Embalming specifically for funeralization, however, was a new practice in the nineteenth century, one that had been popularized only recently when sound recording was introduced to the public. The 1840 translation of *History of Embalming* from French into English provided the first set of widely available and standardized instructions.[18] Around the same time, embalming moved from a practice performed in the private home to a practice performed in the funeral home. In other words, it was professionalizing. States began to license embalmers, and, in 1843, the first patent for an ice-filled corpse preserver was granted. This is significant, not because the technique caught on (it did not until the twentieth century, when the refrigeration of corpses became practicable), but because it indicates increasing interest in the mechanics of preserving the bodies of the dead.[19]

Between 1856 and 1869, eleven major patents were granted for fluids, processes, and media for chemical embalming.[20] This move toward chemical embalming was driven largely by government involvement in the burial industry, an involvement triggered by the massive casualties generated by Civil War battles and, thus, the large numbers of bodies that needed to be transported all over the union at once. Prior to chemical embalming, the two predominant methods for preserving human bodies were storing in containers cooled by ice-based refrigeration and sealing in airtight metal caskets. Both methods presented real problems during the Civil War: maintaining large quantities of ice for the time it took to transport bodies across the country by train was not really practicable. The metal caskets, on the other hand, did not provide for any kind of cosmetic preservation or touch-up and were sometimes not strong enough to resist the buildup of gasses from the corpse, occasionally exploding while in transit.[21] Additionally, with new and ever more gruesome forms of warfare, the Union government was engaging in a rudimentary form of public relations by seeking to ensure that bodies were returned to their families in the best possible cosmetic condition. Once again, the cosmetic preparation of the corpse was a necessary adjunct to its embalming.[22]

Chemical embalming eliminated the need for maintaining large quantities of ice, facilitated cosmetic touching up, and could, to a great extent, stop the decay of the body. Research into embalming fluids was already in process in fields like medicine, chemistry, and anatomy where the preservation of specimens was of great importance. Federal interest in this approach spurred research and experimentation in the chemical embalming of human bodies, and, once officials were satisfied with the result, it was immediately put into practice. Even undertakers who were not themselves skilled in chemical preservation were required to hire someone who was if they hoped to obtain a federal contract. Government interest helped build the industry for the next half century: many of the best-known undertakers of the period got their start working for the federal government.[23]

Initially, there was some public resistance to chemical embalming. Although people had little say in what the federal government did, chemical embalming was frequently met with objections concerning the sanctity of the body (since it involved direct alteration of the body in a way that refrigeration or metal caskets did not). In some cases, opponents of embalming worried that the soul of a chemically embalmed body would not be allowed into heaven.[24] The first subjects of new embalming practices developed by Thomas Holmes and others were elites since the government's goal was to popularize the practice. Mourners of government officials, war heroes, and prominent officers were, therefore, among the first to witness innovations in chemical embalming, thereby offering some high-profile examples to help legitimate the practice. After Lincoln's assassination in 1865, hundreds of thousands of mourners viewed the president's embalmed body on its way from Washington, D.C., back to Springfield, Illinois.[25] But, by this time, chemical embalming had already been well established.

From the end of the Civil War through the remainder of the nineteenth century, the popularity of chemical embalming grew to the point where it eclipsed other approaches to the preservation of the dead. The exteriority of function had overcome the internal consistency of form. The desirability of preserving a corpse's outward appearance overcame concerns about altering the internal composition of the body.

At the core of claims about chemical embalming lay a set of claims about the nature of preservation. When combined with other beautification techniques, chemical embalming opened up a whole field of cosmetic possibilities for the presentation of the dead to the living. Embalming could also provide almost indefinite preservation through altering the state of the body. Ultimately, its effectiveness at preserving the body for transport and at cosmetic modification led to its more widespread acceptance.[26] In other words, while competing methods were concerned with interiority, with preserving the body in its original form, chemical embalming was concerned only with exteriority, with the appearance of the body and its potential to perform its social function.

The implied logic in preserving the voices of the dead through sound recording follows an identical path to this new approach to preserving the dead in general: a disregard for the preservation of the voice in its original form, instead aiming for the preservation of the voice in such a form that it may continue to perform a social function. In the last chapter, I argued that the very nature of sound production is transformed prior to reproduction. *The studio* is an organizing principle of sound reproduction. In many ways, embalming is an analogue of this studio process. Both transform the interiority of the thing (body, sound performance) in order that it might continue to perform a social function after the fact. Like the cosmetic touch-up of corpses, even the most "realist" approaches to sound recording took extensive steps to beautify the product for future ears.

A ca. 1900 article on recording studios made this connection explicitly in describing celebrities' anxieties about recording: "'To obtain the records of celebrities,' says a well known phonographic expert, 'is often a matter of difficulty. In the first place it is a task sometimes to gain their consent. Some of them fear the accusation of seeking notoriety. Others declare that they do not wish their voices to remain beyond the period of their own lives.'"[27] This suggestive passage directly connects the peculiar performance anxiety arising in the studio environs and the audition of the not yet born. Performers had to accommodate themselves to the peculiar practices of studio recording so that, while living, their voices could become (potentially) available to the not yet born. Perhaps the frightening aspect of this process, then, was that, in recording, the performers felt obliged to contemplate their own deaths.

An early reaction to the possibility of sound recording makes the same connection: "we" may now address the ears of the not yet born; future generations will, thus, be able to hear the voices of the dead. On receiving a letter from Edison's assistant Edward H. Johnson announcing the invention of the phonograph, *Scientific American* printed this reaction:

It has been said that Science is never sensational; that it is intellectual and not emotional; but certainly nothing that can be conceived would be more likely to create the profoundest of sensations, to arouse the liveliest of human emotions, than once more to hear the familiar voices of the dead. Yet Science now announces that this is possible, and can be done. That the voices of those who departed before the invention of the wonderful apparatus described in the letter given below are for ever stilled is too obvious a truth; but whoever has spoken or whoever may speak into the mouthpiece of the phonograph, and whose words are recorded by it, has the assurance that his speech may be reproduced audibly in his own tones long after he himself has turned to dust. The possibility is simply startling. A strip of indented paper travels through a little machine, the sounds of the latter are magnified, and our great grandchildren or posterity centuries hence hear us as plainly as if we were present. Speech has become, as it were, immortal.[28]

The phonograph as a way of preserving the dead: its contemporaries thought themselves brilliant and the invention wonderful for contriving such a possibility. Yet this peculiar construct of sound recording did not simply allow speech to live on forever: it essentially embalmed the voice. While this fascination with the voices of the dead might conjure up "the metaphysics of presence," voices of the dead are not present to self-affecting subjects; they are not part of an apparatus of self-awareness (and perhaps this is why Derrida hails sound recording as ending the era of writing).[29] The voices of the dead are present to their auditors only, and even then in modified form. As a cultural analogue of chemical embalming, sound recording preserved the exteriority of the voice while completely transforming its interiority, its insides.

While this philosophical accounting might suggest that sound recording occasions a rethinking of the phenomenology of speech, it would be uselessly anachronistic if I did not also note the condition of sound recording in 1877 when *Scientific American* printed its reaction. Even today, some historians are inclined to take the *Scientific American* article at face value, arguing that the *invention* of phonography made it possible to preserve the voices of the dead. In point of fact, the early phonograph did not preserve the voices of the dead, except for a short time. The machine itself recorded sound onto tinfoil. Given the fragility of the recording medium, the moment the record left the machine, it was essentially destroyed. Early wax cylinders could be taken off the phonograph and put back on, but they did wear out, sooner rather than later. Recordings were ephemera. To write in 1877 that "whoever has spoken or whoever may speak into the mouthpiece of the phonograph, and whose words are recorded by it, has the assurance that his speech may be reproduced audibly in his own tones long after he himself has turned to dust" is to write fiction, or, more accurately, prognostication. Taken as a diagnosis of the cultural power of recording technology in 1877, *Scientific American's* voice fantasy is wildly inaccurate. The article is not an artifact of phonography's accomplishments; it is a program for developing sound recording as a preservative medium. It speaks to the fascination with death as a receding limit—to hear the voices of the dead and to send messages to the future. This is why we cannot accurately claim that sound recording radically altered the cultural status of speech. It would, therefore, be more accurate to effect an inversion of the usual wisdom on the voice and the phonograph: the cultural status of the voice transformed sound recording.

Despite their overtures to the possibility of permanence, early phonograph users were well aware of the fragility of the medium itself. A ca. 1902 article from the *Kansas City Star* describes the process of multiple electroplating over the wax record and eventually replacing the wax record, thereby making a record cylinder of "imperishable metal" that will last forever. The article heralds the possibility of permanence, but that possibility is understood more as a disappointed hope: "For several years the expectations of the scientists as to the durability of the Phonographic process first employed received no encouragement. It was found that the cylinders used would not stand the ravages of time, being necessarily of soft, impressionable material."[30] Although the new process held theoretical promise, it was by no means a simple solution to the unfulfilled promise of permanence.

The short-lived Indestructible Phonographic Record Company was founded as a result of the fallout from patent battles over this new process, and it began manufacturing cylinders made of celluloid in 1906. The company's advertisements emphasized the durability of the cylinders and played on the idea of indestructibility with pictures of a child putting a stick of dynamite into a cylinder or polar bears rolling around on Arctic ice with one in a cylinder (figures 46–47), although the process or the material composition of the cylinders was never explained. More to the point, the company's very name pointed to the fragility of most recordings. Responses from competitors suggested not only that more fragile records and players were the norm, but also that such fragility was a significant aspect of the medium. Dealers wrote letters to the company complaining that competitors were telling their customers that Indestructible records would damage sapphire needles. Although the company responded both in its advertisements—"YOUR phonograph and YOUR reproducer will give excellent results with *Indestructible Records*"—and in special circulars sent to their dealers, it appears that listeners continued to shy away from the cylinders precisely because of their special, "indestructible" status. The company also offered a special stylus for playing its records but cautioned potential customers that this stylus would destroy regular wax records. Ironically,

46

the emergence and disappearance of the Indestructible Phonographic Record Company demonstrated that the phonograph itself, the machine with the promise of communication across the centuries, was designed for fragile, ephemeral wax records until well into the second decade of the twentieth century. By 1909, Indestructible had been absorbed by Columbia Records, and experimentation with other surfaces continued.[31]

I offer the Indestructible Records story here to frame the following discussions of death and permanence. Although early users were fascinated with phonography's potential to reanimate the voices of the dead and direct their own recordings to the ears of the not yet born, this conception of persistence in death was widely available in the late nineteenth century and was not even an accurate description of the machine and its recordings.

Figure 46. "The Indestructible Records"—ca. 1908 advertisement for the Indestructible Phonographic Record Co. (courtesy Archives Center, National Museum of American History)

Figure 47. "Indestructible Phonographic Records—Do Not Wear Out"—1908 advertisement (courtesy Archives Center, National Museum of American History)

Permanence became a wish and a program for sound recording, not simply a fate realized.

His Voice, Mastered

The persistence of the voices of the dead—or at least the voices of absent speakers—is a hallmark of early visual and narrative representations of recording. The case of Nipper, the RCA dog, illustrates the peculiar Victorian culture of death and dying into which sound recording was inserted. Nipper is the dog of *His Master's Voice* fame, standing with an ear cocked at the horn of a gramophone. But there is more to this trademark than is generally acknowledged today. In its earliest versions, Nipper is clearly seen

48

sitting on a shiny surface that reflects both the dog's body and the gramophone, wider at one end than the other, with an edge and sides. Many contemporaries who viewed that picture considered Nipper to be positioned on a coffin (figure 48). Although RCA began cropping the picture higher and higher, eventually eliminating any clear detail about the surface on which the dog sits, the original picture remains at least ambiguous in this respect. In his history of the Nipper trademark, Leonard Petts rejects the coffin reading, but other writers have been less dismissive of it.[32] Friedrich Kittler, for instance, calls Nipper "the dog that started sniffing at the bellmouth of the phonograph upon hearing its dead master's voice, and whose vocal-physiological loyalty was captured in oil by the painter Francis Barraud, the brother of the deceased."[33] "Dogs were given a firm place in the Victorian language of grief," writes John Morley, and John Peters has aptly pointed out the iconographic similarity between Barraud's painting and Victorian images of dogs at empty cradles and in attendance at funerals.[34] There is also a long history of images of dogs showing interest in their masters' musical and vocal performances.[35]

Figure 48. *His Master's Voice* (courtesy Archives Center, National Museum of American History). Notice the shiny surface beneath the dog's feet and the appearance of edges in the lower-right-hand corner and between the dog's right foot and the support for the gramophone's horn (also behind the gramophone and to its left). Many early viewers took the angles and edges of this shiny surface to be the outlines of the top of a coffin.

302 THE AUDIBLE PAST

The history of the painting *His Master's Voice* itself offers some evidence in this respect. Nipper belonged to the painter Francis Barraud's brother Mark Henry, who died young. Francis painted the picture around 1893, before gramophones were commercially available, and the original sound reproduction device in the painting was not a gramophone (which could not record on its own) but a phonograph (which could). When Emile Berliner's agent, William Barry Owen, saw the picture while working in England in 1899, he hired Barraud to replace the phonograph with a gramophone and bought the rights to the painting. On the basis of these facts, the collector Robert Feinstein concludes that the dog could well have been perched on the casket of Mark Barraud and that the Edison Home Phonograph that originally graced the painting could well have been used to record the voice of the painter's brother, who lay in a casket at home for the week prior to burial.[36] To add supporting evidence for his claim, Feinstein makes reference to the widely reported use of phonographs at funerals.

Funereal phonography never really caught on, but there are many cases reported in the early phonograph industry press—usually in the most sensational style possible. When we see a dog listening to a gramophone, we understand that the important issue is the *sound* of the voice, not what was said, since dogs are known for heeding the voices of their masters more often than their words. Stories of phonographic funerals recast Barraud's painting in narrative terms. In almost every case, the event is reported as having significance for sound recording, not for what was said. As in the case of Nipper, it was the persistence of the voice itself that fascinated contemporary writers.

One of the earliest reports was of the Reverend Thomas Allen Horne, a resident of Larchmont, N.Y., preaching his own funeral. The article reported that, in his last year, he listened to the singing of his deceased wife on a phonograph and then recorded his own funeral ceremony, complete with hymns and sermon. After eulogizing himself and his wife, "the voice of the deceased had evidently broken down, and from the instrument the terrible sound of a strong man weeping and unable to restrain himself broke out with realistic force and caused a shudder of horror among those who were present."[37] Other ministers intentionally sought publicity through this spectacle. In one widely reported instance, Henry C. Slade, a minister in Rideout, Kentucky, was said to have recorded his final sermon when he knew that the end was near, secretly instructing his deacons concerning the nature of the funeral ceremony. When the time came, the funeral was a massive public spectacle, with people coming from neighboring towns to hear the "wonder" of the minister's final sermon. An industry magazine reported his sermon:

> The voice of the dead minister spoke, saying:
> "The Lord giveth and the Lord taketh away."
> The voice of him who lay dead in the coffin gave out the hymn, and, half frightened, the mountaineers arose and sang.
> Then the funeral sermon opened.
> Plainly, without effort, the voice told of the early struggles of the dead man, of his hopes, his fears, his troubles, his prayers. It told of his coming to the Pine mountains, of his reception, his striving against great odds.
> And, as the climax of the sermon, the voice adjured them to be constant in well doing.
> Then, suddenly the voice commanded the congregation to rise and sing, and they sang, "Jesus Lover of My Soul."[38]

The portrayal of the event as particularly spectacular suggests that this had not become a common practice. It also suggests how little the religious content of the event mattered to the writer and to the people from neighboring towns who attended the usual event—the "half frightened" mountaineers were clearly responding to the phonographic performance; if they mourned the man, we do not know it from this article. Most other accounts of phonographic funerals, whoever the performer, maintained a similar tone. Although Feinstein notes that a German patent was taken out in 1907 by Elisabeth Hauphoff, who had invented a "phonographic hearse," the idea of the phonographic funeral seems to have been more captivating to the sound-reproduction industry press than it was to the grieving public, again suggesting that it was, above all else, an auditory spectacle.[39]

Although phonographic funerals never caught on, the persistence of the voice after death was a common trope in early phonographic advertising. The image of Nipper on the coffin is completely consonant with conventions of Victorian sentimentality used to describe the phonograph's temporality. An advertisement for the Duplex Phonograph enticed readers with the promise, "Once more you can hear the voice of old Joe Jefferson as, with matchless pathos, he delivers the lines of Rip Van Winkle so familiar to a former generation. For just before his death, this greatest and best loved of American actors left a perfect record, which, reproduced by the Duplex Phonograph, will preserve his living tones for the admiration and delight of thousands yet unborn."[40] An advertisement for the gramophone hailed it as "the only permanent means able to reproduce, in a *natural* quality, a living breath of air and speech—of those who will hereafter pass from this life."[41] Even Edison's original list of uses for sound recording included an auditory version of the Victorian family album as well: "The 'family record'—a registry of sayings, reminiscences, etc., by members of a family in their own voices, and the last words of dying persons." Edison's example takes much from the Victorian family album.[42]

The phonograph industry took great care to make sure that readers understood that Edison's "predictions"—more accurately considered programs—were coming true. When the actor W. J. Florence died, he left behind recordings of his voice, and the official organ of Edison's National Phonograph Company picked up the story and ran with it: "In Boatman's bank, St. Louis, Mo., there assembled since the decease of the great actor, four of his intimate friends, who stood before a phonograph to listen again to his voice. The scene was quaint and odd, and one that the nineteenth century alone could evolve. They listened with feelings of sadness to the quaint humor with which he had amused great audiences during life. . . . [A]ll agreed the record was marvelous, the voices of Jefferson [another famous actor] and Florence being as distinctly recognized as if they were present in person."[43] The recordings did not even need to be performances in the commonly understood sense. Following the sudden death of a stenographer, for example, his clerks were left with a number of untranscribed cylinders: "The feelings of his clerks, who thus wrote from the dictation of their dead employer, hearing his veritable voice, although he was no more, can better be imagined than described."[44] An Australian writer told of making a record of a man playing a clarinet solo before going off to be killed in the war in South Africa. Afterward, the recordist sent the record to the dead man's mother, who bought a phonograph to hear her dead son performing. Again, the woman's response is left to the imagination.[45] The reader who had never heard the voices of the dead could presumably know full well the response of those who had because the presumed common context was a particular culture of death and dying; alternatively, the response may be so personal and idiosyncratic—as it is today—that it slips away into the journalistic sublime.

Although the preservation of the voices of the dead appears in turn-of-the-century writing as an ideal-typical instance of the phonograph's purported power, it was not the only instance. The persistence of the voice could take on other valences as well; phonography was presented as an antidote to *any* silence into which speech could dissipate. In a particularly unpleasant anecdote, the *Phonogram* relates the experience of a well-traveled man by the name of Crampton who contracted a horrible (but unspecified) infection of the tongue:

> When Crampton learned of the surgeon's decree that his tongue must be removed to save his life, he conceived the idea of recording his experiences on Phonograph Cylinders, and of using these records to describe his adventurous career in public lectures, letting the Phonograph speak for him. So for three days before the operation he talked his adventures into a Phonograph's wide mouth; now modulating his voice when he would be pathetic, again raising it to meet the climaxes of his narrative. His articulation was clear, his voice did not tremble, although every word stabbed him with acute pain.
>
> So the Phonograph makes a living for Crampton, dumb and tongueless. His son Sherwood is his manager. Crampton is known as the "Tongueless Lecturer."[46]

Here we have a particularly acute case of the absence of speech, but not the death of the speaker. This account offers a particularly disturbing instance of delegation.[47] Here, the speech is methodically delegated to a machine, in the sense of turning over the activity once and for all — more delgation than Latour even imagined when he coined the term. The pathos of the story lies in Crampton's knowledge that, a few days hence, he will never speak again, that he must delegate his voice to the machine. But, apart from the exact nature of the deadline, this account is structurally similar to the death narratives: the phonograph is the signature on a guarantee specifying the persistence of the voice. In Crampton's story, the account is especially troubling because it abstracts the tongue into the organ of speech. Not only did our sad protagonist lose his ability to speak, his ability to eat, drink, sneeze, and yawn and to perform many other basic activities of self-preservation would be affected. The fetishization of the tongue as the organ of speech speaks, not to Crampton's particular fate, but rather to the possibilities of a phonograph as designated delegate for Crampton in public performance.

This tour of phonographic funerals, dead performers, soldiers, and office managers, and subjects of gruesome surgeries should give a sense of depth to this Victorian fascination with the persistence of the voice in phonography. In each case, sound recording is a kind of embalming — the voice is transformed so that it may continue to perform a function *as* the voice. To return to Nipper and the gramophone, it is easy from this broader context to interpret Barraud's painting of Nipper within a voices-of-the-dead hermeneutic, regardless of whether it actually represented his brother's coffin or even a coffin at all. Even if one remains skeptical about the actual surface on which the dog sits, the implication of the picture is clearly that the absent master's voice has the same effect on the dog as if the master were there. The dog needs nothing more than the exteriority of the voice.

This morbid mentality behind *His Master's Voice* was not lost on contemporary critics. Theodor Adorno saw Nipper as an emblem of self-importance in listeners, as a hinge into identification with the image:

> The dog on records listening to his master's voice off of records through the gramophone horn is the right emblem for the primordial affect which the gramophone stimulated and which perhaps even gave rise to the gramophone in the first place. What the gramophone listener actually wants to hear is himself, and the artist merely offers him a substitute for the sound image of his own person, which he would like to safeguard as a possession. The only reason he accords the record such value is because he himself could also be just as well preserved. Most of the time records are virtual photographs of their owner, flattering photographs—ideologies.[48]

The attitude that Adorno diagnoses here moved well beyond the iconography of Victorian sentimentality into a much larger and murkier field of bourgeois self-understanding. Adorno's man listens to the gramophone to hear the possibility of himself being heard after death: a truly convoluted scenario. But this is essentially the future hailed in *Scientific American:* the fantasy is as much about speaking to the not yet born as it is about hearing the voices of the dead. This very convolution was expressed repeatedly at a cultural level: the captivating possibility of sound recording was in the preservation of sound beyond its immediate moment, extended perhaps to infinity. In other words, it was the context of reproducibility itself that mattered; the specifics of speech and voice itself did not even really matter. The inside of sound was transformed so that it might continue to perform a cultural function.

Messages to the Future

The voices of the dead had their cultural converse in the ears of the not yet born. Beyond the idea of retaining the voices of the recently departed for a final graveside performance or for the ears of loved ones, writers quickly developed a sense of the metahistorical possibilities of sound recording. They hoped that recording would enable transgenerational speech, where any "present" could address itself to an almost infinite range of possible futures. Nietzsche once referred to history as a dialogue of greatness across the ages. Sound recording promised its Victorian beholders at least a serial monologue:

> If thus we could but listen to the voice of the great founders of this mighty commonwealth: Washington, Jefferson, Jackson, Lincoln and others, how easy it would be for us to grasp their great ideas and teachings and follow in their footsteps. But in their time the talking machines had not been thought of. To-day we are in a position to reap the full benefit of the genius of our great inventors.
>
> How salutary and consoling it is for loving children and friends to be able to retain the voices of their dear departed ones for communion in time of trouble, and of pleasure. The voice of that mother whose every thought has been for our welfare, whose last prayer was to call blessings down on us from Heaven; of that father whose stern, unbending, yet loving character first instructed us in the hard realities of life. Death cannot now deprive us of their help, advice and encouragement, if we will but record their voices whilst they live, and treasure them not only in our hearts, but in a certain and lasting form, on the surfaces of phonograph and graphophone cylinders.
>
> . . . The voice, formerly invisible and irretrievably lost as soon as uttered, can now be caught in its passage and preserved practically for ever.
>
> The great speakers, singers, actors of to-day have it in their power to transmit to posterity all the excellencies they are so richly endowed with. Art in its perfection need no longer be lost to succeeding generations, who now shall be able to enjoy all its benefit by setting in motion the wheels of a simple machine. . . .
>
> Death has lost some of its sting since we are able to forever retain the voices of the dead.[49]

Here, sound recording is understood as an extension of the art of oratory—a set of practices that depended heavily on the persona and style of the speaker and relations between speaker and audience. But, in *this* oratory, the construct of audience undergoes a wild permutation—the medium itself is the audience. Phonography marks both a sociospatial network and a sociotemporal network, where one time could potentially speak to (if not with) another. One could easily view this as a pipe dream, a high modernist conceit—since phonographic orators would never hear the response of their future audiences. But this is to forget that any medium requires a modicum of faith in the social relations that constitute it. This is precisely the investment remarked on by the novice radio singer Leon Alfred Duthernoy—the initial silence of his audience terrifies him; the emergent faith that he will reach "tens of thousands" through the medium offers a new level of gratification greater than that of the most adoring audience.[50] The message to the future requires two kinds of faith: that the audience is at the other end of the phonographic network and that the embalming of the voice promises sufficient durability to fulfill a social function indefinitely into the future. True believers can be found everywhere: as Friedrich Kittler writes, "Once technological media guarantee the similarity of the dead to stored data by turning them into the latter's mechanical product, the boundaries of the body, death and lust, leave the most indelible traces."[51] But this permanence, this "indelible trace," was itself based in a kind of faith.

Like radio performances, phonographic messages to the future were contrivances. Although forged sound recordings of McKinley's final speech emerged shortly after his death and the 1910s found Congress experimenting with mass dissemination of important political speeches via disk recordings, rarely did the early phonograph happen to capture a performance—political or otherwise—as it happened. More often, the voices of the dead were taken from the living, whose words were specifically targeted toward the horn of a recording phonograph—and implicitly targeted toward an imagined future audience. Recording was a studio art, designed for ears distant in time and space yet connected through the medium.

One can find many "messages to future generations" among early recordings. A record exists in Alexander Graham Bell's laboratory notebook that some apparently prominent men from England visited his laboratory with the express intent of recording words for future listeners: "Mr. Chamberlain made a speech to the people of the 20th century!!—followed by his two friends. The phonautogram produced was presented to Prof. Langley for preservation in the National Museum."[52] Professor Langley then made an accompanying recording certifying the authenticity of the first. These messages may well have been lost in the twentieth century—if they even made it that far. When asked about the recording, current Smithsonian staff responded that they were unsure whether it was still in the institution's possession and whether it would be playable if it were. O. Henry similarly recorded a message to future generations in the early 1890s, informing listeners such as myself that he hopes we continue to read his books; P. T. Barnum expressed his gratitude to the queen and the people of England.[53] In fact, as Gary Gumpert has bemusedly pointed out, recorded messages to future generations tend to lack any meaningful content whatsoever: the 1980s-vintage talking tombstones of which he writes offer such wisdom as, "Do what's right, come what may."[54]

Gumpert's trope of the talking tombstone suggests a certain continuity between present and past in attitudes toward sound recording—the resonant tomb. The very idea of making recordings for listeners in a distant and unknown future also carries within it a distinctively threefold sense of time: this time is at once (1) a linear, progressive historical time, (2) the internally consistent time on a record, a present cut into fragments, and (3) the almost geologic time of the physical recording itself. Here, the phonograph collides with a larger sense of the connections between time and culture that Matei Calinescu names *bourgeois modernity*. Bourgeois modernity hails a sense of present time as a thing, something measurable, something that can be apprehended and felt, stockpiled, repeated, spent, saved, broken, fragmented, or mended. Yet this sense of the present immediately collides with linear-progressive "historical time, linear and irreversible, flowing irresistably onwards."[55] As Janet Lyon writes, modernity is "subject to the very discontinuities of time that its narratives seek to disguise: different 'times' coexist within the same discrete historical moment, just as surely as homologous 'times' exist across centuries."[56]

The technology of sound reproduction fits oddly into this description of modernity as a form of at once hypertemporalized and detemporalized social consciousness. In bourgeois modernity sound recording becomes a way to deal with time. Sound recording came to embody three conflicted senses of time for its early users. "Bourgeois modern" recording is articulated to a linear-progressive sense of time, where the present inevitably disappears into the future, modernity being assumed to assure the perpetuity of changes, the constancy of upheaval and transformation. But the sound recording itself also embodies fragmented time. It offers a little piece of repeatable time within a carefully bounded frame. A few moments caught on cylinder, disk, or tape that bear some past consistency can be made manifest in the present. As Jacques Attali puts it, sound recordings allow for the stockpiling of other peoples' time. For Attali, this is a matter of property, of owning another moment of labor.[57] For others, it could be distilled into a simple fact: sound recording stores time.[58] In addition, this time is also something more, the retention of a certain sequence, isolation, and repeatability of moments—a fragmented consciousness of time. These two temporalities are then set into play with a third, physical temporality: the decay of the recording itself, the ephemerality of the medium. The bourgeois modernity of sound recording is polyrhythmic: it becomes an interplay of telos and cycle shaped by the physical possibilities and limits of materials; it moves between the ephemerality of moments and the possibility of an eternal persistence.

This distinctive temporality of sound recording is thus forged in the making of the machine itself. Phonographic time was the outgrowth of a culture that had learned to can, to embalm, in order to "protect" itself from seemingly inevitable decay. This sensibility would have to be built into the medium itself, literally. Permanence, or, at least, the persistence of the voice, was a promise that people desired to make to others whom they could only imagine. The phonograph did not introduce a jarring new temporality into the culture; on the contrary, this "bourgeois modern" sensibility was a means by which phonography was introduced.

"The Voices of Dying Cultures": Audio Ethnography and the Ethos of Preservation

Early recording enthusiasts praised sound recording for its preservative promise; early anthropological uses of sound recording expanded the metaphor. While Edison wrote of using the phonograph to preserve the voices of dying persons, the American anthropologists who first used sound recording in their work often explicitly justified it in terms of the phonograph's potential to preserve the voices of dying *cultures*. Alongside the notions of time embedded in Calinescu's construct of bourgeois modernity were another set of attitudes about time that shaped the history of recorded sound and the desire to emphasize recording's preservative function.

Writing about anthropology's conception of modernity, Johannes Fabian argues that cultures outside the anthropologist's own became representative of some kind of collective past, thereby implying that the anthropologist's home culture represented the future of so-called primitive cultures. This denial of coeval existence—that is, coexistence at the same historical moment—results in a relentless "othering" where anthropologists construct themselves as living in a society that is more developed or advanced than the societies that they study; it is a form of primitivism. Fabian writes that ethnology and ethnography "promoted a scheme in terms of which not only past cultures, but all living societies were irrevocably placed on a temporal slope, a stream of Time—some upstream, others downstream."[59] Despite Native Americans and American anthropologists occupying the same geographic space, early ethnography and ethnology cast Native Americans as existing in the collective past of white society. As Philip Deloria writes, early ethnographers depicted their native subjects as living "in a different temporal zone" from modern, white society. Since they existed in the same space, time was used as the measure of cultural difference between native and white cultures.[60]

Although Fabian is writing about American and European anthropology, the "denial of coevalness" was particularly acute in the United States, where nineteenth-century federal policies had—to put it bluntly—genocidal effects on Native American populations as tribes were moved around to make space for white settlements. What had been an explicit program of land clearance in the early nineteenth century was by the 1880s represented as the inevitable force of civilization and modernity: this is the denial of coevalness. Part of this collapse of cultural difference into a progress narrative was embodied in a somewhat earlier shift in federal policy toward Native Americans. Starting during Ulysses Grant's administration, the government pursued what the historian Francis Prucha calls a "peace policy," "a state of mind, a determination that since the old ways of dealing with the Indians had not succeeded, a new emphasis on kindness and justice was in order."[61] Although the final military campaign against native populations would not be until 1890, from this period on the federal government at least tried to appease reformers and humanitarians through pursuing an officially paternalistic stance toward Native Americans. But policies continued explicitly to target Native American cultures, if not Native Americans as persons, often to similar effect as more officially brutal policies. The 1887 Dawes Act, for instance, was popular with reformers because it allotted property to Native Americans and was intended to help them assimilate into white culture. But, in forcing Native Americans to divide up tribal lands into family plots and sell off the rest of their territory to white settlers, it also facilitated the kind of "land clearance" (moving Native Americans out so that whites could move in) that had been a staple of federal policy throughout the nineteenth century. The Dawes Act was essentially about encouraging Native American assimilation by abandoning communal ownership of land in favor of a private-property model. Similarly, many treaties specifically directed cash settlements for Native American lands into funding for schooling, supplies, and infrastructure aimed at integrating Native Americans into white American ways of life. Federal programs were, thus, aimed at getting Native Americans to abandon their tribal lands, their religions, and their economic and cultural ways of life in favor of secular, mainstream American "white" values. By the 1890s, most of the major native nations had been relocated from their earlier tribal grounds—many to the Indian Territory (now Oklahoma).[62]

A comparable change occurs in attitudes among anthropologists during roughly the same period. Beginning with the writings of Lewis Henry Morgan, anthropologists aimed to paint a more sympathetic portrait of Native Americans—"a kinder feeling toward the Indian." But, at the same time, this work was heavily influenced by the evolutionism sweeping the human sciences following the publication of Darwin's *Origin of the Species*. Morgan thus sought to locate Native Americans on an evolutionary ladder, "behind the Aryan family in the race of progress." As Robert Berkhofer writes, nineteenth-century anthropologists used a concept of culture, but it was culture in the singular—not in the plural sense common today. By the 1890s, however, a more relativistic, pluralistic, empirical, and specific tradition was emerging in the work of Franz Boas and his followers. Boasian anthropology did think of culture in the plural sense, focusing on native tribes as cultures rather than as examples of the racial type *Indian*. By World War I, Boas and his students dominated the anthropological profession.[63]

In the writings of the first audio ethnographers, one can see a mixture of these various influences. They retain some vestiges of the evolutionism and primativism of their predecessors—as Johannes Fabian notes of fin de siècle anthropologists, they "did not solve the problem of universal human Time; they ignored it at best, and denied its significance at worst."[64] This sense of the immutable force of modernity can be read as a displaced evolutionism. Rather than portraying their native subjects as occupying a stable place on the cultural-evolutionary ladder, these writers instead cast modernity as the dynamic factor: explicitly political and cultural programs designed to eradicate native cultures were recast as almost unspeakable forces of nature in much American anthropological writing from this period. Deliberately hostile political and military programs and their purportedly kinder descendants liquefied in the imaginations of many anthropologists coming of age in the 1880s and 1890s; together, they welled up into a temporal tidal wave that swept away native cultures and civilizations as if it were an immutable, natural force.

This sense of loss and cultural change was itself a result of broader trends in nineteenth-century interactions between various Native American cultures and white culture. As Fred Hoxie has pointed out, "The academic discovery of Indian beliefs and traditions . . . fit neatly within an old tradition that classified native lifeways as both exotic and backward." They were "defined from the outset as irrelevant to contemporary concerns."[65] Hoxie is suggesting not a lack of intellectual relevance—since he is writing of academics' fascination with Native Americans—but rather a sense of irreducible difference, alterity, between native "tradition" and white, academic "modernity." Fewkes's and others' writing is peppered with this kind of nostalgia for Indian culture as a past existing in the present. Thus, the "peace policy" mixed with evolutionism, sympathy, pluralism, empiricism, foreboding, nostalgia, and a preservationist ethos in the writings of the early audio ethnographers.

Because they are artifacts of this strange mix, early anthropological and folkloric recordings are perhaps the quintessential example of the intertwined phonographic tropes of voices of the dead and messages to the future, except that, in this case, it was voices of a dying *culture* that the anthropologists hoped to save. This effacement of the previous generation's deliberate hostility, combined with a view of modernity as ineffable progress and assimilation, led the Bureau of American Ethnology (BAE) ethnologist John Peabody Harrington to compose the following lines:

Give not, give not the yawning grave its plunder,
Save, save the lore for future ages' joy:
The stories full of beauty and of wonder
The songs more pristine than the songs of Troy,
The ancient speech forever to be banished—
Lore that tomorrow to the grave goes down!
All other thought from our horizon banish,
Let any sacrifice our labor crown.[66]

Harrington's sentimental verse captures nicely the twin affects animating ethnographic recording, a deep affection for Native American culture combined with a sense of impending death for the culture—hence the recurring grave metaphor. The cause of death, however, is conspicuously absent from his poetic effort. The willful destruction of Native American culture undertaken by whites throughout the late eighteenth century and the nineteenth became an impersonal, immutable force of history in ethnographic writing at the turn of the twentieth century.

To many of the anthropologists participating in it, this discussion of cultural death was not metaphoric. Many academics at the time saw Native American cultures as actually dying out. By the time academics were developing a significant body of research on Native American cultures, the federal government had essentially won a centuries-long war against native peoples for control over the continent. In addition to federal policies clearly aimed at eliminating many central elements of Native American culture, the native population had been in steady decline for over a century as a result of warfare, famine, and disease: it was widely held that the native population hit an all-time low in 1890, although this figure is debatable since native population statistics were based on purist and eugenic definitions of race and, therefore, did not account for people of multiracial backgrounds (had they done so, the native population figures would have increased significantly). Although immediately thereafter the population began to rise again, eugenic theories and narrow definitions of *race*—along with mistaken conceptions of Native American culture as historically static—led many turn-of-the-century academics to the conclusion that Native Americans were dying out as a race.[67]

Early ethnographic recordings of Native Americans are, thus, marked by a sense of impending loss and the imperative of preservation as well as the hope for their future use. By considering the processes through which these early recordings were made alongside the incredibly nostalgic language of anthropological mourning that accompanied them, we can catch a glimpse of a long cultural-historical process of crystallization. Sound recording came to take on a whole set of temporal and cultural valences specific to a particular time and place—in this case, the ethos of preservation in the late nineteenth century—that would in turn be reattributed to the machine itself.

An account exists of the ethnologist Frank Hamilton Cushing possessing cylinders of Zuni, Apache, and Navajo music in May 1889. It is not clear whether Cushing himself performed the music or whether the cylinders had documented native performers—this is because he is not known to have written about using sound recording in his work. The credit for bringing sound recording to ethnology is, thus, usually given to Jesse Walter Fewkes, a Harvard-trained zoologist who had turned his interests to the study of Native Americans. Although Fewkes's East Coast manners earned him the nickname "The Codfish" in the field and a great deal of his ethnographic work was collected by assistants, Fewkes proved to be an effective organizer; he was able to make full use of his old-boy network connections. With funding from Mary Hemenway, he purchased an Edisonphone, tested it first on a trip to Maine in the winter of 1889–90, and then took it to the Southwest. Between the two trips, Fewkes recorded over forty cylinders of Passamaquoddy and Zuni music and speech.[68] Those recordings, soon deposited at the Peabody Museum at Harvard, provided the basis for a series of articles that demonstrated to anthropologists the utility of sound recording for their work.

Fewkes's early recording work is remarkable when compared with the approaches to recording dominating other fields at the time. While the sound-recording industry was experimenting with office applications and coin-in-the-slot machines and just beginning to tap into the entertainment market, Fewkes and other early users of sound recording like Alice Cunningham Fletcher and Frances Densmore more or less established the parameters for anthropological understandings of sound recording that have in many ways held up until the present day. When Fewkes's recording work began in 1890, anthropologists had been seriously interested in Native American music for only about ten years. Although accounts of Native American music existed prior to 1880, few, if any, took it seriously as music, preferring instead to call it noise. Theodor Baker's 1882 monograph, *Über Die Musik Der Nordamerikanischen Wilden,* was the first serious and extended scholarly treatment of Native American music. Fletcher, also an ethnologist, was the first American to develop scholarly interest in Native American music and began publishing on the subject in 1884.[69] Fewkes's experiments in phonography thus took place in a still-nascent field searching for a coherent understanding of its object and its approach.

Reading Fewkes's early writings while considering the making and preservation of his phonographic cylinders offers a telling dissonance. Fewkes wrote of sound recording as beneficial for both immediate study and preservation. As anthropologists previously had to transcribe stories or music by hand, they would either rely on their memory for what they heard (which, as Fewkes points out, was probably considerably worse than the memory of their subjects)[70] or ask their subjects to repeat the performance many times so that they could capture it in all its detail. Even then, there was no guarantee that, when reenacted by a white reader, the transliteration would be intelligible to Native American listeners as the performance to which it was supposed to refer: "I doubt very much if the Indians could understand many of the words in some of the vocabularies of other Indians which have been published, if the words were pronounced as they are spelled. The records of the phonograph, although of course sometimes faulty, are as a general thing accurate."[71]

The phonograph's accuracy or lack thereof itself sparked a debate concerning Fewkes's recordings and the relation of recording and transcription. A Harvard psychologist of music, Benjamin Ives Gilman, transcribed some of Fewkes's recordings of Zuni and Hopi music and elicited emphatically hostile reactions from two major comparative musicologists—Carl Stumpf and John Comfort Fillmore. Stumpf was critical of the phonograph itself, claiming that it was not mechanically reliable; Fillmore criticized Gilman's transcriptions. Because Gilman had the recordings and was able to base his transcriptions on repeated listenings, his notation was exceptionally detailed for the period, noting even minor deviations from the half-step intervals basic to Western notions of pitch. Fillmore argued that these deviations were idiosyncratic to the specific performance and were likely the results of an untrained singer. Put simply, Fillmore argued that the transcriptions were too detailed; phonography had facilitated a mode of listening too technical and too focused to be of true ethnological use.[72] Erika Brady writes that this was a result of the "paradigmatic" attitude among many ethnologists at the time: "Particular performances were important only insofar as they could be used to reconstruct a paradigm for song, story, narrative, or myth in a given culture." In other words, the focus on a given text allowed too many variables to be generalized. Another writer referred to transcription from phonographic recordings as pointless "as a singer would make alterations to a tune with each performance."[73]

Despite Fewkes's qualifications and others' subsequent criticisms of phonograph-based ethnography (either that it was too accurate or that it was not yet accurate enough), the practice of ethnographic recording took off. Leaders in the field like Franz Boas advocated the use of recording and the inclusion of music as essential to ethnographic research, although Boas too was critical of the comparative method advocated by Stumpf and Fillmore.[74] The task of repetition and memorization had been delegated from anthropologists and subjects to the machine, which allowed the anthropologists to transcribe at a more leisurely pace and check their work more frequently. Fewkes's famous *Journal of American Folklore* piece takes this approach, and the vast majority of the essay is devoted to explaining the contents of several cylinders to readers. Of course, this approach in part reflected the scarcity of phonographs at the time and the relative impossibility of mass-producing the recordings. If Fewkes wanted a large number of people to know what was on the recordings, he had to write about them. In its delegation of memory to the machine and its media, the practice of anthropological recording connects with the multiple temporality so clearly elaborated in relation to the machine's morbid modernity. Fewkes's article manifests an intermixture of linear, historical progress ("dying cultures"); fragmented, repeatable events (the task of repetition and short-term recall); and the physicality of recordings themselves (explaining recordings because they could not be widely heard).

As Fewkes wrote, "When one considers the changes which yearly come to the Indians, and the probability that in a few years many of their customs will be greatly modified or disappear forever, the necessity for immediate preservation of their songs and rituals is imperative." Here, Fewkes enacted that anthropological mourning that converts a long history of deliberate attacks on native ways of life into an immutable force of progress. He hoped to demonstrate the utility of his recordings as an alternative to the method of anthropological transcription then in use, but the promise of preservation superseded any immediate utility in its significance: "Now is the time to collect material before all is lost. . . . The scientific study of these records comes later, but now is the time for collection of them. Edison has given us an instrument by which our fast-fading aboriginal languages can be rescued from oblivion, and it seems to me that posterity will thank us if we use it to hand down to future students of Indian languages this additional help in their researches."[75] This sense of loss constantly resurfaces in Fewkes's discussions of Passamaquoddy speech and music. He repeatedly writes of how this or that song or ritual is remembered by elders but not by younger members of the tribe, how Western and even biblical elements have infiltrated important stories, and how modes of dress and ornamentation are now becoming mere curiosities.[76] Descriptions of later work offered similar rationales. For instance, A. L. Kroeber had two members of the Mojave tribe spend a month in 1903 making over a hundred phonographic cylinders of Mojave speech, language, and singing. Kroeber cited the work as an opportunity to archive Mojave traditions and also as providing fixed artifacts for careful study since there was no written Mojave language.[77]

The dominant paradigm in this moment of American ethnology was very much focused on the collection of texts and artifacts. This textualism would itself be justified in terms of death imagery—in its ethos, anthropologists were to collect the artifacts of dying cultures before they died out altogether. Although anthropologists retained a certain antimodernism because of their interest in and affection for native cultures, this impulse still enacted a denial of coeval existence since it cast the force of modernity as ultimately undoing native life ways. Justified in terms of the immutable flow of history, the textual approach had the side effect of dehistoricizing Native American cultures. Following Curtis Hinsley, Erika Brady refers to this as the *marbling* or *bronzing* of Native Americans—freezing a dynamic native culture at a single moment in time for future study. Perhaps, as Brady suggests, early resistance to sound recording was in part a resistance to this artifactualization of living cultures.[78] We do know that the desire to artifactualize native cultures—themselves understood as *acutely* ephemeral—was a central motif in early writing about phonographic ethnography.

This combination of death imagery and a textual emphasis is an enduring feature of early phonographic ethnography; the phonograph became a tool of embalming an already supposedly frozen native present for the future. Fewkes's analysis of the Passamaquoddy Snake Dance embodied this mixture of foreboding, nostalgia, and belief in the possibility of preservation. His written description survives today in his journal articles; a reproduction of the recording is available through the Library of Congress Folklife Collection. The liner notes to the recording offer the same sentiment as does Fewkes, with the added certainty of an accomplished history rather than a dreaded future: "When it was sung by Noel Josephs into Fewkes' recording horn, it was part of a tradition which was still flourishing. Today the observances are largely a remnant of the past, and the language in which Josephs sang is virtually unknown."[79] This process of forgetting was already under way when Fewkes arrived with his phonograph.

On listening to the recording, one is forced to question the ideology of transparency suggested by ethnographic recordists. The recording confronts my ears as an artifact of an event, not simply as the event itself. If it is an antidote to total forgetting, it still thrives on the forgotten, on a past that recedes and retreats. Through headphones, listening to a grainy LP record copy of a grainy cylinder, one can hear—or, at least, imagine that one hears—the grain of Josephs's voice. Although I imagine Josephs's voice to be deep and full, the horn and diaphragm of the phonograph seem to have thinned it some. He sings a verse; he sings it again, varying it slightly. And so it goes until the recording runs out in a few brief moments. Hearing it without knowing the language, I understand only its internal rhythms, as they are presented to me in a single, truncated performance. Fewkes offers a transcription of both words and music in his *Journal of American Folk-Lore* article, yet the transcriptions offer no enlightenment in and of themselves. In fact, the written score seems to obfuscate: I hear with my ears some spontaneous variation, some improvisation. I see in the score a fixed melody line and lyrics (although I still have no idea what they mean). But Fewkes, who himself never saw the Snake Dance performed, offers a narrative account of the event (paraphrased below) as told to him by others who had observed or participated:

> The dance begins with the leader moving around the room in a stooping posture, holding a rattle in one hand. While singing and shaking the rattle, he pounds the floor rhythmically with one of his feet. He looks around the room, inciting the listeners to take part. He moves to the middle of the room and takes a person's hand, who takes another, and so on, until a long, coiling line of people, facing in alternate directions, emerges. The dancers start stooped over, but, as the intensity of the song increases, they gradually stand up and coil around the leader, all singing. The dance grows louder and faster, the coil rapidly unwinds, and the dance continues until dancers begin to fall or cannot keep up. Then the chain breaks, and the dancers return to their seats, shouting loudly.[80]

What has the phonograph preserved? Can we even say that it has captured the sound of the song? No: the interiority of the performance has changed. One singer, no doubt instructed to modulate his voice so as not to tax the capabilities of the recorder, standing still and not dancing, without drum or rattle, sang the melody a few times into the phonograph. The recording diaphragm and wax medium captured a specific performance, a performance designed and modified specifically for the purposes of reproducibility.[81] The promise of mediation was made but not fulfilled: the mediation of the live music and the dissolution of that mediation into transparency are at best imagined. The thing itself as we imagine it was never there at the moment of the recording; the recording is less a memory and more a mnemonic. The performance itself was transformed in order to be reproduced. The abstraction happened before the cylinder spun or a word was sung. Although written transcriptions can bear the same mark of abstraction, it takes more effort to mistake a written description of a Snake Dance song for the song itself.

Fewkes and his phonograph are documentarians; they are active participants in the culture that he claims they study from the outside. Even if Fewkes's manner of self-presentation earned him a reputation for distance—a reputation that he sought to promote—a rare glimpse of his subjects' view of him offers a substantially different perspective. In 1898, Fewkes abruptly halted his fieldwork while studying the Hopi tribe's winter rituals: the most widely cited account is that he feared that he might contract smallpox. However, the Hopi told a different story. It was said that, during a highly secret ceremony, where Fewkes was instructed to lock himself in his house and ignore the goings-on outside, he was visited by Masauwu, "a mighty and terrible being [who] wears on his head a bald and bloody mask." Finally, Masauwu "cast his spell on him and they both became like little children and all night long they played around together and Masauwu gave the doctor no rest. And it was not long after that Dr. Fewkes went away but it was not on account of the smallpox."[82] That Fewkes himself made it into his subjects' lore emphasizes the degree to which the supposedly fixed Native American cultures studied by early anthropologists were themselves dynamic and shaped by the encounter. Another instance had Fewkes depicted in a Hopi dance, where a pipe representing the phonograph's horn was placed on a table covered with a blanket. One performer yelled into the pipe, another hid under the blanket and yelled back jibberish, and a third, dressed as an "American," stood by and took notes. Fewkes's colleagues' perception of his distance from his subjects was, thus, itself a cultured, subjective perception. The disappearance of Fewkes's supposed ethnographic distance in the Hopi tales and dances is allegorical for my narrative here—far from being an external reporter of Native American culture and history, the phonograph becomes part of that history.[83]

This intertwined relation becomes even clearer when we turn to the work of other phonographic ethnographers. Although Fewkes's phonographic work, the Hemenway expeditions, and Gilman's transcriptions predate Alice Fletcher's adoption of sound recording, her work provided a foundation for theirs and in turn built on their experiments. The publications by Fewkes and Gilman caused some concern for Fletcher, who was working hard to establish herself in the field of anthropology and whose 1886 lectures on Native American music inspired Mary Hemenway to fund Fewkes's southwest expedition in the first place. Fletcher was initially unimpressed with the phonograph as an aid in musical ethnography. Fewkes's and Gilman's success helped change her mind, and she purchased a graphophone in 1895.[84]

Fletcher's approach to recording was a significant departure from that of Fewkes and others. Whereas others went out into the field to get recordings, Fletcher brought the field to her doorstep. She set up a rudimentary studio in her house at 214 1st Street S.E. in Washington, D.C., and worked with Native Americans who came to Washington on government business or whom she herself invited. As a result, she was the most prolific recordist in anthropology for the next five or six years. Although she would occasionally make cylinders while away from Washington, this was her main approach to phonographic ethnography.[85]

Fletcher's approach yielded a large number of important recordings, including some that are important to Native Americans today (as I discuss below). A good deal of her success rested in extending the logic of Fewkes's initial method through to its conclusion. Whereas Fewkes and his team of researchers sought abstracted performances that would fit on cylinders on site, Fletcher redefined fieldwork. Rather than the anthropologist going into the field—essentially defined as someone else's culture or native territory—and bringing back samples, she was able to use her location in the nation's capital as a way to catch Native Americans on their travels through the anthropologist's own "native" territory. Rather than facilitating a certain form of cultural contact, Fletcher took advantage of existing Native American traffic. Her recordings were even more clearly based on the artifice of the studio and practices specific to sound recording, although they were certainly no more constructed than Fewkes's. In her case, as in Fewkes's, the music was transformed at its most basic level before it crossed into the world of the reproducible. Fletcher's method benefited from the prestige of Washington, D.C., as the nation's capital; it helped persuade her subjects to perform for the machine.[86] Her frank and pragmatic approach may also have been partly a result of her orientation toward Native American culture itself: she was a strong advocate of integration and assimilation. In fact, Fletcher had previously worked for federal agencies attempting to get Native Americans to comply with the Dawes Act. As a result, her recording technique embodied not a pristine native culture about to be touched by modernity, but a living native culture in continuous contact with white culture.

It is worth noting that Fletcher's approach found its analogue in other ethnographic practices. A 1907 *Ladies Home Journal* reports that Native Americans doing business in Washington, D.C., could stop and have their portraits taken at the Bureau of American Ethnology; others had "life masks" made of their faces. Of course, the purpose was to document native physiognomy, and the National Museum's Laboratory of Physical Anthropology added to these projects its own different approach—measuring Native Americans' bodies in minute detail. Like Fewkes's writings, the *Journal* expressed a certain degree of nostalgia, mixed with fascination and dread: "After the Indian has become extinct, as such, his picturesque features and graceful form will thus be preserved to art."[87] Often employing people with no training in ethnography, musicology, or linguistics, the BAE made great use of sound recording as a way of creating texts for others' later interpretation.[88] Despite its shoestring budget, the BAE did manage to hire several people who would go on to become famous ethnographers. One such person was Frances Densmore.

Unlike her predecessors, Densmore commenced fieldwork after the phonograph was well established. She arrived for her job at the BAE in 1908 with a small phonograph, but the bureau immediately replaced it with a Columbia Graphophone: "Home recording was at the height of its popularity and this machine was made to meet the demand." It was also, in her estimation, particularly well suited to recording Native American music. Densmore's fieldwork was more traditional than Fletcher's in that she went out to the reservations and did her collecting on site. But her approach was clearly informed by the same studio ethos:

The ideal place for recording Indian songs is a detached building which is not so isolated as to give an impression of secrecy, nor so conveniently located that Indians will linger around the door. The building should be near the agency and trading post, so the Indians can attend to business if they wish to do so. This was important in the old days when they often came 25 miles or more on horseback. Such an ideal "office" is rare, but the Superintendents of the reservations have always given me the best facilities at their disposal. I have recorded in an agent's parlor and in his office on a Saturday afternoon, and also at a Protestant mission. I have even recorded in a school laundry, with the tubs pushed back against the wall, and in an agency jail that was not in use at the time. A tar-paper shack was my office for more than a month on the Dakota prairie where the temperature in similar shacks was 116°—there was no shade for miles around.[89]

For Densmore, the nature of the facilities may not always have been very good, but it was essential to have a facility. Although she occasionally recorded outside, as when Henry Thunder, a Winnebago, refused to sing unless he was in a grove and could see for miles around in all directions, this was to further the studio ethos: Thunder only wanted to be sure that nobody would hear him sing. Densmore's approach refined the artifice of recording, and she explicitly understood it as artifice, as a document of the music rather than as the music itself: "The singer is shown how to sit in front of the horn, and to sing into it from the proper distance. . . . He is also told that he must sing in a steady tone and not introduce the yells and other sounds that are customary to Indian singers. *The recording is not intended to be realistic, but to preserve only the actual melody"* (emphasis added).[90] Here it is stated clearly: sound recording is an exercise in disciplined abstraction. Its goal is not to capture the music as it is but to present the music as it might be for study and careful analysis. Although Densmore characterized Native American music as primarily rhythmic and minor in tone, it was the melody line, the "Anglo" preoccupation, that defined the quality of the recording. Again, the music underwent a transformation in order to be reproduced, in order to be contemplated and studied away from the place and time of its performance.[91]

Yet, without some kind of standardization or guide to recording speed, the accuracy supposedly gained by using a phonograph instead of manually transcribing could be partially lost. Densmore reported that Fewkes once told her that his first phonograph, operated with a foot treadle like a sewing machine, led to problems in recording because, "if he became interested in the singing, he moved the treadle faster, increasing the speed and raising the pitch. Sometimes he moved the treadle slower, with the opposite effect." Even when early researchers like Alice Fletcher and Frances Densmore had the good sense to begin their recordings with pitch pipes, the possible variance in speed is still so great that sometimes it cannot be determined whether the pitch is A or C.[92]

Taken together, the practices of (and artifacts left behind by) Fewkes, Fletcher, and Densmore suggest that ethnographic recording was an extension of the preservative ethos emerging at the turn of the twentieth century. These practices of recording created sound events designed to be reproduced later and elsewhere, even though the method was justified in terms of saving tradition in the "here and now." Recording was a mechanical form of preservation—change the insides so that the outsides can continue to perform a social function. But this instance of the preservative ethos was, like the others, more of a program for sound recording than a description of it. Even if recordings are consistent and the pitch standard, the mere fact of recording was no guarantee of preservation. Although, as we have seen, both casual users and serious ethnographers understood the phonograph to be a medium of possible permanence even in its earliest years (although others, like Fletcher and Densmore, were more immediately interested in transcription and analysis), that permanence was largely imagined for early recordings. It was a Victorian fantasy. Edison's tinfoil recordings could not be played once they had been taken off the cylinder;

the wax cylinders of the 1880s and 1890s were recorded at nonstandard speeds, sometimes at varying speeds in the same recording. The records themselves also required fastidious care; preservation needed to be conscious. The permanence of sound recordings was an imagined future. The message to future generations—whether some banal piece of advice or the fragments of an eroding tradition—presumed institutional and technological frameworks that did not yet exist (such as archives, more stable wax cylinders, better climate control, and an indexing system). Even where those frameworks existed in nascent form, researchers had to learn to treat recordings like they were history. As in their collection, recordings' preservation required a dedicated project and newly emergent institutional contexts.

Permanence as a Project

In his history of folk-song collecting, D. K. Wilgus argues that most early scholarly recordings were made largely for the purpose of later transcription, that few were made purely to preserve folk songs for posterity. This hypothesis is illustrated by the work of John Lomax, one of the first (and certainly the most notable) fieldworkers to collect songs in English by phonograph. Wilgus treats Lomax's *Cowboy Songs,* released in 1910, as a classic in the field. In gathering the material that was eventually excerpted in *Cowboy Songs,* Lomax made over 250 recordings. Forty-seven of those recordings had made their way to the Library of Congress by 1958, when Wilgus was writing, but the rest were lost or broken. Wilgus notes that many other researchers had a similar retention rate. This was because it was the transcriptions that were considered the "primary analytical basis for work in folklore or anthropology," not the recordings themselves.[93]

Far from providing a neutral field on which to register and replay the effects of sound, the materials of cylinder recordings had their own temporality—they were fragile and not well suited to long-term preservation. This was the fate of the majority of Fewkes's recordings. Engineers' listening notes taken from a file on Fewkes's recordings at the American Folklife Center of the Library of Congress also mark sound recording's ephemerality. Although Fewkes did keep rudimentary notes on his cylinders, the recordings still require a great deal of guesswork on the part of more recent listeners. Having transferred the Fewkes cylinders to tape, an engineer evidently attempted to catalog what the center had in its possession.

Interspersed with occasional descriptions of entire performances are notes like the following:

> *no. 2* Two bands. Virtually inaudible. Talking, male voice?
>
> *no. 5* talking. "Very important" audible. 1st take at the same speed as the 2nd take of no. 4. 2nd take returns to the speed of 160 rpm as before. Virtually inaudible. Whistle. "In order to take." . . . "preserving" (2nd take clearer, perhaps not recorded in the field.);
>
> *no. 24* Cylinder cracked and mildewed. only hints of audio information. (Dead bugs in wrapper. Not all copied. Too hard on needle.);
>
> *no. 44* Talking by Fewkes with remarks by informant. Recorded introduction: "Passamaquoddy tribe." . . . "Goodbye Mr. Phonograph" by informant at end, then whistle. 2 takes. ("Now Mr. Phonograph, let's try it again. We have come to record some of your songs." "Stories, I have obtained today." "Our phonograph was invented by Thomas Alva Edison." "That shows that the phonograph can be . . . for a very long time.")[94]

These comments are interesting when contrasted with Fewkes's own regarding the cylinders. He represented himself, not only as having a clear sense of what was on each cylinder, but also as being able to hear distinctly the material recorded on each one.[95] No doubt the cylinders were in much better condition when Fewkes first listened to them; moreover, having heard much of the music as it was recorded, Fewkes could treat them more as a mnemonic—a reminder of something he had already heard. Decades later, Library of Congress engineers had neither of Fewkes's advantages; the "permanence" did not lie in the medium itself.

Far from being a transparent echo of the past, its perfectly preserved remainder, the process of preservation and the historicity of the medium itself—right down to the aging of the wax and where the cylinders were stored—shape the history that remained audible after the fact. As Anthony Seeger and Louise Spear discovered, they could not always be sure of what they heard in old recordings: "Sometimes what sounded like a drum was not a drum at all, but a crack in the cylinder. The close examination of a recording which was described by earlier technicians as 'man sings with drum' showed that the drum-like noise was produced by the thump of the needle hitting a crack. . . . A similar misconception also resulted when a patch of mold on one part of a cylinder produced a sound interpreted as a rattle."[96] This scenario can easily be read as the medium overtaking the content; the physical characteristics of the recording surface enter into the recording itself. But that would assume the possibility of an unmediated message, presumably where there were no cracks in the recording surface and no other surface noise. To put it another way, the records' state of disrepair suggests that the tendency of the medium is, far from permanence, toward simply a different temporalization, a different historicity. The material form of recorded sound—the record itself—is still another form of ephemerality.

In addition to the material of records and the ability to reproduce them, there was also the difficult problem of preserving phonographic history. Recordings for future generations needed archives: the speaking dead needed a cemetery for their resonant tombs. The largest early soundrecording archives date from the founding of the major record companies. For instance, the Columbia Phonograph Company, founded in 1889, had an archive by 1890.[97] Especially once the means of mass production were in place, the retention of a master recording meant that it was possible to make further runs of popular recordings. But corporate archives were only as good as the corporations' motivations. Beyond a certain point, companies lacked a compelling reason to preserve recordings that did not hold commercial promise, and, if a company went out of business, its archive would not necessarily be maintained, and the continued preservation of the recordings housed therein became more a matter of accident than of intention.[98] The first archives existed not for purposes of preserving history or communing with the not yet living, but rather for very basic commercial purposes: keeping the prototype of a product at hand.

The phonographic archive as initially conceived, however, was not commercial. It was meant rather to serve prevailing notions of history and historicity. In 1891, for example, the *Phonogram* ran a piece promising that

> . . . persons will soon be sent to foreign countries to collect the voices of all the living kings, queens, statesmen, composers, artists and novelists, and, if possible of the latter, extract spoken versions from some of their great works, a space will be set apart in our own and in foreign national museums, for the "Phonograph Cabinet," and this rare and valuable collection of phonograms will be duplicated and preserved for future generations.
>
> We have lost Tennyson, Longfellow, Whittier, Bryant and hundreds of other celebrities, now passed away, but Tennyson's prayer for "the touch of a vanished hand and the sound of a voice that is still," will be answered, for though the body may be turned to dust, the phonograph will have made and preserved an exact picture of the sounds it uttered while living.[99]

Two months later, the same magazine ran a feature written by one Dr. J. Mount Bleyer, who had collected over six hundred recordings to begin a phonographic library, the goal of which was to preserve the thoughts and speeches of great people for future generations. The phonograph was to be the voice's "champion against time."[100] The idea of some kind of institutionalized preservation was quite appealing at the time, although no clear sense of the significance of this preservation emerges other than the general desirability of still having voices when bodies have disappeared.

Apart from formal, private attempts to build an archive, there were also organizations that hoped to use recordings as a kind of living history, as when a church society placed a recording in the cornerstone of its new church in Findlay, Ohio, telling of its various church-building enterprises in the voice of the current pastor. The thought was that, when the church would come down in a few decades or a hundred years, "the people then living will be permitted to hear a living voice that has been boxed up for a hundred years or more; all of which is very instructive."[101] Once again, a vague optimism surrounds the enterprise, the sense that somehow this will be an intrinsic good.

Such general optimism, without a clear sense of purpose or utility beyond preservation as an intrinsic good, was not limited to industry periodicals. Philip Mauro, an attorney with an interest in the establishment of a "historical graphophonic collection to be preserved either in the National Museum or in the Congressional Library" through the use of "an indestructible sound-record [this referring to a kind of record, and not the company]," wrote to Alexander Graham Bell that "a national collection . . . should include at least our President, Cabinet Officers, Speaker of the House, and Justices of the Supreme Court." His letter did not include any sense of what such a collection might teach future listeners.[102]

The sound archives that could and actually did preserve recordings for future generations were themselves part of the anthropological impulse toward preservation. They derived their justification from the ethics of the disciplines of anthropology, musicology, and linguistics. Beyond sharing the temporal sensibility of their contemporaries in the phonograph industry and elsewhere, academic and government researchers had the added justification of systematic study and research. They also had the institutional impetus and resources to carry out the project in something approaching a systematic fashion—and certainly less idiosyncratically than hobbyists and church groups.

The Phonogramm-Archiv at the Austrian Academy of Sciences was the first sound-recording archive aimed at preservation and study, followed shortly thereafter by similar archives in Berlin, Paris, and London.[103] Perhaps the most influential of these early archives was the Berliner Phonogramm-Archiv. Carl Stumpf, a psychologist of music and a founding figure in the field of comparative musicology, commissioned Otto Abraham, a medical doctor, and Erich von Hornbostel to build a collection of recordings that would aid in the comparative psychological research in which Stumpf was engaged. Von Hornbostel went considerably further, branching off into explicitly musicological concerns.[104] In a 1905 address on the problems of comparative musicology, von Hornbostel discussed the use of electroplating (discussed above) as a precondition for the archiving of recordings and cited scientific comparison as the primary reason for building and preserving recording archives. The process of archiving, in turn, required systematic collection methods, and von Hornbostel points to American researchers as an example.[105] The majority of his address was devoted to the specifics of comparative-musicological analysis of pitch, rhythm, and other "measurable" musical characteristics. Interestingly, von Hornbostel also speculated that sound recording would assist in the process of translating non-Western musics into a more intelligible form for Western ears and eyes. Sound recording could make it possible to study music in slow motion, so to speak, by slowing down the playback mechanism. As the playback slows, the music decomposes to the point where "individual measures, even individual notes resound on their own" and non-Western music becomes susceptible to Western notation.[106] But his conclusion gestures back to history as a final rationale for recording:

The more extensive the data that we submit for comparison, the sooner we may hope to be able to explain a posteriori the archetypal beginning of music from the course of its development. . . .

The danger is great that the rapid dissemination of European culture will destroy the remaining traces of ethnic singing and saying. We must save whatever can be saved before the airship is added to the automobile and the electric express train, and before we hear "tararabumdieh" in all of Africa and, in the South Seas, that quaint song about little Kohn.[107]

For von Hornbostel, the popular culture of the moment existed in a zero-sum relation with living non-Western cultures.

Although elsewhere von Hornbostel is careful to maintain that non-Western musical cultures are not simply ancestors of Western music and are themselves living traditions, modernity, in the guise of "European culture," is here presented as an immutable force that will eventually overtake them. The preservationist ethos is, thus, in some sense anterior to the scientific justification for the archive: we must collect the world's music and fast, or we will never know our history—or anyone else's. The triple temporality of sound recording reappears in von Hornbostel's work as logical steps in an argument: we must preserve the voices of dying cultures so that we have them (linear-historical time); we must then preserve the recordings themselves so that we can keep them (geologic time), so that we may then break them down and study them at our leisure (fragmented time).

Almost a quarter of a century later, the Library of Congress Archive of Folk Song was founded on grounds similar to those on which the Phonogramm-Archiv was founded: "The time has come when the preservation of this valuable old material is threatened by the spread of the popular music of the hour."[108] The sense of impending loss here exists in analogical counterpoint to the voice that confronts a quizzical Nipper. Social-theoretical questions aside, there was a clear individual-society analogy at work: where phonography appeared to offer the possibility of preserving the voices of people no longer living, it also offered the chance to preserve the sounds of entire cultures no longer in existence. As one article described the process, anthropological recording preserved "the voices of a dying nation."[109]

No doubt this reasoning was ideological and a product of its time; but it is also more than ideology pure and simple. Today, the Archives of Traditional Music at Indiana University (founded in 1945) holds over seven thousand cylinder recordings of Native American music—one of the largest collections of Native American music in the world. According to the collection's archivist, Marylin Graf, many of the people who come to listen to these recordings are themselves educated Native Americans seeking to hear forgotten aspects of their cultures. The archive provides tapes to individuals, cultural centers, libraries, and schools for the purposes of teaching about Native American music and cultures.[110]

Dennis Hastings, an archivist for the Omaha tribe, wrote in 1984 that, when he stumbled on the Omaha cylinder collection at the Library of Congress, he was probably the first Omaha to hear the recordings in eighty years: "The . . . recordings . . . were not really thought about by the Omaha during the last century. They were made as reference notes rather than as documents to come back to the Tribe, and nobody knew they still existed."[111] The cylinders were originally made as anthropological documents, but, through the institutionalization of preservation and the careful research of an interested tribe member, the recordings finally did come back to the Omaha tribe in the early 1980s. Hastings writes that the recorded music helped reanimate forgotten tribal knowledge and spur the reinvigoration of living traditions.

For Hastings and others, sound recordings used in this fashion literally become a way of filling up a missing history. For some of these songs and rituals, there is no other past than what can be heard. The recordings are traces, fragments of events, fabricated at the behest of whites and with the cooperation of performers who often anticipated the absence of the events being recorded. Yet this is not simply a case of extended mnemonics. Taking nothing away from the work of contemporary archivists who make this material accessible or the people who make use of it to help revitalize their tribal traditions, it is clear that the recordings in existence are the result of one particular moment in a much larger and unequal sphere of cultural interchange. It is the anthropologists who delegated the function of memory to the cylinders; the tribespeople cooperated, and, although some were interested in preserving native culture in cylinder form, not all were.[112] In fact, it is hard to imagine that the Native Americans who cooperated in making these recordings thought of themselves with the same pathos and tragic sense employed by their anthropologist contemporaries who wrote about them—even when they did imagine preserving some aspect of the music and tradition for future generations.

That the descendants of these native subjects are now rediscovering these cylinders and using them for their own purposes is no doubt a good thing; however, we should not be so sanguine about the context in which the cylinders were recorded. Again, this is not simply to assail the work of early anthropologists like Fewkes or Fletcher but to insist on the contradictory nature of these recordings' continued existence. Although one cannot miss in their writings these anthropologists' respect for their subjects, the very terms of their discussions of native cultures were organized by forces and events much larger than the anthropological encounter. Set against the history of the U.S. government's essentially genocidal policies toward Native Americans during the nineteenth century and the early twentieth, the ethos of preservation assumes, when applied to recordings of Native American music, the status of a bizarre self-fulfilling prophecy. The recordings were collected to "preserve music, ritual, and languages that federal policy at the time of their recording had intended to drive into the ground within a generation."[113] The cylinder collections represent systematized cultural fragments solicited and preserved by one set of institutions while another set systematically destroyed the culture from which the fragments were taken. The work of anthropological cultural stewardship coincided with the decimation that necessitated the stewardship in the first place.

On several levels, then, that late-nineteenth-century culture of death (in both its literal and its metaphoric form) shaped the possibilities of sound recording. The embalmed corpse helped make sound recording what it is today. Although framed by early users as a matter of technological possibility, the question of sound recording's permanence, of the voices of the dead, articulated the mechanical aspects of sound recording to larger ethical, cultural, institutional, and political conditions. The desire to preserve the voices of the dead led to modifications in the technology and the growth of institutions dedicated specifically to the kind of historicity at first only envisioned for the practice of sound recording. Preservationism itself was, in turn, tied to elements of nineteenth-century culture like canning and embalming that were not at first glance immediately related to the problem of recording and reproducing sound. Because these new technologies were explained through the heuristic of bourgeois modernity, sound recordings themselves took on a triple temporality for their early users: the geologic decay of the medium; the linear sense of historical time and an immutable and inevitable break with the past; and a cyclic notion of time based on the fragment, the sonic element of an event. Until the establishment of sound-recording archives, until people making recordings learned to preserve them, and until the recordings themselves were preservable, the scheme of permanence pervading sound-recording discourse was essentially hyperbole, a Victorian fantasy. Repeatability from moment to moment was not the same thing as preservation for all time. The latter turned out to be a *program* for recorded sound.

When one traces recordings back to their so-called sources, one finds the intersection of cultural forces that made initial and subsequent moments of reproducibility desirable and possible. There was no "unified whole" or idealized performance from which the sound in the recording was then alienated. To whom we attribute the possibility and the desire to record or listen is entirely context dependent. Recording is a form of exteriority: it does not preserve a preexisting sonic event as it happens so much as it creates and organizes sonic events for the possibility of preservation and repetition. Recording is, therefore, discontinuous with the "live" events that it is sometimes said to represent (although there are links, of course). Like the body embalmed, recorded sound continues to be able to have a social presence or significance precisely because its interior composition is transformed in the very process of recording. This unique transformation of the interior to facilitate the functioning of the exterior is one of the defining characteristics of sound recording's so-called modernity. If the past is, indeed, audible, if sounds can haunt us, we are left to find their durability and their meaning in their exteriority.

Conclusion: Audible Futures

I have no idea whether it is actually a Chinese curse, but "may you live in interesting times" has a special significance for people who write about culture, communication, and modernity. It seems that writers who consider these phenomena together always conclude that they live in interesting—or at least disorienting—times. One result of this pose is a striking polemical similarity between claims for *modernity* and its supplement, *postmodernity*. Marx and Engels famously wrote of modernity in 1848, "All that is solid melts into air." Jean-François Lyotard would echo back their sentiments in 1979 as proof of a postmodern condition.[1]

Today, a whole procession of commentators hails emergent digital storage and transmission technologies as indices of a new age of subjectivity. Too many authors to name spent some time in the 1990s taking advantage of the concurrence of proliferating digital technologies in the First World and the approaching turn of the millennium. They adopted the millennial rhetoric that accompanied both phenomena to declare that, once again, we really *are* undergoing another historical transformation of the subject. Those of us interested in the history of the senses have been called with an especial intensity to take heed of new digital technologies and their purported promise to transform the sensory landscape. I could reasonably conclude *The Audible Past* in this millennial fashion—noting that the digital recording and transmission of sound is part of a new set of transformations today. I could go further and argue that, because we are moving into a new sonic age, the salient features of the previous age have come into clearer relief, that we can finally write histories of our audible pasts. But that would be to disown the work on which this book is based and to retract some of the central theses offered in the preceding pages.

Instead, this conclusion discusses some of the philosophical background that lies behind the preceding pages.[2] As we write books, scholars make many choices that we do not think of as choices; to us, they feel like imperatives. But behind those imperatives are concrete philosophical and political positions. We choose how to periodize our histories; we choose how to describe our human subjects; we choose the level of abstraction and concreteness for our objects of study. In fact, we choose what counts as concrete or abstract. These choices come with a good deal of weighty baggage. Any field has its default or "commonsense" choices. To use a term from the sociology of knowledge, every field has its *doxa*. As a concept, *doxa* is different from *orthodoxy*. The latter implies conscious conformance to a set of positions. *Doxa* refers to views of social life that are treated as "self-evident."[3] A variety of theoretical and political perspectives can be doxic, but doxic views tend to be treated as self-evident by those who hold them. Many of the pieties that we find in writing on technology, sound, communication, and culture in fact protect authors' intellectual decisions from scrutiny because they conform to doxa of one sort or another.

In these remaining pages, I scrutinize some of these pieties about the "impact" of new technologies, the self-evidence of historical periods, and the centrality of the speaking and listening subject for theories of communication. Behind each piety lies a doxic position worth questioning. Of course, heresy is not an end in and of itself. We must all be able to defend the value of our ideas. So this conclusion aims to present some of the central themes in sound studies as embodying *choices,* not imperatives, for scholars. In it, I apply some scrutiny to a few of our orthodoxies about sound, technology, culture, communication, and human nature. In doing so, I hope to persuade readers of the value of some of the choices that I made in writing *The Audible Past.*

To consumers and members of institutions in capitalist societies, new technologies often appear as imperatives. This is, after all, the language of advertising. Today, it is "You *need* a new computer" (or a personal data assistant, or a cell phone, etc.); decades ago, it was "You *need* a headset." Scholars often write about new digital communication technologies as if their mere presence demands that social life and social thought be remade. But this is advertising talk masquerading as academic discourse. If there is some social magic in the digital transmission and storage of sound, it is not to be found in the brute fact of the technology itself. Instead, we would have to ask the same questions of CD, DVD, or MP3 players, hard-disk recorders, wireless telephones, and digital-audio workstations that we asked of the telephone, the phonograph, and the radio. Why *these* technologies, *now?* What social forms, what social relations, do they encapsulate? If they are part of a reorganization of sound, then where is that shifting boundary between sound and not-sound this time? If the *Audible Past* is about changing relations among listening and speaking subjects, bodies, sounds, ideas, an emergent middle class, and other dimensions of modernity's maelstrom, then we should not excuse the audible present from a similar examination.

Many practices that appear as emergent in these pages have become pervasive and banal facts of modern life. Few people wonder at the voice at the other end of the telephone or at the repetitions of recorded music unless these technologies break down. Archivists take for granted the necessity of preserving sounds and debate the proper media for long-term storage. Radio receivers are standard equipment on cars. Recorded voices offering commands, reminders, entreaties, and warnings suffuse our public spaces, private homes, and many of our everyday technologies. It is probably not an exaggeration to say that a tympanic device—a microphone or a speaker of some sort—can be found in almost every inhabited building in the United States. In 1982, it was estimated that one-third of all Americans heard programmed music (best known by its brand name, Muzak) every day of the year. That figure has doubtlessly increased since then.

Some have suggested that, as recorded or transmitted images and sounds become forms of data, we will have to consider them as part of larger social and technological complexes, rather than as separate media. But sound technologies, including the human voice and the human ear, have always been parts of larger social and technological complexes. It is not the breaking down of borders between sound and not-sound that should fascinate us but rather the continuous constitution and transformation of the two. If sound is a little domain carved out from the vast, vibrating world, then sound technologies are carved out from the vast world of social and material life. If there is a question raised by the mere existence of digital sound technology, it is not a question of the impact of digital technology on sound culture. It is a question of what happened to the sonic and the digital that led to their mutual entanglement.

Today's sound media—whether analog or digital—embody and extend a panoply of social forms. It does not matter whether the machine in question uses magnetic particles, electromagnetic waves, or bits to move its information: sound technologies are social artifacts all the way down. The secret to a hard-disk recorder may, indeed, lie somewhere inside the hard drive and the microprocessor, but only if we consider those technologies as social artifacts that in turn lead us beyond themselves into other fields of practice. They are parts of networks or assemblages. The ear on Bell and Blake's phonautograph marked the collision of acoustics, physiology, otology, the pedagogy of the deaf, the state's relation to the poor, and Western Union's research agenda. Why should we expect any less of our microprocessors and hard drives?

An expansive field, littered with expected and unexpected connections, awaits in the analysis of these "new" technologies. Even a technology as apparently simple as the portable stereo with headphones (best known by the brand name Walkman) requires a vigorous cultural analysis of the modern cityscape.[4] But it also shows the degree to which a social practice almost two centuries old—the isolation in a world of sounds first developed by medical doctors in the early nineteenth century—can be articulated in new ways. The same is true for other contemporary practices. If we want to understand the politics of music file sharing on the Internet, we also need to understand the long history of music piracy, which takes many other forms—most recently cassette tape recordings—and the history of software piracy. We also have to look at the changing contexts and practices of musical listening, the uneven distribution of Internet connections, and many other phenomena that will not become apparent until we actually begin doing the research. To understand the home studio, we would have to understand the changing status of the private home. To understand the aural dimensions of virtual reality, we need to consider audio engineers' century-long obsession with creating what we would now call *virtual* acoustic spaces in recordings.[5] To understand even the simplest sonic or musical practice, we have to open it out into the social and material world from which it comes. There are likely as many unexpected connections in the audible present as there were in the audible past. But our speculation should be aided by research, for reality is often stranger and more fascinating than anything we can make up.

We should wonder less at the purportedly revolutionary aspects of new sound technologies and more at their most banal dimensions. It is those elements that seem most obvious, least likely to draw our critical attention, that may tell us the most about the central components of sound culture in our own moment. All this is to say that we cannot assume that, by their existence alone, digital transmission and storage media herald a new age, a fundamental transformation in modern sound culture. It may be the case that they do mark some deeper structural changes, or at least lead us to them. But, in order to determine whether they do, we must move beyond the technologies themselves. And, once we have done the research, the choice of periodization is, ultimately, in our hands as scholars. We create periods to explain the stuff of our histories; periods are not simply latent in our objects of study.

A discussion of periodization is perhaps less glamorous than a discussion of new technologies, but it is more important for understanding the stakes in writing sound history. To posit a break between past and present, and thereby authorize a historical study, is an immensely powerful rhetorical move. A claim for interesting times elides the writer's own work of periodizing the past and thereby narrating its relation to the present. By claiming that they live on the cusp of a historical shift, authors grant themselves the necessary purchase on the past to make grand claims about it at the same time as they excuse themselves from dealing with contemporary political or philosophical questions.

Our millennial interlocutors are in good company. "The owl of Minerva spreads its wings only with the falling of dusk," wrote Hegel, and social thought has been haunted by his formulation ever since.[6] He has been read as saying that a social formation can be understood only once it is in decline, but the larger implication is that theory is taken out of the messy world of contemporary politics. Hegel's point was that philosophy "always comes on the scene too late to give [instruction]. As the thought of the world, it appears only when actuality is already there cut and dried after its process of formation has completed." Because philosophy is restricted to "cut and dried" matters, life "cannot be rejuvenated but only understood."[7] What a tragic view of social thought! The past may be nothing but traces and the future nothing but hopes and fears, but we dwell together in the present with those traces, hopes, and fears. We use them to build and maintain the world. Philosophy is for nothing if it is not for rejuvenating life.

The claim for interesting times has worked well with audiences for over a century. In a way, however, this millenarian claim is a classic misdirection ploy. As it has come to be used, Hegel's "owl of Minerva" argument actually promotes the premise that intellectuals can have no purchase on societies that they themselves occupy.[8] This amounts to aborting the primary mission of social thought after Vico: we understand the world because we have made it, and, through understanding the world, we can come to change it. There is no doubt that talking about the present—that placing ourselves within the purview of our object of study—is, indeed, a risky business, but, in a world with "too much evidence and not enough argument," intellectuals must believe in and value the possibility of generalization.[9]

While assertions that we live in "interesting times" conveniently separate the historical from the contemporary, they also perform a useful smoke-and-mirrors trick. Claiming that we live on the edge of an epochal shift simultaneously accomplishes and disowns some of the work of periodization. To say that we live in a revolutionary age makes it sound like the historian has no choice but to periodize as he or she did. Only rarely do authors making epochal claims for the present explicitly examine the prior question of why one would want to posit a break between the present and the past or, for that matter, two moments in the past. Historians sometimes so impress us through their declarative periodization that we might forget—if they are lucky—that periodization is ultimately their responsibility. Make no mistake: periodization is an interpretive exercise. It requires both analytic and artistic sense. To periodize, a historian must make uncomfortable commitments on issues of continuity and change. Historical periods do not exist objectively outside the writing of history. People, practices, memories, sensibilities—all these things can be said to exist outside the historical text. But sorting the stuff of history into groups, naming and delimiting periods, *that* is the historian's responsibility. Periodization is not even necessarily an epistemological matter; rather, it is an issue of motivation and desire: What does it accomplish to divide one historical period from another? Why emphasize continuity at one moment and change at another? An honest answer lays bare the motivations behind a history and also the author's hopes for its significance. So here they are.

The story offered in these pages—of an "Ensoniment," a modern organization of sound—promotes a conception of nature (and human nature) as malleable, as something to be shaped and transformed. In positing a break between modern and nonmodern ways of sound and hearing, I am arguing for the possibility of future transformations. To posit this break is to recognize the significant shifts in Western social and cultural life that occurred between the Middle Ages and the twentieth century: the rise of capitalism, urbanization, the development of new political forms, new cultural sensibilities, and new philosophies. All these shifts are worthy of consideration by historians of the senses. By grounding my historical narrative in human action rather than in the inherent capacities of technologies or sense organs, I have argued that we are, ultimately, in control of our destiny, right down to the most basic aspects of human existence. The power to transform ourselves is a fundamental condition of human existence. Granted, this power resides in collective activity (sometimes over very long periods), but it exists nonetheless. Perhaps a statement on human nature seems too grand for a history of sound technology. Perhaps a statement on human malleability seems too easy in an age where social construction has moved from challenging thesis, to scholarly common sense, to utter banality. But conceptions of human nature drive all histories—especially histories of the senses—so it seems only reasonable to reflect on them.

The Audible Past elevates the question of possibility as itself a central historical problem. Practice is a ground of historical contest, but so too is *possibility*. How and under what conditions did it become possible to manipulate sound in new ways? How and under what conditions did new practices of listening become possible? Possibility is both a conceptual problem and a material issue: a practice or an event must be both thinkable and potentially able to be accomplished. Because I focus on a wide range of technological and cultural possibilities, I have offered a speculative, episodic history. Instead of a clear chronology and a unified movement of history over a single coherent temporality, I have argued that the history of sound contains multiple temporalities and a variety of intersecting chronologies. The history of sound contains many smaller chains of events; history is the amalgamation of these chains. For this book, I considered sound reproduction as the path into sound history; there are certainly other stories that can be told.

I could have easily started from long historical continuities in ways of hearing and listening and then used those continuities to explain the developments in communication, physics, physiology, and all the other cultural domains considered in this book. After all, there are continuities to be found. As I discussed in the introduction, the move toward positing transhistorical constructs like biology and physics is a frequent foundation in the scholarship pertaining to the history of sound. With alarming frequency, historical arguments about sound begin with some version of the audiovisual litany as a basis for explaining the cultural function and significance of sound and hearing. The move to posit a prehistorical phenomenology of hearing is usually criticized as a form of essentialism; it is usually defended as a form of humanism. Both the tired accusation and the tired defense actually conceal what is at stake in positing our fundamental faculties as static bases from which history flows: all histories embed stories about the future in their stories about the past.

My decision to write a history of sound in modernity goes beyond epistemology to a more fundamental question of the goals and possibilities for change in human society. It is not that continuity arguments are fundamentally pessimistic about the future or that discontinuity arguments such as my own are fundamentally optimistic about the future. One can easily imagine all sorts of configurations of historical description and temperament toward the future. It is that histories of the senses have tended to posit long continuities based on the audiovisual litany. The audiovisual litany, in turn, carries with it deeply conservative ruminations about the shape of human societies. Let us linger on this argument for a moment because it is so easy to cast the audiovisual litany as a kind of humanism and thereby forget that there are political and philosophical divisions within humanism.

In locating the transhistorical basis of historical formations in static sensory capacities, the audiovisual litany claims that it has named all that we can become. Walter Ong is the clearest advocate of a return to past forms of social organization since he makes his nostalgia programmatic. Ong hopes for some of the mystery of orality to return so that people can once again hear the word of God.[10] But the audiovisual litany also comes into play in theories of communication more generally, and it always carries with it a certain hostility toward large-scale societies. Whether we are considering Plato's *Phaedrus* or Jürgen Habermas's *Structural Transformation of the Public Sphere,* the idea of face-to-face conversation embodied in mutual spoken dialogue is a central tenet of Western philosophy.[11]

Sound theory offers a unique purchase on the philosophical privilege of speech because of its sometimes excessive literalism when dealing with the faculties of speech and hearing or the physics of sound. The theorist R. Murray Shafer's ideal acoustic community is limited by the powers of the human voice: "In his model Republic, Plato quite explicitly limits the size of the ideal community to 5,040, the number that can be conveniently addressed by a single orator." Schafer lambastes the "lo-fi landscape of the contemporary megalopolis."[12] His ideal sound culture is one limited to what he calls a *human scale*—the spatiality of the unamplified human voice. For Schafer, the human is the small. This definition of humanity reduces it to the scale of a single human being and confuses cacophony with social disorder or, worse, inhumanity. Schafer's definition of a "hi-fi" soundscape conceals a distinctly authoritarian preference for the voice of the one over the noise of the many.[13]

I have moved the speaking subject away from the center of sound theory as a means of countering this aesthetics and politics of the singular. It is not the voice that orients our theories of sound; the voice is a *special case* of sound: as sound goes, so goes the voice. As the drawings of sound-reproduction networks reproduced in chapter 5 illustrate, voices and speech exist within whole social networks of sound. They are not, by their nature or by divine decree, the point of orientation for those networks. They represent a field of possibility, themselves objects of historical contest. Sound is always defined by the shifting borders that it shares with that vast world of not-sound phenomena. Sound is not the result of transhistorical interior states of the body or the subject.

The emphasis on sound's exteriority in the preceding pages drives toward a larger philosophical point: regardless of situation, before we can consider the experience of sound, before we can talk about the so-called inherent interiority of sound and being in the world, we must consider the constitution of sound as a thing and the listening subject as a social and physical being. Our bodies must be able to transform physical vibrations into perceptible sounds, and we must know how to hear and listen to those sounds. Sound in itself is always shaped by and through its exteriors, even as it acts on and within them. Sound reproduction as we know it depends on a whole set of phenomena that we would not necessarily assume to have anything to do with sound. Capitalism, cities, industries, the medicalization of the human body, colonialism, the emergence of a new middle class, and a host of other phenomena turn out to be vital elements of the history of sound—and sound turns out to be a vital element of their history. To think the terms *sound* and *modernity* together is to conceive of sound as a variable inside a history made of variables. Considered apart from modernity, sound has been treated as a nonhistorical or transhistorical substance, immune to human action and human practice, with a fixed and predetermined relevance for historiography—in short, a constant.

To deprive the speaking subject of its presumptive privilege is not necessarily to signal the death of the Enlightenment subject or the humanist subject as such. It is to suggest, instead, a more thoroughgoing humanism, a more sophisticated enlightenment, one that can move beyond the idealized voice of the one—a god-like voice in a human guise. Communication is a collective endeavor, not reducible to a model of two people talking. One could cast Schafer's and others' preference for the few over the many as a form of nostalgic elitism. Even to entertain a suggestion that we recover an ancient or medieval way of life requires that one identify with the elites of that former society. It follows that one of the criticisms of Habermas (and others who privilege face-to-face speech as the basis of communication theory) has been that he occludes labor and social stratification in his idealized communication contexts.[14] But there is still a more fundamental issue.

As far as I can tell, large-scale societies are necessary for realizing the idealistic political programs of the Enlightenment: liberty, democracy, equality for all people, and the ability to "pursue happiness." In order for people to be free to pursue these goals, they need to be born into a society large enough that the circumstances of their birth do not necessarily determine their life chances. This is by no means to celebrate contemporary urban life with its vast inequities and deleterious effects on its ecological surroundings. It is simply to argue that small-scale societies do not necessarily afford their subjects very much freedom, equality, and flexibility. Obviously, these are much larger political-theoretical questions: entire careers are devoted to the relation between theories of politics and theories of communication. For now, it is enough to note the connection between theories of sound and theories of society and to insist that theories of sound take responsibility for the theories of humanity and society that they articulate.[15]

More often than not, good sound—and good communication—emerges from the many, and from the many it travels to one or many. Even the cry of a beloved child presupposes the prior interaction of two parents. It is this sociality, this "manyness," that should be valued in both theories of sound and theories of communication. Of course, sociality does not guarantee goodness, but it is a better theoretical starting point than the authoritative voice of a single, isolated individual. Modern political theory is littered with metaphoric equations between the voice of the one (or a collective group acting as if it had a single voice) and agency. This book repeatedly demonstrates that the presence of voice and sound is not necessarily indicative of the agency of the speaker—even when that speaker's voice is really heard, whatever that might mean in a given situation. The agency always resides in the social relation making possible the moment of sonic communication. By emphasizing the voice of the individual and the desirability of its persistence in dialogue, we forget all the other—possibly mute—people who enable and structure even the most fleeting moment of dialogue. The persistence of the voice as a metaphor for political, cultural, or sonic agency both leaves aside the potential agencies of listening and promotes a fundamental hostility to large-scale society.

In recent years, the philosophical privilege of dialogue, the centrality of the voice, and static theories of our senses and faculties—all the aspects of the audiovisual litany—have come under heavy criticism. Critics have argued that the metaphysics of presence ignores the more fundamental and constitutive play of difference. Others have argued that attention to speech ignores the conditions under which it becomes possible or reduces the voice to intelligible speech. Still others have claimed that dialogue itself is nothing more than a special case of dissemination, a privileged form of turn taking.[16] This book is clearly a contribution to that literature, but with a sonic twist. As it appears in Ong's eschatology or Schafer's ideal society, the audiovisual litany carries with it assertions about what we necessarily are, and, by extension, it is an argument about the limits of what we can become. If we excuse the senses and the faculties themselves from history and instead begin by assuming that certain configurations of activity belong to certain senses or faculties by right of origin, we pose a very limited set of possibilities for the organization of human activity in the future. We appeal to a fundamentally theological construct for the purposes of constructing social theory and thereby severely circumscribe what it means to be fully human.

The question of the body is a short step from questions of the voice and the human subject. As I discussed in the introduction, the body has become a major preoccupation in social theory. But the body can mean different things. It can be treated as the universal ground of all experience, as in transcendental phenomenology. It can be treated as a set of positively identifiable phenomena and functions, as in medical science. It can be treated as the site of social and cultural difference, as in feminist and some poststructuralist thought. And it can be treated as itself an artifact of social life, as the locus of enacted and lived subjectivity. Writers who privilege speech as the metaphoric or real locus of agency and subjectivity do so on the basis of a universal body that is the ground for all possible experience.

Throughout this book, I have called into question that notion of a unitary body, from my criticism of the audiovisual litany, to my analysis of the physical positions of people's bodies in visual depictions of sound technologies. Ears excised from cadavers, doctors' disgust for their patients, and the popularization of embalming are all central motifs of sound history because the sounding or listening subject is coterminous with a sounding or listening body, and that body is itself an object of cultural struggle and historical transformation. If this seems like a strangely violent or gory history, if all this talk of power and human relations seems somehow secondary to the history of sound technologies and human hearing, if this book seems like anything less than a love letter to the phenomenon of sound reproduction, it is only because, as Walter Benjamin admonishes us, there is no document of civilization that is not also a document of barbarism. Our bodies are not givens, the grounds from which we enter social life; they are the domains through which we are constituted (and constitute ourselves) as social beings. The human body is suffused with the struggles of history. It is no wonder that body parts play a major role in a history of sound.

There are real stakes in the way we choose to describe the relations among sound, hearing, bodies, and subjects. This is well illustrated by the profound hostility toward the deaf in the equation of voice with agency and silence with its absence. *Silence* has become a metaphor for exclusion in a great deal of political theory, to the point that it has become a verb describing the act of exclusion. *Deafness,* meanwhile, often appears as a metaphor for the refusal of intersubjectivity or respect. To offer but one banal example, three recent books on difficulties in international relations have the phrase *dialogue of the deaf* in their titles.[17] A full catalog would have to include the wide abuse of the cliché *turned a deaf ear toward . . .* to mean "refused intersubjectivity" in narratives of political and interpersonal conflict. In the audiovisual litany, speaking is generally a good thing, silence is a bad thing, and deafness is an antisocial thing. If these were caricatures of a gender, a race, or an ethnic group, they would long since have disappeared from serious scholarship. A history of sound that does not presumptively privilege a speaking subject allows us to reconsider this prejudice as itself a historical artifact, not as a reasonable philosophical position to be taken seriously on its own terms. With the notable exception of our colleagues in the "hard" sciences, scholars of speech, hearing, and sound seem largely ignorant of the cultural work on deafness. For sound and communication scholars to take seriously the problem of deafness, we will—once again—need to leave behind the audiovisual litany and the idea that audible speech and hearing are such defining faculties that to be without them is to be less than fully human.

Hostility toward the deaf among the hearing has a long history. When the first schools for the deaf began in sixteenth-century Spain, it was widely believed that, because they could not speak, deaf children had no souls. The connection between hearing and full humanity likely stemmed from Saint Augustine's literalization of the dictum, "Faith comes by hearing." (It is worth noting that Augustine's preference for the auditory is also manifest in Ong's construct of orality.) Secularized versions of this prejudice followed deaf people into the twentieth century. Even respected sound theorists repeat the lie that the blind are humane and compassionate but that the deaf are rageful and antisocial.[18] Scholars and activists have challenged these caricatures of the deaf—and their calamitous effects on the rights, the culture, and the education of the deaf—for almost two centuries now. It is time for scholars of sound to recognize this work, engage its spirit and substance, and let go of the Augustinian baggage that we have been carrying around with our theories of speech and hearing.

The cultural study of deafness is in many ways complementary to the cultural study of hearing; both lead us to new ways of thinking about the plasticity of the human subject. As Oliver Sacks and many others have shown, the study of deafness "shows that much of what is distinctively human in us—our capacities for language, for thought, for communication, and culture—do not develop automatically in us.... [These things] are a *gift*—the most wonderful of gifts—from one generation to another." The study of deafness "shows us that the brain is rich in potentials we would scarcely have guessed of, shows us the almost unlimited plasticity and resource of the nervous system, the human organism, when it is faced with the new and must adapt."[19] The idea of a history of sound suggests the possibility of *other* social organizations (and scholarly assessments) of our sonic faculties, where the hearing might not be—in the first instance—philosophically privileged as more human than the deaf.

As I argued in chapter 1, Western culture has fetishized and generalized certain aspects of deafness as a condition in its development of sound-reproduction technologies, all the while maintaining a strong stigma against deafness itself. Sound-reproduction technologies were connected with an ongoing project to make the deaf like the hearing. They wound up making the hearing more like the deaf. By 1878, it was possible to think that ears were (at least potentially) imperfect versions of a tympanic mechanism that could be mimed and amplified. Hearing demanded supplementation, so we are now surrounded by media that hear for us. Of course, there are many more connections to be made between deaf culture and sound culture. This is only one.

By providing alternatives to our most comfortable ways of thinking about and describing sound and hearing, *The Audible Past* casts the relative status of speech, hearing, silence, and deafness as matters worthy of philosophical, historical, political, and ethical reflection. Otherwise, the terms *voice, silence,* and *deafness* inhabit philosophical and political discourse as nothing more than clichés; they become nothing more than inanimate answers to long-forgotten political and cultural questions. They live on in our doxa as fossilized metaphors.

Questions raised by thematizing sound and sound culture reach far beyond their immediate domains. To make that intellectual stretch, we must first recognize that there *is* a domain of significant and connected questions surrounding the social life of sound in all its manifestations. We have the difficult task of breaking down the intellectual ghetto walls surrounding those practices long associated with sound: speech, music, acoustic design, soundscape studies, studies of sound technologies and media, deaf studies. Questions derived from these fields belong at the very center of the human sciences because—like any field that considers the senses—the study of sound ultimately deals with the meaning of humanity itself.

At the same time, the centrality of sound to social thought requires us to admit larger philosophical concerns into the study of sound. The fundamental problem of social thought is to explain the relation between the personal and the social. For too long, scholars of sound have ignored the two-way connection between the personal and the social, opting instead to deduce social reality from ossified descriptions of human experience. It is no accident that Marx begins one of his earliest published discussions of communism with the argument that the history of the senses is essential to the history of society. Similarly, Benjamin wrote that, "during long periods of history, the mode of human sense perception changes with humanity's entire mode of existence."[20] This is not to say that sensory constructivism automatically grants its user a progressive politics—far from it. But it is to argue that a hopeful vision of the future requires a certain level of constructivism—a commitment to the idea that people can become more and other than what they have been in the past. It requires a sense that, although our possible futures are conditioned by our pasts, they are not in any simple way determined by our pasts.[21] This is why sensory history needs social theory, and this is why the deliberate work of periodization and the eschewal of transcendental continuities must accompany such histories. A fully historicist history of sound suggests that large-scale social transformation is possible, right on down to the level of the individual subject.

As I have suggested throughout this conclusion, histories always raise questions about the relation between the past and the present. Can the claims in the history be extended into the present? What do they portend for the future? Some of the horizons for sound's possibility traced in *The Audible Past* are still in force today, but others may be undergoing new transformations. The history of sound conditions its present shape, but origins do not guarantee the present or the future.

For the past few years, NPR (National Public Radio, an American network) has run a series called "Lost and Found Sound," as part of a project called "Quest for Sound." The journalist Jay Allison, coproducer of the series, explains the goals of the project:

In homes around America, we have preserved ourselves, on magnetic tape, vinyl, digital bits, acetate, and wax—the homemade audio ephemera of this century. We at NPR are putting out a call for it. We want yours. . . .

Quest for Sound is a call to listeners to send in their home recordings of the last one hundred years to be shaped into stories that capture the rituals and sounds of everyday life.

We're asking you for your favorite sonic artifacts. If you have audio treasures to send us, call us first at our National Quest for Sound Hotline. We want to hear what we all decided was worth saving about ourselves, the ordinary and fabulous, the joyous and miserable, the ancient and the modern.

We want recorded letters sent home from the war, debate club practice tapes, pen-pal audio files from the Internet, personal recordings of historical events, your unique collection of doorbell sounds. What else is out there? You tell us.

We want your sounds and the stories that go with them . . . the childhood voices of famous men, the recorded letters of lovers, mysterious dialogues on forgotten cassettes found by the side of the road.[22]

The stories of voices alienated from once-whole bodies—so beloved by theorists of sound—are nowhere to be found. Gone are the fears of sound technologies as imposters or surrogates for real interaction. Absent is the sense that recordings distort or misrepresent a prior reality. Perhaps "Lost and Found Sound" promotes the doxa of sound reproduction that we first encountered in discourses on sound fidelity and audio preservation: that, in hearing the recording, we will get back to the thing itself before it was recorded. But, then again, Allison asks contributors for their stories to go with the sounds that they submit. The context is not given; it is something that contributors and listeners establish together.

"Lost and Found Sound" is a document of audile modernity in all its subtlety and depth. There are plenty of voices in the recordings that have been collected, but the voice carries no presumptive privilege. Spoken words dwell alongside loon calls, steam train whistles, foghorns, Morse code signals, and the sounds of falling Douglas fir trees. In some cases, the recordings are the only remaining traces of otherwise forgotten events. People have been collecting sounds for over a century, says Allison, and it is time to hear the shared archive that they have built in their attics and dining rooms. The recordings collected for "Lost and Found Sound" are artifacts left from the past; they are also fragments of everyday life. They bear all the biases that come with any archive, to be sure. No doubt the upper-middle-class, highly cultured listeners of NPR will produce a very particular version of twentieth-century everyday life. But the recordings' collective existence testifies to the expansion of sound reproduction in twentieth-century American life. Sound reproduction exists inside lived history even as it documents that history: speaking into the reel-to-reel recorder is a cherished memory from Allison's childhood; he fondly recalls his fascination with the recorder and the ritual of speaking into it. As objects invested with memory or nostalgia, recordings, recorders, and voices stand together as equals. Today, we could tell similar stories about telephones and radios.

Few people wonder at the fact of sound reproduction today. Instead, we are more likely to ask sound reproduction to tell us something about the present, the past, or the future. We are more likely to invest sound-reproduction technologies—and reproduced sounds—with the emotional energy that we put into other artifacts from everyday life. Many are ephemeral; a few carry with them cherished feelings or memories. In this way, "Lost and Found Sound" is only one iteration of an expansive cultural sensibility. The World Soundscape Project is a more academic version of "Lost and Found Sound." Its goal is to document "vanishing" sounds and to promote its conservationist approach to acoustic ecology. The two archival institutions most heavily covered in *The Audible Past,* the Library of Congress and the Smithsonian, have joined together in a project called "Save Our Sounds." The project aims to raise $750,000 (to match a $750,000 federal grant) to locate and preserve fragile, one-of-a-kind recordings in their collections and to conduct a campaign to promote their curatorial work.

One finds a similar orientation among the musicians who use turntables and samplers to construct a musical present from the twentieth century's vast archive of recorded music. Countless hip hop producers, DJs, and audio collage artists have claimed that the use of prerecorded material to make new sound art is a self-consciously political and historiographic project. In fact, that orientation goes beyond musicians—many record collections carry their own implied commentaries on twentieth-century history. As I discussed at the end of chapter 6, Native Americans' appropriation of century-old ethnographic recordings also carries with it a sensibility about an audible past and contemporary sound culture. All these practices ask in concrete ways what the long-established fact of sound reproduction can tell people about who they are, where they came from, and where they we are going. In this way, archivists and DJs have a common cultural project.

Sound leaves its traces, and our interest in those traces is a fact of modern life. The call to turn our attention to a continuously constructed audible past is part of the present. We contemplate the history that people have made through shaping and reshaping the experience of sound. The banality of that power over and through sound is a defining feature of modern life. *The Audible Past* is a story about how that power came to be.

Nobody can predict the future with any certainty, but this much is sure: changes in the form and consistency of sensory experience are bound up in much larger social and cultural transformations. This book has explored one such conjuncture between sensory transformation and social transformation. It happened before; it could happen again. In the meantime, give NPR a call if you find a recording of a lost sound; they might want to play it on the radio.

Notes

Hello!

1. In both the Bell and the Edison cases, the inventors had a partially functional device before the moment of their "famous first."
2. Oliver Read and Walter L. Welch, *From Tin Foil to Stereo: Evolution of the Phonograph* (New York: Herbert W. Sams, 1976), 4; Michael Chanan, *Repeated Takes: A Short History of Recording and Its Effects on Music* (New York: Verso, 1995), 2.
3. Marshall Berman, *All That Is Solid Melts into Air: The Experience of Modernity* (New York: Penguin, 1992).
4. Alan Burdick, "Now Hear This: Listening Back on a Century of Sound," *Harper's Magazine* 303, no. 1804 (July 2001): 75.
5. For the sake of readability, I have largely kept with the standard practice of using light and sight metaphors for knowledge. Replacing all these with sonic metaphors would be largely a formalist exercise and of dubious value in helping readers understand my argument.
6. For a full discussion of the status of vision in modern thought and the idea that vision is central to the categories of modernity, see Martin Jay, *Downcast Eyes: The Denigration of Vision in Twentieth-Century French Thought* (Berkeley and Los Angeles: University of California Press, 1993); and David Michael Levin, ed., *Modernity and the Hegemony of Vision* (Berkeley and Los Angeles: University of California Press, 1993). See also Marshall McLuhan, *The Gutenberg Galaxy: The Making of Typographic Man* (Toronto: University of Toronto Press, 1962); Michel Foucault, *The Birth of the Clinic: An Archaeology*

of Medical Perception, trans. A. M. Sheridan Smith (New York: Pantheon, 1973); and Walter Ong, *Orality and Literacy: The Technologization of the Word* (New York: Routledge, 1982). Ong's work and the phenomenology of listening are discussed below.

7 Although one can hope that this, too, is changing. In addition to some of the scholars of sound cited elsewhere in this introduction, see, e.g., Laura Marks, *The Skin of the Film: The Senses in Intercultural Cinema* (Durham, N.C.: Duke University Press, 1999); David Howes, *The Varieties of Sensory Experience: A Sourcebook in the Anthropology of the Senses* (Toronto: University of Toronto Press, 1991); and Alain Corbin, *The Foul and the Fragrant: Odor and the French Social Imagination* (Cambridge, Mass.: Harvard University Press, 1986), and *Time, Desire, and Horror: Toward a History of the Senses,* trans. Jean Birrell (Cambridge: Blackwell, 1995).

8 In addition to the works discussed below, see Kaja Silverman, *The Acoustic Mirror: The Female Voice in Psychoanalysis and Cinema* (Bloomington: Indiana University Press, 1988); Amy Lawrence, *Echo and Narcissus: Women's Voice in Classical Hollywood Cinema* (Berkeley and Los Angeles: University of California Press, 1991); and Claudia Gorbman, *Unheard Melodies: Narrative Film Music* (Bloomington: Indiana University Press, 1987). Anahid Kassabian, *Hearing Film: Tracking Identifications in Contemporary Hollywood Film Music* (New York: Routledge, 2001), provides an interesting alternative approach.

9 See C. Wright Mills, *The Sociological Imagination* (New York: Oxford University Press, 1959), 50–75; Peter Novick, *That Noble Dream: The "Objectivity Question" and the American Historical Profession* (New York: Cambridge University Press, 1988); Georg C. Iggers, *Historiography in the Twentieth Century: From Scientific Objectivity to the Postmodern Challenge* (Hanover, N.H.: Wesleyan University Press, 1997); and Bonnie Smith, *The Gender of History* (Cambridge, Mass.: Harvard University Press, 1998).

10 See, e.g., Hadley Cantril and Gordon Allport, *The Psychology of Radio* (New York: Harper and Bros., 1935); Rudolf Arnheim, *Radio,* trans. Margaret Ludwig and Herbert Read (London: Faber and Faber, 1936); and Hanns Eisler and Theodor Adorno, *Composing for the Films* (New York: Oxford University Press, 1947). For a contemporary example, see David Michael Levin, *The Listening Self: Personal Growth, Social Change, and the Closure of Metaphysics* (New York: Routledge, 1989).

11 A significant share of the English-lanugage literature appears in my notes and bibliography.

12 Karl Marx, *Economic and Philosophic Manuscripts of 1844,* trans. Martin Milligan (New York: International, 1968), 140–41.

13 Jonathan Crary, *Techniques of the Observer: On Vision and Modernity in the Nineteenth Century* (Cambridge, Mass.: MIT Press, 1990), 16.

14 These questions recur constantly in the classic texts, such as Max Horkheimer and Theodor Adorno, *Dialectic of Enlightenment* (New York: Continuum, 1944); Walter Benjamin, "The Storyteller," and "The Work of Art in the Age of Mechanical Reproduction," in *Illuminations,* trans. Hannah Arendt (New York: Schocken, 1968); and Theodor Adorno, "The Curves of the Needle," trans. Thomas Levin, *October,* no. 55 (winter 1990): 49–56.

15 Stephen Kern, *The Culture of Time and Space, 1800–1918* (Cambridge, Mass.: Harvard University Press, 1983), passim; Donald M. Lowe, *History of Bourgeois Perception* (Chicago: University of Chicago Press, 1982), 9, 111–17; and Marshall McLuhan, *Understanding Media: The Extensions of Man* (New York: McGraw-Hill, 1964), 265–83, 297–307.

16 Claude S. Fischer, *America Calling: A Social History of the Telephone* (Berkeley and Los Angeles: University of California Press, 1992), 5; Jacques Attali, *Noise: The Political Economy of Music* (Minneapolis: University of Minnesota Press, 1985), 87; John Durham Peters, *Speaking into the Air: A History of the Idea of Communication* (Chicago: University of Chicago Press, 1999), 160; Susan Douglas, *Listening In: Radio and the American Imagination from Amos 'n Andy and Edward R. Murrow to Wolfman Jack and Howard Stern* (New York: Times Books, 1999), 9.

17 D. L. LeMahieu, *A Culture for Democracy: Mass Communication and the Cultivated Mind in Britain between the Wars* (Oxford: Clarendon, 1988), 81.

18 Douglas Kahn, "Histories of Sound Once Removed," in *Wireless Imagination: Sound, Radio, and the Avant-Garde,* ed. Douglas Kahn and Gregory Whitehead (Cambridge, Mass.: MIT Press, 1992), 2.

19 Technological determinism is, more or less, the premise that technology determines the conduct and form of cultural life. For criticisms of technological determinism from perspectives sympathetic to my own, see Jennifer Daryl Slack, *Communication Technologies and Society: Conceptions of Causality and the Politics of Technological Intervention* (Norwood, N.J.: Ablex, 1984); Raymond Williams, *Television: Technology and Cultural Form* (Middletown, Conn.: Wesleyan University Press, 1992); and Carol Stabile, *Feminism and the Technological Fix* (New York: St. Martin's, 1994). From a different angle, Martin Heidegger points out that there are actually four kinds of causality when we consider technology: material, form, use, and that which shapes the material into a particular form for a particular use (see *The Question Concerning Technology and Other Essays,* trans. William Lovitt [New York: Harper Torchbooks, 1977], 6–12 and throughout).

20 Bruno Latour, *We Have Never Been Modern,* trans. Catherine Porter (Cambridge, Mass.: Harvard University Press, 1993), 3–8; Jody Berland, "Cultural Technologies and the Production of Space," in *Cultural Studies,* ed. Lawrence Grossberg, Cary Nelson, and Paula Treichler (New York: Rout-

ledge, 1992); J. Macgregor Wise, *Exploring Technology and Social Space* (Thousand Oaks, Calif.: Sage, 1997), xvi, 54–55, 68; Gilles Deleuze and Felix Guattari, *A Thousand Plateaus: Capitalism and Schizophrenia,* trans. Brian Massumi (Minneapolis: University of Minnesota Press, 1987), 4, 90, 503–5.

21 Jonathan Sterne, "Sound Out of Time/Modernity's Echo," in *Turning the Century,* ed. Carol Stabile (Boulder, Colo.: Westview, 2000), 9–30.

22 This list is drawn from Berman, *All That Is Solid Melts into Air,* 5–12, 16; Matei Calinescu, *Five Faces of Modernity: Modernity, Avant-Garde, Decadence, Kitch, Postmodernism* (Durham, N.C.: Duke University Press, 1987), 42; Zygmunt Bauman, *Modernity and Ambivalence* (Cambridge: Polity, 1991), 5; and Henri Lefebvre, *Introduction to Modernity,* trans. John Moore (New York: Verso, 1995), 168–238.

23 Johannes Fabian, *Time and the Other: How Anthropology Makes Its Object* (New York: Columbia University Press, 1983).

24 *Oxford English Dictionary,* s.v. "aural," "auricular."

25 Michael Taussig, *Mimesis and Alterity: A Particular History of the Senses* (New York: Routledge, 1993), xvi–xviii. See also Ian Hacking, *The Social Construction of What?* (Cambridge, Mass.: Harvard University Press, 1999).

26 Johannes Müller, *Elements of Physiology,* trans. William Baly, arranged from the 2d London ed. by John Bell (Philadelphia: Lea and Blanchard, 1843), 714.

27 Michel Foucault, *Discipline and Punish: The Birth of the Prison,* trans. Alan Sheridan (New York: Vintage, 1977), 136. Foucault has a similar discussion in *The History of Sexuality,* vol. 1, *An Introduction,* trans. Robert Hurley (New York: Vintage, 1978), 139.

28 Jean-Marc Gaspard Itard, *The Wild Boy of Aveyron,* trans. George Humphrey and Muriel Humphrey (New York: Meredith, 1962), 26–27; Douglas Keith Candland, *Feral Children and Clever Animals: Reflections on Human Nature* (New York: Oxford University Press, 1993).

29 Karl Marx, *Capital,* vol. 1, *The Process of Capitalist Production* (New York: International, 1967), 168.

30 Marcel Mauss, "Body Techniques," in *Sociology and Psychology: Essays,* trans. Ben Brewster (Boston: Routledge and Kegan Paul, 1979), 104, 121.

31 Gilles Deleuze, *Foucault,* trans. Sean Hand (Minneapolis: University of Minnesota Press, 1988), 109.

32 Alain Corbin, *Village Bells: Sound and Meaning in the Nineteenth-Century French Countryside* (New York: Columbia University Press, 1999), 254–83.

33 "Man himself must first of all have become *calculable, regular, necessary,* even in his own image of himself, if he is to be able to stand security for *his own future*" (Friedrich Nietzsche, *On the Genealogy of Morals and Ecce Homo,* trans. Walter Kauffman [New York: Vintage, 1967], 58). Nietzsche makes this

comment in a discussion of promises and contracts. For him, a sense of the human calculability is inextricably tied to the contemplation of an interrelated past, present, and future. For our purposes, it is enough to note that Nietzsche's self-calculating subject who can make a promise is a short step from Zygmunt Bauman's subject who contemplates the relation between continuity and change (see Bauman, *Modernity and Ambivalence*).

34 Michel Chion, *Audio-Vision: Sound on Screen,* trans. Claudia Gorbman (New York: Columbia University Press, 1994), 94.

35 Maurice Merleau-Ponty, "The Primacy of Perception and Its Philosophical Consequences," trans. James M. Edie, in *The Primacy of Perception and Other Essays on Phenomenological Psychology, the Philosophy of Art, History, and Politics,* ed. James M. Edie (Evanston, Ill.: Northwestern University Press, 1964), 20. For a discussion of the temporal and ephemeral character of perception, see ibid., 35.

36 The following summarizes an argument that I develop more fully in an essay in progress entitled "The Theology of Sound."

37 This list is most clearly elaborated in Walter Ong, *The Presence of the Word: Some Prolegomena for Cultural and Religious History* (Minneapolis: University of Minnesota Press, 1981). See also Ong, *Orality and Literacy,* 30–72; Attali, *Noise;* Lowe, *History of Bourgeois Perception;* Marshall McLuhan and Edmund Carpenter, "Acoustic Space," in *Explorations in Communication: An Anthology,* ed. Edmund Carpenter and Marshall McLuhan (Boston: Beacon, 1960); Rick Altman, "The Material Heterogeneity of Recorded Sound," in *Sound Theory/Sound Practice,* ed. Rick Altman (New York: Routledge, 1992); John Shepherd, *Music as a Social Text* (Cambridge: Polity, 1991); Barry Truax, *Acoustic Communication* (Norwood, N.J.: Ablex, 1984); Eric Havelock, *Preface to Plato* (Cambridge, Mass.: Harvard University Press, 1963), and *The Muse Learns to Write: Reflections on Orality and Literacy from Antiquity to the Present* (New Haven, Conn.: Yale University Press, 1986), esp. 1–29; and Bruce R. Smith, *The Acoustic Culture of Early-Modern England: Attending to the O-Factor* (Chicago: University of Chicago Press, 1999), 3–29.

38 Caryl Flinn, *Strains of Utopia: Gender, Nostalgia, and Hollywood Film Music* (Princeton, N.J.: Princeton University Press, 1992), 7. Although it is still quite influential, this Romantic notion of music has been widely criticized in the past few decades. See, e.g., Janet Wolff, "The Ideology of Autonomous Art," in *Music and Society: The Politics of Composition, Performance, and Reception,* ed. Richard Leppert and Susan McClary (New York: Cambridge University Press, 1987).

39 Susan Handelman, *Slayers of Moses: The Emergence of Rabbinic Interpretation in Modern Literary Theory* (Albany: State University of New York Press, 1982),

15–21. Handelman's book offers an extended discussion of the spirit/letter distinction as it is manifested in Western metaphysics and hermeneutics. See also Peters, *Speaking into the Air*, 36–51, 66–74; Jacques Derrida, *Of Grammatology*, trans. Gayatri Chakravorty Spivak (Baltimore: Johns Hopkins University Press, 1976), 323 n. 3, *Dissemination*, trans. Barbara Johnson (Chicago: University of Chicago Press, 1981), 61–171, and *The Postcard: From Socrates to Freud and Beyond*, trans. Alan Bass (Chicago: University of Chicago Press, 1987); Plato, "Phaedrus," in *Collected Dialogues*, ed. Edith Hamilton and Huntington Cairns (Princeton, N.J.: Princeton University Press, 1961), 475–525; and Havelock, *Preface to Plato*.

40. Ong, *Presence of the Word*, 6. Pre-Christian is an important modifier here since it treats rabbinic thought as an incomplete prelude to Catholic Christianity.

41. Ibid., 6, 11, 12, 288–89, 324. In *Orality and Literacy* (perhaps at the request of his editors), Ong largely removed the religious content of the distinction, treating it instead as a purely secular academic discovery (see pp. 1, 6).

42. Ong, *Orality and Literacy*, 75, 77, 123, 129, 166–71.

43. Jacques Derrida, *Speech and Phenomena and Other Essays on Husserl's Theory of Signs*, trans. David B. Allison (Evanston, Ill.: Northwestern University Press, 1973), 77.

44. Derrida, *Of Grammatology*, 323 n. 3. For a reading of Derrida's critique of Christian metaphysics as an instance of a "heretical" rabbinic hermeneutics, see also Handelman, *Slayers of Moses*.

45. Rick Altman, "Four and a Half Film Fallacies," in Altman, ed., *Sound Theory/Sound Practice*, 37, 39.

46. Ong, *Presence of the Word*, 111 (quotation), and *Orality and Literacy*, 32 (on sound existing only as it goes out of existence).

47. James Lastra (*Sound Technology and American Cinema: Perception, Representation, Modernity* [New York: Columbia University Press, 2000], 133) criticizes nominalism on the basis that it treats technological reproduction as a false mediation of a real event. He is actually criticizing bad nominalism since a fully developed nominalism would treat *all* events in their uniqueness. See, e.g., Gilles Deleuze, *Spinoza: Practical Philosophy* (San Francisco: City Lights, 1988), 122–30.

48. Don Ihde, *Listening and Voice: A Phenomenology of Sound* (Athens: Ohio University Press, 1976), 58. Steven Feld has also roundly criticized the notion of orality as a universal construct of sound culture. See his "Orality and Consciousness," in *The Oral and the Literate in Music*, ed. Yoshiko Tokumaru and Osamu Yamaguti (Tokyo: Academia Music, 1986).

49. Chanan, *Repeated Takes*, 12 (on Schaeffer); John Corbett, *Extended Play:*

Sounding Off from John Cage to Dr. Funkenstein (Durham, N.C.: Duke University Press, 1994), 37. Andrew Goodwin has criticized Corbett's idea of visual lack in *Dancing in the Distraction Factory: Music Television and Popular Culture* (Minneapolis: University of Minnesota Press, 1992), 49.

Musique concrète uses found recorded sounds to create an original piece through montage. It is a partial forerunner of sampling.

50 Truax, *Acoustic Communication,* 120. See also R. Murray Schafer, *The Soundscape: Our Sonic Environment and the Tuning of the World* (Rochester, N.Y.: Destiny, 1994), 90–91.

51 Stuart Hall, "The Narrative Construction of Reality: An Interview with Stuart Hall," *Southern Review* 17 (March 1984): 8.

52 In addition to the discussion of the transcendental phenomenology of listening presented above, see Derrida, *Speech and Phenomena,* 70–87; Briankle Chang, *Deconstructing Communication* (Minneapolis: University of Minnesota Press, 1996), 187–220; and John Durham Peters, "The Gaps of Which Communication Is Made," *Critical Studies in Mass Communication* 11, no. 2 (1994): 117–40, and *Speaking into the Air,* 33–108.

53 On genealogy as demystification, see Michel Foucault, "Nietzsche, Genealogy, History," in *Language, Counter-Memory, Practice: Selected Essays and Interviews,* ed. Donald Bouchard, trans. Donald Bouchard and Sherry Simon (Ithaca, N.Y.: Cornell University Press, 1977), 139–40; and Nietzsche, *On The Genealogy of Morals and Ecce Homo,* 25–26.

54 Charles Burnett, "Sound and Its Perception in the Middle Ages," in *The Second Sense: Studies in Hearing and Musical Judgement from Antiquity to the Seventeenth Century,* ed. Charles Burnett, Michael Fend, and Penelope Gouk (London: Warburg Institute, 1991), 48, 49; Penelope Gouk, "Some English Theories of Hearing in the Seventeenth Century: Before and After Descartes," in ibid., 96.

55 On this point, see also Douglas Kahn, *Noise, Water, Meat: A History of Sound in the Arts* (Cambridge, Mass.: MIT Press, 1999), 9; and Friedrich Kittler, *Gramophone, Film, Typewriter,* trans. Geoffrey Winthrop-Young and Michael Wutz (Stanford, Calif.: Stanford University Press, 1999), 24–25. Felicia Frank's *The Mechanical Song: Women, Voice, and the Artificial in Nineteenth-Century French Narrative* (Stanford, Calif.: Stanford University Press, 1995) offers a full study of the female voice in nineteenth-century French literature, arguing that new attitudes connecting voice and artifice became prevalent during that period.

56 Here I am using *articulation* to convey a sense of cultural connection as suggested by Stuart Hall, "On Postmodernism and Articulation," *Journal of Communication Inquiry* 10, no. 2 (1986): 45–60; and Lawrence Grossberg, *We Gotta Get Out of This Place: Popular Conservatism and Postmodern Culture*

(New York: Routledge, 1992), 52–61. I discuss articulation further in chapter 4.

57 Fabian, *Time and the Other,* esp. 37–69. See also Robert F. Berkhofer Jr., *The White Man's Indian: Images from Columbus to the Present* (New York: Vintage, 1978); Francis Paul Prucha, *The Great Father: The United States Government and the American Indians,* abridged ed. (Lincoln: University of Nebraska Press, 1986); Calinescu, *Five Faces of Modernity,* 26–32; Edward Said, *Culture and Imperialism* (New York: Vintage, 1993), 15–19; and Rita Felski, *The Gender of Modernity* (Cambridge, Mass.: Harvard University Press, 1995), 13.

58 Corbin's *Village Bells* offers an interesting history of structured environmental sounds—church bells. He chronicles struggles over the use and meaning of the bells and shows that those struggles are connected with anxieties over modernity, subjectivity, and the emerging French state.

59 See, e.g., Joan Scott, *Gender and the Politics of History* (New York: Columbia University Press, 1988), 5; and Pierre Bourdieu and Loïc J. D. Wacquant, *An Invitation to Reflexive Sociology* (Chicago: University of Chicago Press, 1992), 132–35.

60 Hayden White, "The Burden of History," in *Tropics of Discourse: Essays in Cultural Criticism* (Baltimore: Johns Hopkins University Press, 1978), 50.

1. Machines to Hear for Them

1 Although, throughout this book, I invoke the names of famous individuals, I do not want to give the impression that this is a "great man" history. On the contrary, following Pierre Bourdieu, I consider Bell, Blake, Berliner, and all the others as "epistemic individuals": their actions are instances and particular locations of social activity, rather than being purely matters of personal biography. The goal is to use the artifacts that they left behind to establish a history of activities and ideas, not to establish the merits or characteristics of particular people. See Pierre Bourdieu, *Homo Academicus,* trans. Peter Collier (Stanford, Calif.: Stanford University Press, 1988).

2 *Oxford English Dictionary,* s.v. "tympanic," "tympanum." *Tympanum* has a considerably longer history, but the adjectival form is significant here because it indicates that the term has come to have some mobility. *Tympanum* also refers to an architectural form, the recessed part of a pediment, often adorned with sculpture.

3 The two predominant techniques for early film sound—sound on disk and sound on film—are also tympanic. Sound on disk was essentially based on a modified version of the gramophone; sound on film used light pulses in a manner analogous to the telephone's use of magnetism. For a discussion of

the working of these technologies, see, e.g., Stephen Neale, *Cinema and Technology: Image, Sound, Colour* (Bloomington: Indiana University Press, 1985), 71–76; and James Lastra, *Sound Technology and American Cinema: Perception, Representation, Modernity* (New York: Columbia University Press, 2000), 92–122.

4 Emile Berliner, "The Gramophone: Etching the Human Voice" (paper read before the Franklin Institute, 16 May 1888, EB [unsorted papers]), 4; Edward Wheeler Scripture, *The Elements of Experimental Phonetics* (New York: Scribner's, 1902), 17; Thomas L. Hankins and Robert J. Silverman, *Instruments and the Imagination* (Princeton, N.J.: Princeton University Press, 1995), 135.

5 Melville Bell, *A New Elucidation of the Principles of Speech and Elocution* (1849), quoted in Robert V. Bruce, *Bell: Alexander Graham Bell and the Conquest of Solitude* (Boston: Little, Brown, 1973), 19.

6 John Durham Peters, "Helmholtz, Edison, and Sound History," in *Memory Bytes: History, Technology, and Digital Culture,* ed. Lauren Rabinovitz (forthcoming); a revised English-language version of John Durham Peters, "Helmholtz und Edison: Zur Endlichkeit der Stimme," *Zwischen Rauschen und Offenbarung: Zur kulturellen und Medien-geschichte der Stimme,* ed. Friedrich A. Kittler, Thomas Macho, and Sigrid Weigel, trans. Antje Pfannkuchen (Berlin: Akademie, in press).

7 George Bernard Shaw, "Preface to *Pygmalion,*" in *Bernard Shaw: Collected Plays with Their Prefaces* (New York: Dodd, Mead, 1972), 4:664. See also Shaw, *Pygmalion,* in ibid.

8 Friedrich Kittler, *Gramophone, Film, Typewriter,* trans. Geoffrey Winthrop-Young and Michael Wutz (Stanford, Calif.: Stanford University Press, 1999), 27.

9 Alexander Graham Bell, *The Telephone: A Lecture Entitled Researches in Electric Telephony by Professor Alexander Graham Bell, Delivered before the Society of Telegraph Engineers, October 31st, 1877* (New York: Society of Telegraph Engineers, 1878), 20.

10 Alexander Graham Bell quoted in Charles Snyder, "Clarence John Blake and Alexander Graham Bell: Otology and the Telephone," *Annals of Otology, Rhinology, and Laryngology* 83, no. 4, pt. 2, suppl. 13 (July–August 1974): 30.

11 Scripture, *Elements of Experimental Phonetics,* 27.

12 Snyder, "Clarence John Blake and Alexander Graham Bell," 11.

13 See Alexander Graham Bell, *Memoir upon the Formation of a Deaf Variety of the Human Race* (New Haven, Conn.: National Academy of Sciences, 1883).

14 "We recall that the Deaf were considered more radically deprived of life than the blind, for blindly still we dwell in language" (Avital Ronell, *The*

Telephone Book: Technology—Schizophrenia—Electric Speech [Lincoln: University of Nebraska Press, 1989], 324).

15 Douglas C. Baynton, "'Savages and Deaf-Mutes': Evolutionary Theory and the Campaign against Sign Language in the Nineteenth Century," in *Deaf History Unveiled: Interpretations from the New Scholarship,* ed. John Vickery Van Cleve (Washington, D.C.: Gallaudet University Press, 1993), 92–112. See also Richard Winefield, *Never the Twain Shall Meet: Bell, Gallaudet, and the Communications Debate* (Washington, D.C.: Gallaudet University Press, 1987); and Douglas C. Baynton, *Forbidden Signs: American Culture and the Campaign against Sign Language* (Chicago: University of Chicago Press, 1996).

16 Baynton, "Savages and Deaf-Mutes," 108, 100. See also Jack R. Gannon, *Deaf Heritage: A Narrative History of Deaf America* (Silver Spring, Md.: National Association of the Deaf, 1981), esp. 75–79; and Bruce, *Bell,* 409–12.

17 Lennard Jeffries, *Enforcing Normalcy: Disability, Deafness, and the Body* (New York: Verso, 1995), 32, 81, and passim. Jeffries provides a detailed critique of the category *disability,* especially as it is applied to deafness.

18 Kittler, *Gramophone, Film, Typewriter,* 22.

19 Bruce, *Bell,* 121; Bell, *The Telephone,* 22. See also Michael Gorman, *Simulating Science: Heuristics, Mental Models, and Technoscientific Thinking* (Bloomington: Indiana University Press, 1992).

20 S. Morland, *Tuba Stentoro-Phonica: An Instrument of Excellent Use, as Well at Sea, as at Land; Invented and Variously Experimented in the Year 1670 and Humbly Presented to the Kings Most Excellent Majesty Charles II in the Year 1671* (London: Printed by W. Godbid and Sold by M. Pitt, 1672), 6–7.

21 Kittler, *Gramophone, Film, Typewriter,* 24–25; Hankins and Silverman, *Instruments and the Imagination,* 130–32.

22 Kittler, *Gramophone, Film, Typewriter,* 47 (see also 71).

23 "The Invention of the Euphon, and Other Acoustic Discoveries of C. F. Chladni . . . ," *Philosophical Magazine* 2 (October 1798): 391; "Chladni's Experiments on the Resonant Figures on Glass; Philosophical Transactions; Berthoud on Time-Pieces," *Journal of Natural Philosophy, Chemistry, and the Arts* 3 (April 1799–March 1800): 185; T. H. Huxley, *On the Physical Basis of Life* (New Haven, Conn.: Charles C. Chatfield, 1870), 112–20; Hankins and Silverman, *Instruments and the Imagination,* 130–32.

24 Adam Politzer, *History of Otology,* trans. Stanley Milstein, Collice Portnoff, and Antje Coleman (Phoenix: Columella, 1981), 250.

25 Thomas Young, *A Course of Lectures on Natural Philosophy and the Mechanical Arts* (London: J. Johnson, 1807), 1:369. See also Robert Silverman, "Instrumentation, Representation, and Perception in Modern Science: Imitating

Human Function in the Nineteenth Century" (Ph.D. diss., University of Washington, 1992), esp. 96–189.

26 Hankins and Silverman, *Instruments and the Imagination,* 133.
27 Scott quoted in ibid., 135.
28 Edouard-Léon Scott de Martinville, *Le Problèm de la parole s'écrivant alle-même: La France, l'Amérique* (Paris, 1878), cited in Hankins and Silverman, *Instruments and the Imagination,* 137; and in Thomas Y. Levin, "For the Record: Adorno on Music in the Age of Its Technological Reproducibility," *October,* no. 55 (winter 1990): 36. Levin cites Scott in support of his thesis that, just as early cinema was heralded as a transparent reproduction of images that would supersede national languages, the prehistory of sound recording articulated an "analogous discourse of democratization and univocal, natural signs" (36).
29 Jacques Derrida, *Of Grammatology,* trans. Gayatri Chakravorty Spivak (Baltimore: Johns Hopkins University Press, 1976), 25–26; Briankle Chang, *Deconstructing Communication* (Minneapolis: University of Minnesota Press, 1996), 187–220; Lastra, *Sound Technology and American Cinema,* 73–91.
30 Arnold Pacey's discussion of "virtuosity values" is apropos here. Pacey argues that scientific and technical research often seeks technological innovation for its own sake, rather than for some specific end toward which it can be put. Thus, even though Berliner would probably have had no idea what to do with phonautograms (it is not known for certain whether he was familiar with Bell's plans for them), he could herald their existence with the hope that a use would later be found. See Arnold Pacey, *The Culture of Technology* (Cambridge, Mass.: MIT Press, 1983), esp. 78–96.
31 Berliner Gramophone Co., Philadelphia Division, "The Gramophone," EB, scrapbook 2.
32 Bell, *The Telephone,* 22; Berliner, "The Gramophone," 17; General Electric Co., "Radio Signals Recorded by the General Electric Visual and Photograph Recorder," GHC, series 93, box 389; Hankins and Silverman, *Instruments and the Imagination,* 135.
33 Edward Wheeler Scripture, *Researches in Experimental Phonetics: The Study of Speech Curves* (Washington, D.C.: Carnegie Institution of Washington, 1906), 3.
34 Ibid., 4.
35 Derrida, *Of Grammatology,* 17.
36 Theodor Adorno, "The Form of the Phonograph Record," trans. Thomas Y. Levin, *October,* no. 55 (winter 1990): 59; see also Levin, "For the Record," 38–41.
37 Digital audio technologies allow for the reversal of that formulation, and it is now possible to subject auditory phenomena to the orderings of images.

38 *Oxford English Dictionary*, s.v. "phonograph"; Bruce, *Bell*, 351; William Elterich to Emile Berliner, n.d. (probably late 1890s), EB, scrapbook 2.
39 Bruce, *Bell*, 112. Bell and Blake met in 1871.
40 Ibid.; Alexander Graham Bell, "Early Telephony" (address before the Telephone Society of Washington, Washington, D.C., 3 February 1910, AGB, box 383, speech file "Early Telephony"), "Prehistoric Telephone Days" (n.d., AGB, box 385, speech file "Prehistoric Telephone Days"), and *The Telephone;* Clarence J. Blake, "Sound and the Telephone" (paper read before the British Society of Telegraph Engineers, London, 8 May 1878) (an offprint of this talk can be found at the Library of Congress), and "The Use of the Membrana Tympani as a Phonautograph and Logograph," *Archives of Opthamology and Otology* 5 (1878): 108–113. This last article provides the most complete instructions for the construction of an ear phonautograph.
41 Blake, "Sound and the Telephone," 5–7.
42 Snyder, "Clarence John Blake and Alexander Graham Bell," 13.
43 S. Morland, *Tuba Stentoro-Phonica*, 7.
44 P. Kennedy, *Ophthalmographia; or, A Treatise of the Eye in Two Parts* (London: Bernard Lintott, 1713), H2.
45 William R. Wilde, *Practical Observations on Aural Surgery and the Nature and Treatment of Diseases of the Ear with Illustrations* (Philadelphia: Blanchard and Lea, 1853), 60–61.
46 Martin Jay, *Downcast Eyes: The Denigration of Vision in Twentieth-Century French Thought* (Berkeley and Los Angeles: University of California Press, 1993); and David Michael Levin, ed., *Modernity and the Hegemony of Vision* (Berkeley and Los Angeles: University of California Press, 1993).
47 Anton von Tröltsch, *Treatise on the Diseases of the Ear, Including the Anatomy of the Organ,* trans. D. B. St. John Roosa, 2d American ed., from the 4th German ed. (New York: William Wood, 1869), vii (quotation), 25–26. (The first German edition, from which the quoted passage is taken, appeared in 1862.) See also Dennis Pappas, "Anton Friedrich von Troltsch (1829–1890): The Beginning of Otology in Germany," *Ear, Nose, and Throat Journal* 75, no. 10 (October 1996): 50–51.
48 Snyder, "Clarence John Blake and Alexander Graham Bell," 3–4.
49 Joseph Toynbee, *The Diseases of the Ear: Their Nature, Diagnosis, and Treatment* (London: John Churchill, 1860), 1–2 (quotation), 7–9.
50 N. Rüdinger, *Rüdinger Atlas of the Osseous Anatomy of the Human Ear, Comprising a Portion of the Atlas of the Human Ear,* translated, with notes and an additional plate, by Clarence Blake (Boston: A. Williams, 1874), 18.
51 Snyder, "Clarence John Blake and Alexander Graham Bell," 5, 12. See also Gustav Brühl and Adam Politzer, *Atlas and Epitome of Otology,* trans. and

ed. S. MacCuen Smith (Philadelphia: W. B. Saunders, 1903); and Hermann Helmholtz, *The Mechanism of the Ossicles of the Ear and the Membrana Tympani*, trans. Albert H. Buck and Norman Smith (New York: William Wood, 1873). Brühl and Politzer's book is both an anatomical and a diagnostic manual, containing illustrations from Politzer's collection. It is a testimony to the significance of Politzer's collection of models and bones in the development of European otology. Its translation and mass dissemination in the United States are also indicative of the greater degree of institutionalization and professionalization of otology in the United States by the turn of the twentieth century as well as of the professionalization of medicine in general.

52 Sylvan Stool, Marlyn Kemper, and Bennett Kemper, "Adam Politzer, Otology, and the Centennial Exhibition of 1876," *Laryngoscope* 85, no. 11, pt. 1 (November 1975): 1898–1904; Neil Weir, "Adam Politzer's Influence on the Development of International Otology," *Journal of Laryngology and Otology* 110, no. 9 (September 1996): 824, 826, 828.

53 Elisabeth Bennion, *Antique Medical Instruments* (Berkeley and Los Angeles: University of California Press, 1979), 99–101.

54 Georg Békésy and Walter Rosenblinth, "The Early History of Hearing—Observations and Theories," *Journal of the Acoustical Society of America* 20, no. 6 (November 1948): 727–28.

55 Soyara de Chadarevian, "Graphical Method and Discipline: Self-Recording Instruments in Nineteenth-Century Physiology," *Studies in the History and Philosophy of Science* 24, no. 2 (1993): 279. See also Hankins and Silverman, *Instruments and the Imagination*, 113–48.

56 Alexander Munro, *Observations on the Organ of Hearing in Man and Other Animals* (Edinburgh: Adam Neill, 1797); Neil Weir, *Otolaryngology: An Illustrated History* (Boston: Butterworths, 1990), 63.

57 Peter Degravers, *A Complete Physico-Medical and Chirurgical Treatise on the Human Eye, Second Edition, Corrected and Considerably Enlarged, to Which Is Now Added a Treatise on the Human Ear* (Edinburgh: Printed for the Author, 1788), 267.

58 Charles Bell, *The Nervous System of the Human Body: Embracing the Papers Delivered to the Royal Society on the Subject of Nerves* (Washington, D.C.: Stereotyped by D. Green, for the Register and Library of Medicine and Chirurgical Science, 1833), 21 (see also 20, 21, 333, 37, 51, 71). See also Charles Bell, *Ideas of a New Anatomy of the Brain: A Facsimile of the Privately Printed Edition of 1811* (London: Dawsons of Pall Mall, 1966).

59 Jonathan Crary, *Techniques of the Observer: On Vision and Modernity in the Nineteenth Century* (Cambridge, Mass.: MIT Press, 1990), 89–90.

60 Johannes Müller, *Elements of Physiology*, trans. William Baly, arranged from

the 2d London ed. by John Bell (Philadelphia: Lea and Blanchard, 1843), 587–88 (see also 709).

61 Ibid., 710, 712.
62 Ibid., 711.
63 Crary, *Techniques of the Observer*, 92.
64 I appear to be in the good company of another Jonathan: Crary (ibid., 92) notes homologies between Müller's account of vision and theories of photography. In "Helmholtz, Edison, and Sound History," John Peters argues that the telegraph was the inspiration for the physiological instrumentation that in turn begat modern electronic media.
65 Müller, *Elements of Physiology*, 744–71.
66 Hermann Helmholtz, *On the Sensations of Tone as a Physiological Basis for the Theory of Music,* 2d English ed., trans. from the 4th German ed. by Alexander J. Ellis (New York: Dover, 1954), 7.
67 Helmholtz, *Sensations of Tone,* 3–4. On Helmholtz's elaboration of Müller's theory, see Stephan Vogel, "Sensation of Tone, Perception of Sound, and Empiricism: Helmholtz's Physiological Acoustics," in *Hermann von Helmholtz and the Foundations of Nineteenth-Century Science,* ed. David Cahan (Berkeley and Los Angeles: University of California Press, 1993), 282–83.
68 Helmholtz, *Sensations of Tone,* 129 (first quotation), 142 (second quotation), 166 (discussion of Scott and Politzer).
69 Robert Walser, *Running with the Devil: Power, Gender, and Madness in Heavy Metal Music* (Hanover, N.H.: Wesleyan University Press, 1993), 41–44. For Helmholtz's definition of *lower partial* and *upper partial,* see *Sensations of Tone,* 23. For a summary of the effects of different kinds of upper partials for timbre, see ibid., 118–19. Helmholtz writes that the entire first part of *On the Sensations of Tone* is essentially an explanation of upper partials (ibid., 4). See also Theodore Gracyk, *Rhythm and Noise: An Aesthetics of Rock* (Durham, N.C.: Duke University Press, 1996), 111–14; and Steve Waksman, *Instruments of Desire: The Electric Guitar and the Shaping of Musical Experience* (Cambridge, Mass.: Harvard University Press, 1999), 137–38.
70 Peters, "Helmholtz, Edison, and Sound History" (forthcoming). Although Helmholtz's synthesizer (discussed later in this chapter) is quite distant from the twentieth-century instrument, it foreshadows the latter—creating a sound with a distinctive timbre through the combination of more basic tones. Thaddeus Cahill's telharmonium is, perhaps, a missing link between Helmholtz's synthesizer and the mid-twentieth-century variety. See Reynold Weidenaar, *Magic Music from the Telharmonium* (Metuchen, N.J.: Scarecrow, 1995). See also Hankins and Silverman, *Instruments and the Imagination,* 203.
71 Peters, "Helmholtz, Edison, and Sound History" (forthcoming).

72 Helmholtz, *Sensations of Tone*, 43 (see also 43–44, 129, 149).
73 Ibid., 148.
74 Peters, "Helmholtz, Edison, and Sound History" (forthcoming). Peters is quoting Brian Winston, *Media, Technology, and Society: A History: From the Telegraph to the Internet* (New York: Routledge, 1998), 38.
75 Berliner, "The Gramophone," 2.
76 For a full account, see Paul Starr, *The Social Transformation of American Medicine* (New York: Basic, 1982), esp. 93–144.
77 Snyder, "Clarence John Blake and Alexander Graham Bell," 4–5.
78 See Ruth Richardson, *Death, Dissection, and the Destitute* (New York: Routledge and Kegan Paul, 1987; and Suzanne Shultz, *Body Snatching: The Robbing of Graves for the Education of Physicians* (Jefferson, N.C.: McFarland, 1993). Shultz discusses the incident on pp. 85–89. See also Roger French, "The Anatomical Tradition," in *Companion Encyclopedia to the History of Medicine,* ed. W. F. Bynum and Roy Porter (London: Routledge, 1993), 81–84, 99–100.
79 See Ulrich Beck, *Risk Society: Towards a New Modernity,* trans. Mark Ritter (Newbury Park, Calif.: Sage, 1992).
80 See Richardson, *Death, Dissection, and the Destitute,* 207–10; and Shultz, *Body Snatching,* 90–94.
81 Richardson, *Death, Dissection, and the Destitute,* 266.
82 On the passage of anatomy acts in the United States, see Shultz, *Body Snatching,* 78–94; and Linden F. Edwards, *The History of Human Dissection* (Fort Wayne, Tex.: Fort Wayne and Allen County Public Libraries, 1955), 18–20. On class politics in nineteenth-century America, see, e.g., Alexander Saxton, *The Rise and Fall of the White Republic: Class Politics and Mass Culture in Nineteenth Century America* (New York: Verso, 1990).
83 Catherine F. Mackenzie, *Alexander Graham Bell: The Man Who Contracted Space* (Boston: Houghton-Mifflin, 1928), 71. Ronell writes, "This kind of conduct tends to border on illegality, but it turned out that because the ears were a missing pair, they legally assisted Bell" (*The Telephone Book,* 333).
84 Walter Benjamin, "Theses on the Philosophy of History," in *Illuminations,* trans. Harry Zohn (New York: Schocken, 1968), 256.
85 Kittler, *Gramophone, Film, Typewriter,* 75.
86 For a discussion of the subsumption of the function of the mouth and speech under a larger rubric of sound production for reproduction, see chapter 5. For an alternate reading of the historical relation between automata and other sound-reproduction technologies, see Michael Taussig, *Mimesis and Alterity: A Particular History of the Senses* (New York: Routledge, 1993), 212–25; Hankins and Silverman, *Instruments and the Imagination,* 178–220; Kittler, *Gramophone, Film, Typewriter,* 24–29; and Lastra, *Sound Technology*

and American Cinema, 16–60. Hankins and Silverman and Lastra both provide a more in-depth analysis than I can here.

87 National Phonograph Co., *The Phonograph and How to Use It, Being a Short History of Its Invention and Development Containing also Directions Helpful Hints and Plain Talks as to Its Care and Use, Etc.* (New York: National Phonograph Co., 1900), 13–14.

88 Quoted in Snyder, "Clarence John Blake and Alexander Graham Bell," 30.

89 Francis Bacon, *The Advancement of Learning and the New Atlantis* (Oxford: Clarendon, 1974), 294.

90 Derek J. de Solla Price, "Automata and the Origins of Mechanism and Mechanistic Philosophy," *Technology and Culture* 5 (1964): 9. See also Jamie C. Kassler, "Man—a Musical Instrument: Models of the Brain and Mental Functioning before the Computer," *History of Science* 22 (1984): 59–92.

91 Kassler, "Man," 62.

92 Price, "Automata," 23.

93 René Descartes, *Treatise of Man* (1664), trans. Thomas Steele Hall (Cambridge, Mass.: Harvard University Press, 1972), 2–4. Descartes clearly wants to argue for his own version of the body over the conventional one, but he is careful simply to pose his version as an alternative.

94 René Descartes, *Discourse on Method and Meditations on First Philosophy,* 4th ed., trans. Donald A. Cress (Indianapolis: Hackett, 1998), 26, 31. On Descartes's "unfulfilled promise," see Descartes, *Treatise of Man,* 2 n. 2.

95 Arthur W. J. G. Ord-Hume, *Clockwork Music: An Illustrated History of Mechanical Musical Instruments from the Musical Box to the Pianola, from Automaton Lady Virginal Players to Orchestrion* (New York: Crown, 1973), 18, 25.

96 Hankins and Silverman, *Instruments and the Imagination,* 182–85.

97 *The History and Analysis of the Supposed Automaton Chess Player of M. de Kempelen, Now Exhibiting in This Country, by Mr. Mälzel; with Lithographic Figures, Illustrative of the Probable Method by Which Its Motions Are Directed* (Boston: Hilliard, Gray, 1826), esp. 3–9; Hankins and Silverman, *Instruments and the Imagination,* 186–98; Erasmus Darwin, *The Letters of Erasmus Darwin,* ed. Desmond King-Hele (Cambridge: Cambridge University Press, 1981), 63. See also Lastra, *Sound Technology and American Cinema,* 31–32.

98 J.B. [John Bulwer], *Philosophicus* (1648), 48, copied "verbatim" from the original by Professor J. C. Gordon, 6 February 1886, AGB, box 256, folder "Subject, Phonograph, Miscellany." It appears as though this quote was sent to Bell.

99 Bell, "Prehistoric Telephone Days," 17, 19, 21–22. See also Bell, "Early Telephony," 7–8.

100 Sir Charles Wheatstone, *The Papers of Sir Charles Wheatstone* (London: Tay-

lor and Francis, 1879), 62–63; Müller, *Elements of Physiology,* 699–700 (Müller documents extensive experimentation with human glottises removed from the body and with the glottises of living dogs; the discussion of an artificial glottis comes at the end of this section, pp. 687–700); Hankins and Silverman, *Instruments and the Imagination,* 213–17; Ord-Hume, *Clockwork Music,* 19.

101 Helmholtz, *On the Sensations of Tone,* 120–29, 163; Hankins and Silverman, *Instruments and the Imagination,* 198–209.

102 Berliner, "The Gramophone," 2. It is not clear whether *tympanum membrani* was a phrase in common use at the time or whether this is an error on Berliner's part.

103 König's manometric flame was also based on the diaphragm principle, although it is not commonly thought of as a predecessor of other sound-reproduction technologies.

104 Berliner, "The Gramophone," 4.

105 Ibid., 4.

106 Bell quoted in Bruce, *Bell,* 252.

107 Columbia Phonograph Co., *The Latest and Best Talking Machines: The Perfected Graphophone with Clockwork or Electric Motor* (Washington, D.C.: Columbia Phonograph Co., 1895), n.p., Smithsonian Institution, National Museum of American History, Mechanisms Division, phonograph box 1.

108 Hermann Helmholtz, "On the Interaction of Natural Forces," in *Popular Lectures on Scientific Subjects,* trans. E. Atkinson (New York: Appleton, 1885), 154.

109 Hankins and Silverman, *Instruments and the Imagination,* 223.

110 "New Remedy for Deafness: Dr. Leech Believes That the Phonograph Will Cause a Cure," *New York Times,* 23 May 1892. See also "Miracle of the Phonograph: Edison's Invention Applied to the Scientific Cure of Deafness," *New York Times,* 12 October 1895; and "Possibilities of the Phonograph," *New York Times,* 13 December 1891.

111 Grant Eldridge, "The Human Telephone," *Buffalo (N.Y.) Times,* 24 January 1897, National Museum of American History, Medical Sciences Division, folder "Hearing."

112 Kenneth Berger, *The Hearing Aid: Its Operation and Development* (Livonia, Mich.: National Hearing Aid Society, 1974), 24.

113 On the concept of the diagram, see Gilles Deleuze, *Foucault,* trans. Sean Hand (Minneapolis: University of Minnesota Press, 1988), 34; Michel Foucault, *Discipline and Punish: The Birth of the Prison,* trans. Alan Sheridan (New York: Vintage, 1977), 205.

114 On Foucault's reading of his own work in relation to Weber's notion of ideal types, see his "Questions of Method: An Interview with Michel Fou-

cault," in *The Foucault Effect: Studies in Governmentality,* ed. Graham Burchell, Colin Gordon, and Peter Miller (Chicago: University of Chicago Press, 1991), 8–9.

115 Wiebe Bijker (*Of Bicycles, Bakelites, and Bulbs: Toward a Theory of Sociotechnical Change* [Cambridge, Mass.: MIT Press, 1995]) offers one noninstrumental theory of technological change that does not fall into technological determinism.

116 Edward Wheeler Scripture reports that, as late as 1898, Nagel and Samojloff "used the ear in the head of a freshly killed animal as a manometric capsule" in a manometric flame (*Elements of Experimental Phonetics,* 29).

117 Blake, "Sound and the Telephone," 7.

118 Snyder, "Clarence John Blake and Alexander Graham Bell," 21.

2. Techniques of Listening

1 The language of modern advertising is unmistakable in the Brandes ad; it presumes all sorts of knowledge on the part of the reader. For a discussion of the emergence of advertising language, see Richard Ohmann, *Selling Culture: Magazines, Markets, and Class at the Turn of the Century* (New York: Verso, 1996), esp. 176.

2 "You *need* a headset," advertisement for Brandes headphones, GHC, series 49, box 301.

3 See Susan Smulyan, *Selling Radio: The Commercialization of American Broadcasting, 1920–1934* (Washington, D.C.: Smithsonian Institution Press, 1994), 13–20.

4 I have used R. T. H. Laennec, *A Treatise on the Diseases of the Chest and on Mediate Auscultation,* 3d ed., trans. John Forbes (New York: Samuel Wood; Collins and Hannay, 1830). The difference between the second and the third editions rests in only minor revisions.

5 Michel Foucault, *The Birth of the Clinic: An Archaeology of Medical Perception,* trans. A. M. Sheridan Smith (New York: Pantheon, 1973).

6 Nor is it a definitive history of acoustic technique. Emily Thompson's "Mysteries of the Acoustic: Architectural Acoustics in America, 1800–1932 (Ph.D. diss., University of Pennsylvania, 1992) considers the development of architectural acoustics; her concerns are largely with changing notions of engineering and the development of the profession as such.

7 Michel Foucault, "Questions of Method," in *The Foucault Effect: Studies in Governmentality,* ed. Graham Burchell, Colin Gordon, and Peter Miller (Chicago: University of Chicago Press, 1991), 75, 81.

8 See, e.g., Katherine Ott, *Fevered Lives: Tuberculosis in American Culture since 1870* (Cambridge, Mass.: Harvard University Press, 1996), 6, 21, 25; and

Audrey B. Davis, *Medicine and Its Technology: An Introduction to the History of Medical Instrumentation* (Westport, Conn.: Greenwood, 1981).

9 Marcel Mauss, "Body Techniques," in *Sociology and Psychology: Essays,* trans. Ben Brewster (Boston: Routledge and Kegan Paul, 1979), 104–5.

10 See, e.g., ibid., 112–19.

11 Ibid., 108.

12 Aristotle, *Nicomachean Ethics,* trans. Martin Ostwald (New York: Bobbs-Merrill, 1962), 151–52; Martin Heidegger, *The Question concerning Technology and Other Essays,* trans. William Lovitt (New York: Harper Torchbooks, 1977), 12; Jacques Ellul, *The Technological Society* (New York: Vintage, 1964), 4, 79–133.

13 Pierre Bourdieu, *The Logic of Practice,* trans. Richard Nice (Stanford, Calif.: Stanford University Press, 1990), 52–65; Pierre Bourdieu and Loïc J. D. Wacquant, *An Invitation to Reflexive Sociology* (Chicago: University of Chicago Press, 1992), 13–14, 120–40. To describe the way a particular society organizes practices of listening, John Mowitt has coined the suggestive phrase *structure of listening*—meant as a play on Raymond Williams's *structure of feeling*. I have chosen not to use the term here mainly because of the connotative meaning of the word *structure,* but Mowitt's notion gels well with my discussion of habitus. See John Mowitt, "The Sound of Music in the Era of Its Electronic Reproducibility," in *Music and Society: The Politics of Composition, Performance, and Reception,* ed. Richard Leppert and Susan McClary (New York: Cambridge University Press, 1987), 175.

14 Laennec's failed attempt to develop an acoustic lexicon for the stethoscope will be discussed below.

15 A whole range of writers have contested the notion that hearing is necessarily an imprecise sense. See, e.g., Barry Truax, *Acoustic Communication* (Norwood, N.J.: Ablex, 1984), 15–24; and Don Idhe, *Listening and Voice: A Phenomenology of Sound* (Athens: Ohio University Press, 1974), 6–8, 58. Although they do not extend the argument to a direct consideration of the relation between audition and techniques of rationality and instrumental perception, their arguments enable the move that I make here.

16 Rudolf Arnheim, *Radio,* trans. Margaret Ludwig and Herbert Read (London: Faber and Faber, 1936); Hadley Cantril and Gordon Allport, *The Psychology of Radio* (New York: Harper and Bros., 1935). See also Hanns Eisler and Theodor Adorno, *Composing for the Films* (New York: Oxford University Press, 1947); and Theodor Adorno, "On the Fetish-Character in Music and the Regression of Listening," in *The Essential Frankfurt School Reader,* ed. Andrew Arato and Eike Gebhardt (New York: Continuum, 1982), 270–99, and *Introduction to the Sociology of Music,* trans. E. B. Ashton (New York: Seabury, 1976).

17 *Oxford English Dictionary,* s.v. "audile." Aesthetic preference and convenience also motivate my choice of *audile* as my adverb and adjective: *aural* is too easily confused with *oral* when the word is spoken, and *auscultatory* seems a bit baroque and is too closely associated with the specific case of medical listening.

18 Truax, *Acoustic Communication,* 15–24; Steven Feld, "Aesthetics as Iconicity of Style (Uptown Title); or, (Downtown Title) 'Lift-up-over Sounding': Getting into the Kaluli Groove," in *Music Grooves,* by Charles Keil and Steven Feld (Chicago: University of Chicago Press, 1994), 115–31; Michel Foucault, *The History of Sexuality,* vol. 1, *An Introduction,* trans. Robert Hurley (New York: Vintage, 1978). On audile sensibilities, see also Christopher Small, *Music-Society-Education* (London: John Calder, 1977); and John Miller Chernoff, *African Rhythm and African Sensibility: Aesthetics and Social Action in African Musical Idioms* (Chicago: University of Chicago Press, 1979).

19 James H. Johnson, *Listening in Paris: A Cultural History* (Berkeley and Los Angeles: University of California Press, 1995), 284. For a more flattering account of the importance of musical sound in modern listening, see Simon Frith, *Performing Rites: On the Value of Popular Music* (Cambridge, Mass.: Harvard University Press, 1996), 99–122.

20 Foucault, *Birth of the Clinic,* xiv.

21 Ibid., xii and passim (I discuss Foucault's dogmatic emphasis on vision below); Stanley Joel Reiser, *Medicine and the Reign of Technology* (Cambridge: Cambridge University Press, 1978), 1–44; Jacalyn Duffin, *To See with a Better Eye: A Life of R. T. H. Laennec* (Princeton, N.J.: Princeton University Press, 1998), 6–7.

22 *Oxford English Dictionary,* s.v. "auscultation." Although the entry dates the earliest medical use of the term in English to 1833, it is likely that the term was already in common medical usage when John Forbes first translated Laennec's *Treatise on Mediate Auscultation* into English.

23 Laennec on Buisson (1802) quoted in Duffin, *To See with a Better Eye,* 43 (see also 153).

24 For more discussion of the history of the physiology of hearing, see chapter 1.

25 Duffin, *To See with a Better Eye,* 302.

26 See Paul Starr, *The Social Transformation of American Medicine* (New York: Basic, 1982), 17–24 and passim.

27 Laennec, *Treatise on Mediate Auscultation,* 5. This narrative became the subject of popular lore throughout the nineteenth century (Davis, *Medicine and Its Technology,* 90).

28 Duffin, *To See with a Better Eye,* 121.

29 As discussed in chapter 1, Young used a stylus to trace the vibrations of sound.
30 Ibid., 122–24.
31 Ibid., 153–55.
32 Davis, *Medicine and Its Technology*, 97–102.
33 Duffin, *To See with a Better Eye*, 129.
34 Martin Jay, *Downcast Eyes: The Denigration of Vision in Twentieth-Century French Thought* (Berkeley and Los Angeles: University of California Press, 1993), 83–148. On light as a metaphor for truth in Greek, medieval, and early-modern thought, see Hans Blumenberg, "Light as a Metaphor for Truth: As the Preliminary Stage of Philosophical Concept Formation," in *Modernity and the Hegemony of Vision*, ed. David Michael Levin (Berkeley and Los Angeles: University of California Press, 1993), 30–62.
35 This also served as a useful preemptory critique of François Double's work on listening directly to the patient's body.
36 Laennec, *Treatise on Mediate Auscultation*, 25–26.
37 The physical examination "was the result of a recasting at the level of epistemic knowledge (*savoir*) itself, and at the level of accumulated, refined, deepened, adjusted knowledge (*connaissances*)" (Foucault, *Birth of the Clinic*, 137).
38 Malcolm Nicolson, "The Introduction of Percussion and Stethoscopy to Early Nineteenth-Century Edinburgh," in *Medicine and the Five Senses*, ed. W. F. Bynum and Roy Porter (New York: Cambridge University Press, 1993), 135.
39 Distilled from Laennec, *Treatise on Mediate Auscultation*, 27–28; John Hughes Bennett, *Clinical Lectures on the Principles and Practice of Medicine* (New York: Samuel S. and William Wood, 1860), 49–50; and Austin Flint, *A Manual of Percussion and Auscultation; of the Physical Diagnosis of Diseases of the Lungs and Heart, and of Thoracic Aneurism* (Philadelphia: Henry C. Lea, 1876), 72–74.
40 Foucault, *Birth of the Clinic*, 164.
41 For a discussion of the separation of the senses and the construction of hearing as an object of knowledge, see chapter 1.
42 Flint, *Manual of Percussion and Auscultation*, 73–74.
43 Reiser, *Medicine and the Reign of Technology*, 41; Davis, *Medicine and Its Technology*, 104.
44 For the most part, stethoscope construction in the nineteenth century was not heavily based on the principles of acoustics. Although the instrument embodied a basic principle of acoustics, none of the more advanced thinking concerning acoustics was applied until the turn of the century. Additionally, starting in the 1890s, inventors sought to incorporate new sound-

reproduction technologies into the stethoscope—most notably, sound recording and electric amplification (Davis, *Medicine and Its Technology,* 109).

45 Ibid., 104, 259.
46 Flint, *Manual of Percussion and Auscultation,* 71.
47 The phrase comes from Langdon Winner, *Autonomous Technology: Technics-out-of-Control as a Theme in Political Thought* (Cambridge, Mass.: MIT Press, 1977), 315. Winner is very ambivalent about the *license to forget:* for him, it is "the source of the colossal passivity in man's dealings with technical means," but "the benefits in terms of health, mobility, material comfort, and the overcoming of the physical problems of production and communication are well known" (ibid.). I consider this process of forgetting less a source of passivity than a basic precondition for human tool use or any other kind of repeatable behavior.
48 See Pierre Bourdieu, *Distinction: A Social Critique of the Judgment of Taste,* trans. Richard Nice (Cambridge, Mass.: Harvard University Press, 1984), 169–225.
49 Laennec, *Treatise on Mediate Auscultation,* 4.
50 Duffin, *To See with a Better Eye,* 79.
51 Norbert Elias, *The Civilizing Process: Sociogenetic and Psychogenetic Investigations,* trans. Edmund Jephcott, ed. Eric Dunning, Johan Goudsblom, and Stephen Mennell, rev. ed. (Malden: Blackwell, 2000), 420. Elias's thesis that the repression and management of drives goes hand in hand with ever-increasing rationalization clearly bears the influence of both Freud and Weber, although one need not accept the larger claims of either writer for Elias's insight to have some explanatory utility.
52 Duffin, *To See with a Better Eye,* 110–13.
53 Ruth Schwartz Cowan, *More Work for Mother: The Ironies of Household Technology from the Open Hearth to the Microwave* (New York: Basic, 1983), 76–77; Starr, *Social Transformation of American Medicine,* 72–77.
54 Laennec, *Treatise on Mediate Auscultation,* 8: "It is only in an hospital that we can acquire, completely and certainly, the practice and habit of this new art of observation."
55 Duffin, *To See with a Better Eye,* 126.
56 Ibid., 126–27; Charles E. Rosenberg, "And Heal the Sick: Hospital and Patient in the Nineteenth Century," *Journal of Social History* 4 (1977): 428–71.
57 Duffin, *To See with a Better Eye,* 79, 104.
58 Foucault, *Birth of the Clinic,* 163.
59 Davis, *Medicine and Its Technology,* 88.
60 This was also a result of the clinicization of medicine and the shaping of the doctor-patient relation through its rationalization and institutionalization.

61 Reiser, *Medicine and the Reign of Technology*, 1–7.

62 Ibid., 24.

63 Ibid., 19 (see also 10–19). Dissection-based pedagogy, which had existed throughout the Middle Ages, gradually came into greater favor as the scientific view gained currency in medicine.

64 Of course, the taking of medical history continues down to the present day, but it exists as only one of an array of techniques to be used in medical diagnosis. It is a residual technique.

65 Elias, *The Civilizing Process*, 420.

66 Laennec (*Treatise on Mediate Auscultation*, 24) acknowledged Hippocrates' experiment with immediate auscultation but "considered it as indeed it is, one of the mistakes of that great man."

67 Leopold Auenbrugger, "On Percussion of the Chest," trans. John Forbes in 1824 from the 1761 Latin ed., with an introduction by Henry Sigerist, *Bulletin of the History of Medicine* 4 (1936): 379. Forbes's translation of Auenbrugger first appeared as part of John Forbes, *Original Cases with Dissections and Observations Illustrative of the Use of the Stethoscope and Percussion in the Diagnosis of Diseases of the Chest; Also Commentaries on the Same Subjects Selected and Translated from Auenbrugger, Corvisart, Laennec, and Others* (London: Printed for T. and G. Underwood, Fleet Street, 1824).

68 Reiser, *Medicine and the Reign of Technology*, 21; Duffin, *To See with a Better Eye*, 32.

69 Reiser, *Medicine and the Reign of Technology*, 21–22; see also Henry Sigerist, introduction to Auenbrugger, "On Percussion of the Chest," 374. Reiser (*Medicine and the Reign of Technology*, 37) notes that this prejudice remained in residual form into Laennec's time: one of the early objections to the stethoscope was that it would cause physicians to be classed with surgeons as craftsmen.

70 Forbes, *Original Cases*.

71 Laennec, *Treatise on Mediate Auscultation*, 22–23.

72 Duffin, *To See with a Better Eye*, 33–35.

73 Davis, *Medicine and Its Technology*, 89; Reiser, *Medicine and the Reign of Technology*, 29–30.

74 Foucault, *Birth of the Clinic*, 137, 141, 143, 144.

75 Freud's "talking cure" is sometimes cast as a return to subjectivity, but even here speech is interesting, no longer for its semantico-referential content, but as a manifestation of the patient's interior psyche. In that sense, Freud's emphasis on exterior speech as an index of interior condition was not a full departure from empiricist medicine. Obviously, there are significant differences between psychoanalytic method and the methods of diagnosis discussed here. On the talking cure, see, e.g., Josef Breuer and Sigmund Freud,

Studies in Hysteria, trans. A. A. Brill (New York: Nervous and Mental Disease Publishing Co., 1936).

76 Reiser (*Medicine and the Reign of Technology,* 167–68) argues that the emergent predominance of laboratory medicine led to the decline of physical examination since the latter was seen as less precise and more time-consuming; the eclipse of mediate auscultation by other diagnostic methods should be understood in this context. Of course, as with patient histories, the stethoscope remains in current use, but its general importance in accurate diagnosis has greatly diminished.

77 Duffin, *To See with a Better Eye,* 43.

78 J.-B.-P. Barth and Henri Roger, *A Practical Treatise on Auscultation,* trans. Patrick Newbigging (Lexington, Ky.: Scrugham and Dunlop, 1847), 1, cited in Reiser, *Medicine and the Reign of Technology,* 31.

79 Reiser, *Medicine and the Reign of Technology,* 43–44.

80 Of course, other sounds could conceivably mislead a physician and lead to misdiagnosis, but they would not deceive in the same way that speech could since speech was ascribed a level of intentionality that other bodily sounds were not.

81 Ibid., 36.

82 Foucault, *Birth of the Clinic,* 162.

83 Laennec, *Treatise on Mediate Auscultation,* 38.

84 I discuss automata in chapter 1. See also Friedrich Kittler, *Gramophone, Film, Typewriter,* trans. Geoffrey Winthrop-Young and Michael Wutz (Stanford, Calif.: Stanford University Press, 1999), 25.

85 Laennec, *Treatise on Mediate Auscultation,* 51 (quotation), 58.

86 Duffin, *To See with a Better Eye,* 147.

87 Laennec, *Treatise on Mediate Auscultation,* 36–37.

88 Duffin, *To See with a Better Eye,* 157.

89 Ibid., 134. See also Foucault, *Birth of the Clinic,* 160; Reiser, *Medicine and the Reign of Technology,* 29.

90 Reiser, *Medicine and the Reign of Technology,* 33.

91 Duffin, *To See with a Better Eye,* 138.

92 Foucault, *Birth of the Clinic,* 165.

93 Lisa Cartwright, *Screening the Body: Tracing Medicine's Visual Culture* (Minneapolis: University of Minnesota Press, 1995), xi.

94 Laennec, *Treatise on Mediate Auscultation,* 163 (quotation), 162–63.

95 Duffin, *To See with a Better Eye,* 203–4. Duffin points out that *specificity* and *sensitivity* are somewhat anachronistic terms for Laennec's thinking, but they are at least heuristically useful for understanding how he built his semiology.

96 Flint, *Manual of Percussion and Auscultation,* 14.

97 Charles Sanders Peirce, *Philosophical Writings of Peirce,* ed. Justus Buchler (New York: Dover, 1955), 107.
98 Laennec, *Treatise on Mediate Auscultation,* 566, 569, 571–72.
99 Reiser, *Medicine and the Reign of Technology,* 28. Prior to the innovations of Josef Skoda later in the century, Laennec's teachings did not undergo significant innovation or revision.
100 The analytic language of sound remains incomplete to this day. While all sorts of aspects of visual phenomena can be described in abstract language (e.g., shape, color, texture, size), apart from the specialized technical languages of engineers, musicians, and others who work with sound (none of which has achieved the kind of general currency that abstractions of visual phenomena have in everyday language use) there exists no commonly used equivalent to describe the texture, shape, density, timbre, or rhythm of sound.
101 Ibid., 39–40; Josef Skoda, *Auscultation and Percussion,* trans. W. O. Markham (Philadelphia: Lindsay and Blakiston, 1854), passim.
102 Flint, *Manual of Percussion and Auscultation,* 34.
103 See Tom Turino's discussion of sound and indexicality in "Signs of Imagination, Identity, and Experience: A Peircian Semiotic Theory for Music," *Ethnomusicology* 43, no. 2 (1999): 221–55.
104 Flint, *Manual of Percussion and Auscultation,* 32–33.
105 Bennett, *Clinical Lectures,* 55–56.
106 Peirce, *Philosophical Writings,* 102.
107 John Forbes, translator's introduction to Laennec, *Treatise on Mediate Auscultation,* vi.
108 Bennett, *Clinical Lectures,* 52; Flint, *Manual of Percussion and Auscultation,* 31 (see also 69–70).
109 Foucault, *Birth of the Clinic,* 64–85.
110 Flint, *Manual of Percussion and Auscultation,* 33.
111 Bennett, *Clinical Lectures,* 51.
112 Davis, *Medicine and Its Technology,* 90.
113 Flint, *Manual of Percussion and Auscultation,* 69–70.
114 Forbes, translator's introduction to Laennec, *Treatise on Mediate Auscultation,* vii.
115 Davis, *Medicine and Its Technology,* 108.
116 Reiser, *Medicine and the Reign of Technology,* 38.

3. Audile Technique and Media

1 Harold Innis, *The Bias of Communication* (Toronto: University of Toronto Press, 1951), 59, 167–69; Menahem Blondheim, *News over the Wires: The*

Telegraph and the Flow of Public Information in America, 1844–1897 (Cambridge, Mass.: Harvard University Press, 1994).

2. Daniel Czitrom, *Media and the American Mind: From Morse to McLuhan* (Chapel Hill: University of North Carolina Press, 1982), 3–29; James Carey, "Technology and Ideology: The Case of the Telegraph," in *Communication as Culture: Essays on Media and Society* (Boston: Unwin Hyman, 1988), 203. See also Steven Lubar, *Infoculture* (Boston: Houghton Mifflin, 1993), 73–100.

3. Michael Warner, following Jürgen Habermas, considers the print culture of colonies and, by extension, institutions such as newspapers and the post office central to the formation of modern media culture. See Michael Warner, *The Letters of the Republic: Publication and the Public Sphere in Eighteenth-Century America* (Cambridge, Mass.: Harvard University Press, 1990); and Jürgen Habermas, *The Structural Transformation of the Public Sphere: An Inquiry into a Category of Bourgeois Society,* trans. Thomas Burger with the assistance of Frederick Lawrence (Cambridge, Mass.: MIT Press, 1991). For a history of postal communication that argues for the mail's significance in modern media history, see Richard R. John, *Spreading the News: The American Postal System from Franklin to Morse* (Cambridge, Mass.: Harvard University Press, 1995).

4. Aeschylus, *The Agamemnon,* trans. Louis MacNeice (London: Faber and Faber, 1936), 13.

5. For a fuller technical history of telegraphy, see Lazlo Solymar, *Getting the Message: A History of Communications* (New York: Oxford University Press, 1999), 7–98.

6. For a fuller critique of this position with respect to cinema, see Rick Altman, "Four and a Half Film Fallacies," in *Sound Theory Sound Practice,* ed. Rick Altman (New York: Routledge, 1992), 37–38.

7. C.M. [Charles Morrison], "An Expeditious Method of Conveying Intelligence," *Scots' Magazine* 15 (1753): 73, cited in E. A. Marland, *Early Electrical Communication* (New York: Abelard-Schuman, 1964), 18–19. Morrison, a Scottish surgeon, apparently did not want his neighbors to think that he was a wizard—hence the anonymity of the letter.

8. Norbert Wiener, *The Human Use of Human Beings: Cybernetics and Society* (New York: Doubleday, 1954), 14–15.

9. Charles Bright received an English patent in 1855 for an acoustic telegraph, which involved the use of dots and dashes similar to Morse's code, the dot ringing the bell clearly and the dash giving a longer, more muffled sound. This telegraph was an innovation on Cook and Wheatstone's needle telegraphs. Although the mechanical principle was the same, the code was different. For an account of technical developments in telegraphy between

C.M.'s letter and early sounders, see Marland, *Early Electrical Communication*, 19–116. For a discussion of Bright's acoustic telegraph, see ibid., 118–20.
10 Samuel F. B. Morse to the secretary of the Treasury, 27 September 1837, quoted in Alfred Vail, *The American Electric Magnetic Telegraph: With the Reports of Congress and a Description of All Telegraphs Known, Employing Electricity or Galvanism* (Philadelphia: Lea and Blanchard, 1845), 70. As with all first-person accounts of invention, we should take Morse's dates and realizations with a grain of salt. He was likely trying to establish precedence over Cook and Wheatstone even at this early point.
11 From a longer description of the telegraph's function in Vail, *Electric Magnetic Telegraph*, 25–26.
12 See Friedrich Kittler, *Gramophone, Film, Typewriter,* trans. Geoffrey Winthrop-Young and Michael Wutz (Stanford, Calif.: Stanford University Press, 1999), 1–20.
13 Lubar, *Infoculture*, 81–82.
14 Ibid., 79; Lewis Coe, *The Telegraph: A History of Morse's Inventions and Its Predecessors in the United States* (Jefferson, N.C.: McFarland, 1993), 67–69. Coe argues that, although Morse code fell out of international favor, it remained the faster and more efficient of the two codes.
15 See Czitrom, *Media and the American Mind*, 5–7.
16 Ibid., 7.
17 Ibid., 5.
18 Wireless telegraphy also initially made use of the printer, including Marconi's original apparatus and several of its predecessors, probably for the same reasons that Morse found them useful to begin with: they automatically generated a permanent record. On the technological evolution of wireless telegraphy, see Lubar, *Infoculture,* 102–11. See also Hugh G. J. Aitken, *The Continuous Wave: Technology and American Radio, 1900–1932* (Princeton, N.J.: Princeton University Press, 1985).
19 John Wilson Townsend, *The Life of James Francis Leonard, the First Practical Sound-Reader of the Morse Alphabet* (Louisville: John P. Morton, 1909), 18–20. Townsend also makes a brief reference to a Cincinnati operator, George Durfee, who listened to the telegraph.
20 Ibid., 21, 24–25, 34–37.
21 R. W. Russell, *History of the Invention of the Electric Telegraph, Abridged from the Works of Lawrence Turnbull, M.D., and Edward Highton, C.E. with Remarks on Royal E. House's American Printing Telegraph and the Claims of Samuel F. B. Morse as an Inventor* (New York: Wm. C. Bryant, 1853), 44–45.
22 *American Telegraph Magazine,* 31 January 1853, quoted in Russell, *History of the Invention of the Electric Telegraph,* 54.

23 Robert Sabine, C.E., *The History and Progress of the Electric Telegraph with Description of Some of the Apparatus,* 2d ed. (New York: D. Van Nostrund, 1869), 58–59. See also telegraphy guides such as R. S. Culley, *A Handbook of Practical Telegraphy,* 8th ed. (London: Longmans, Green, 1885), 250, which treats the sounder as standard operating apparatus.

24 Marshall McLuhan, *Understanding Media: The Extensions of Man* (New York: McGraw-Hill, 1964), 22–32, 254–57.

25 Carolyn Marvin, *When Old Technologies Were New: Thinking about Electric Communication in the Late Nineteenth Century* (New York: Oxford University Press, 1988), 93–94.

26 See Erving Goffman, *Presentation of Self in Everyday Life* (Garden City, N.Y.: Doubleday Anchor, 1959), 106–40; Edward Hall, *The Hidden Dimension* (London: Bodley Head, 1966); and Anthony Giddens, *The Constitution of Society: Outline of the Theory of Structuration* (Berkeley and Los Angeles: University of California Press, 1984), 122–26.

27 Giddens, *The Constitution of Society,* 123; John Thompson, *The Media and Modernity: A Social Theory of the Media* (Stanford, Calif.: Stanford University Press, 1995), 87–100.

28 Marvin, *When Old Technologies Were New,* 9–62, 94–95.

29 "Kate: An Electro-Mechanical Romance," in *Lightning Flashes and Electric Dashes: A Volume of Choice Telegraphic Literature, Humor, Fun, Wit, and Wisdom, Contributed to by All of Principal Writers in the Ranks of Telegraphic Literature as Well as Some Well-Known Outsiders, with Numerous Wood-Cut Illustrations* (New York: W. J. Johnston, 1877), 56–57.

30 Marvin, *When Old Technologies Were New,* 84.

31 Ibid., 10.

32 Charles L. Buckingham, e.g., was writing by 1890 that modern life would be unthinkable without telegraphy. See Charles L. Buckingham, "The Telegraph of Today," in *The Telegraph: An Historical Anthology,* ed. George Shiers (New York: Arno, 1977), 163.

33 It should, however, be noted that the geographic development of telephony, phonography, and radio was uneven. All these media started out as local phenomena and spread only gradually.

34 As with mediate auscultation and sound telegraphy, some sounds made possible by the technology were considered "interior" and, therefore, proper objects of attention. Other sounds of the technology, especially those that drew attention to the process of mediation, were coded as "exterior" sounds and were supposed to be ignored. Through this coding of sounds, the process of sonic mediation, so central to audile technique, was effectively erased. This is considered at length in chapter 5.

35 Audrey B. Davis, *Medicine and Its Technology: An Introduction to the History of Medical Instrumentation* (Westport, Conn.: Greenwood, 1981), 107–8; Stanley Joel Reiser, *Medicine and the Reign of Technology* (Cambridge: Cambridge University Press, 1978), 41.

36 George L. Carrick, "On the Differential Stethoscope and Its Value in the Diagnosis of Diseases of the Lungs and Heart," *Aberdeen Medical and Chirurgical Tracts* 12, no. 9 (1873): 902.

37 Alexander Graham Bell, "Experiments in Binaural Audition," *American Journal of Otology* (July 1880): 3, 4, 5, NMAH, Division of Mechanisms, phonograph box 1 (this is an offprint signed "with compliments of the author"). See also Audrey B. Davis and Uta C. Merzbach, *Early Auditory Studies: Activities in the Psychology Laboratories of American Universities* (Washington, D.C.: Smithsonian Institution Press, 1975).

38 Bell, "Experiments in Binaural Audition," 5.

39 Relatively few writers have made much of Horkheimer and Adorno's discussion of the detail. Instead, discussions of Frankfurt school writers have considered the idea of the *fragment*—either at the level of the text itself of *Dialectic of Enlightenment* (since the original title and the German subtitle was *Philosophical Fragments*) or in Walter Benjamin's more mystical version of the fragment (see, e.g., Walter Benjamin, *Illuminations,* trans. Hannah Arendt [New York: Schocken, 1968], 262–64).

40 See Max Horkheimer and Theodor W. Adorno, *Dialectic of Enlightenment* (New York: Continuum, 1944), 125.

41 American Telephone Booth Co., "Booths of Special Design Made to Match Office Furniture" (1893), NWA, telephone box 1, folder 2.

42 Ultimately, it is the social form that counts, not the actual construction of a booth: the proliferation of cellular phones in recent years has demonstrated that people can set up their private acoustic spaces almost anywhere and with little assistance from the ambient environment.

43 This sense of the possibility of owning communication space is also a tributary formation in the shape of broadcast licensing. See Thomas Streeter, *Selling the Air: A Critique of the Policy of Commercial Broadcasting in the United States* (Chicago: University of Chicago Press, 1996), 219–55.

44 Richard Leppert, *The Sight of Sound: Music, Representation, and the History of the Body* (Berkeley and Los Angeles: University of California Press, 1993), 15–41, esp. 39–40.

45 Leppert, ibid., 24–25. See also James H. Johnson, *Listening in Paris: A Cultural History* (Berkeley and Los Angeles: University of California Press, 1995); David Nasaw, *Going Out: The Rise and Fall of Public Amusements* (New York: Basic, 1993), 19–33, 120–34; Lizabeth Cohen, *Making a New Deal:*

Industrial Workers in Chicago, 1919–1939 (New York: Cambridge University Press, 1990), 99–158; and Miriam Hansen, *Babel and Babylon: Spectatorship in American Silent Film* (Cambridge, Mass.: Harvard University Press, 1991). This is clearly a significant issue in sound history, but, since these other authors have considered it at some length, I have chosen to focus on other matters here.

46 Of course, we also find the countervailing tendency in modern life: certain spaces and certain sounds preempt people's entitlement to private acoustic space. A particularly obvious and pervasive case would be programmed music, or Muzak—a topic that I explore in my "Sounds Like the Mall of America: Programmed Music and the Architectonics of Programmed Music," *Ethnomusicology* 41, no. 1 (winter 1997): 22–50. There are also many contexts like sporting events where unruly audience behavior is still encouraged within limits, but that is a whole other issue.

47 I realize that this is a tremendous simplification of social contract theory. Rousseau developed notions of moral development and general will, whereas earlier writers had emphasized the social contract as a defense against the "state of nature." In any case, the loose narrative presented above is purely for heuristic purposes and is not meant to suggest that audile technique is structured according to any kind of formal social contract theory.

48 Many nineteenth-century teachers of medicine also attempted to construct models that would produce sounds identical to those heard through the stethoscope, thereby offering reproductions of the sounds of the body *outside* the body. These sound-reproduction apparatus were themselves miniature automata. As Audrey Davis puts it, these devices were constructed to "duplicate the physical conditions that produced the chest sounds." As an example, she cites Flint's instructions for making models: "Murmurs may be produced by forcing fluids into rubber tubes and bags of varying size. Valvular-like sounds are produced by forcible tension of pieces of linen or muslin held below the surface of a liquid. Musical notes may be caused by a stream of liquid acting on a vibrating body in a closed cavity" (*Medicine and Its Technology*, 93).

49 Reiser, *Medicine and the Reign of Technology*, 43–44.

50 Davis, *Medicine and Its Technology*, 109.

51 MacDowell Associates, *Sound: Can It Be Put to Work? How? By Whom?* 2, WBA, acoustics box 1, folder 3. For a history of architectural acoustics, see Emily Thompson, "Mysteries of the Acoustic: Architectural Acoustics in America, 1800–1932" (Ph.D. diss., University of Pennsylvania, 1992).

52 On the commodification of music, see Alan Durant, *Conditions of Music* (Albany: State University of New York Press, 1984); Evan Eisenberg, *The Recording Angel: The Experience of Music from Aristotle to Zappa* (New York:

Penguin, 1987), 11–34; and Michael Chanan, *Musica Practica: The Social Practice of Music from Gregorian Chant to Postmodernism* (New York: Verso, 1994), and *Repeated Takes: A Short History of Recording and Its Effects on Music* (New York: Verso, 1995).

53 William Howland Kenney, *Recorded Music in American Life: The Phonograph and Popular Memory, 1890–1945* (New York: Oxford University Press, 1999), 4.
54 Ibid. This aspect of recording is a fundamental condition for large-scale musical-cultural phenomena. See, e.g., Gilbert Rodman, *Elvis after Elvis: The Posthumous Career of a Living Legend* (New York: Routledge, 1996).
55 Johnson, *Listening in Paris,* 284.
56 "Your Telephone Horizon" (1912), NWA, box 21, folder 1.
57 Norbert Elias, *The Civilizing Process: Sociogenetic and Psychogenetic Investigations,* trans. Edmund Jephcott, ed. Eric Dunning, Johan Goudsblom, and Stephen Mennell, rev. ed. (Malden: Blackwell, 2000), 118–19.
58 Kenney, *Recorded Music in American Life,* 7. See also Claude S. Fischer, *America Calling: A Social History of the Telephone* (Berkeley and Los Angeles: University of California Press, 1992), 5; and Susan Douglas, *Listening In: Radio and the American Imagination from Amos 'n Andy and Edward R. Murrow to Wolfman Jack and Howard Stern* (New York: Times Books/Random House, 1999), 7–8, 21.

4. Plastic Aurality: Technologies into Media

1 Andre Millard, *America on Record: A History of Recorded Sound* (New York: Cambridge University Press, 1995), 30; Russel Sanjek, *American Popular Music and Its Business: The First Four Hundred Years,* vol. 2, *From 1790 to 1909* (New York: Oxford University Press, 1988), 364; Steven Lubar, *Infoculture* (New York: Houghton Mifflin, 1993), 170. The important issue for the present account is Edison's sense of ownership, although, interestingly, the historical record on Edison's reaction to the Johnson/Tainter meeting is somewhat unclear. Oliver Read and Walter Welch (*From Tin Foil to Stereo: Evolution of the Phonograph* [New York: Herbert W. Sams/Bobbs-Merrill, 1976], 38) cite an 1888 *Electrical World* article claiming that Edison was ill and unable to attend the meeting; they also suggest that Volta made no meaningful contributions to sound recording other than a variable-speed turntable. Andre Millard (*Edison and the Business of Innovation* [Baltimore: Johns Hopkins University Press, 1990], 65–66) cites an undated note to Johnson and suggests that it contains evidence that Edison was simply angry that someone else had improved the phonograph and that he planned to develop the machine himself (although, since Millard does not actually quote that document, its

exact contents remain a mystery). Clearly, the two accounts are not entirely compatible since, if the Volta work was of no major consequence, Edison would not have been upset or concerned.

2 Edward Johnson to A. G. Bell, 11 September 1885, AGB, box 255, folder "Phonograph—Correspondence." Judging by the date on the letter, Johnson was essentially covering for himself and his boss. He could from then on argue that his impression at the meeting was that the graphophone simply was not "good enough" and that the new Edison machine grew from a separate line of research and development.

3 R. W. Russell, *History of the Invention of the Electric Telegraph, Abridged from the Works of Lawrence Turnbull, M.D., and Edward Highton, C.E. with Remarks on Royal E. House's American Printing Telegraph and the Claims of Samuel F. B. Morse as an Inventor* (New York: Wm. C. Bryant, 1853), 54–55.

4 Alexander Graham Bell to Melville Bell, 26 February 1880, AGB, box 256, folder "Photophone, Miscellany."

5 Of course, a literary notion of authorship is even more intensely articulated to the concept of the lone creative individual than is invention. But it is equally a mystification of the creative process to conceive of authors "birthing" their writings.

6 Inventors were far from the only people who described their work in this way. Felicia Frank translates the title of the second chapter of Villers de l'Isle-Anam's novel *L'Eve future* (which features Edison prominently as a character) as "Phonograph's Papa." See Felicia Frank, *The Mechanical Song: Women, Voice, and the Artificial in Nineteenth-Century French Narrative* (Stanford, Calif.: Stanford University Press, 1995), 144. See also Thomas A. Watson, *The Birth and Babyhood of the Telephone* (reprint, n.p.: American Telephone and Telegraph Co., Information Department, 1934).

7 For criticisms of technological determinism from perspectives sympathetic to my own, see Raymond Williams, *Television: Technology and Cultural Form* (Middletown, Conn.: Wesleyan University Press, 1992); Jennifer Daryl Slack, *Communication Technologies and Society: Conceptions of Causality and the Politics of Technological Intervention* (Norwood, N.J.: Ablex, 1984); and Carol Stabile, *Feminism and the Technological Fix* (New York: St. Martin's, 1994).

8 This definition is adapted from John Nerone's approach to the question of what a newspaper is in *Violence against the Press: Policing the Public Sphere in U.S. History* (New York: Oxford University Press, 1994), 13–17.

9 Jonathan Sterne, "Television under Construction: American Television and the Problem of Distribution, 1926–62," *Media, Culture, and Society* 21, no. 4 (July 1999): 504.

10 Lawrence Grossberg, *We Gotta Get Out of This Place: Popular Conservatism and Postmodern Culture* (New York: Routledge, 1992), 54. See also Stuart Hall,

"On Postmodernism and Articulation," *Journal of Communication Inquiry* 10, no. 2 (1986): 45–60.

11 In addition to the extant political-economic histories of individual sound media, Jacques Attali's *Noise: The Political Economy of Music* (trans. Brian Massumi [Minneapolis: University of Minnesota Press, 1985]), Michael Chanan's *Musica Practica: The Social Practice of Music from Gregorian Chant to Postmodernism* (New York: Verso, 1994) and *Repeated Takes: A Short History of Recording and Its Effects on Music* (New York: Verso, 1995), Sidney Finkelstein's *Composer and Nation: The Folk Heritage in Music* (New York: International, 1989), Max Horkheimer and Theodor W. Adorno's *Dialectic of Enlightenment* (New York: Continuum, 1944), and much of Adorno's writing on music as well as Walter Benjamin's "Work of Art in the Age of Mechanical Reproduction" (in *Illuminations,* trans. Hannah Arendt [New York: Schocken, 1968]) and Alan Durant's *Conditions of Music* (Albany: State University of New York Press, 1984) all make impressive but incomplete attempts to connect the history of music with a more general political economy. A truly political-economic history of music and sound has yet to be written.

12 James Carey, *Communication as Culture: Essays on Media and Society* (Boston: Unwin Hyman, 1988), 201–30. See also Daniel Czitrom, *Media and the American Mind: From Morse to McLuhan* (Chapel Hill: University of North Carolina Press, 1982), 3–29; and Lubar, *Infoculture,* 73–100.

13 Other institutional and intellectual contexts for the technical development of sound technologies, such as research into deafness and hearing, are covered in earlier chapters.

14 Both electrical engineering and research and development would become more professionalized in the following decades.

15 Read and Welch, *From Tin Foil to Stereo,* 3–5, 11–24; Robert V. Bruce, *Bell: Alexander Graham Bell and the Conquest of Solitude* (Boston: Little, Brown, 1973), 90–97, 104–20, 215–36; Millard, *America on Record,* 17–18.

16 Millard, *America on Record,* 23.

17 See Millard, *Edison and the Business of Innovation.*

18 Sumner Tainter and Chichester Bell to Alexander Graham Bell, 29 November 1881, AGB, box 25, folder "Bell, Chichester, Alexander."

19 Chichester A. Bell to Alexander Graham Bell, 22 December 1886, ibid.

20 Alexander Graham Bell, entry dated 31 May 1881, typescript copied from Home Notes, 3:55–61, AGB, box 256, folder "Phonograph, Miscellany."

21 Bruce, *Bell,* 82–83, 293, 340. Bell and Hubbard originally met because Hubbard's daughter Mabel was deaf and he hired Bell to teach her to speak. Teacher and pupil would later marry.

22 Read and Welch, *From Tin Foil to Stereo*, 39.

23 George Crossette, "Chichester Alexander Bell," *Cosmos Club Bulletin*, May 1966, 3, AGB, box 25, folder "Chichester Alexander Bell."

24 Read and Welch, *From Tin Foil to Stereo*, 25; Thomas Edison to Alexander Graham Bell, n.d. [1879], AGB, box 122, folder "Thomas Edison."

25 Bruce, *Bell*, 281–87; Read and Welch, *From Tin Foil to Stereo*, 273

26 See the discussion of the "cult of invention" during the first two decades of the twentieth century (with respect to radio) in Susan Douglas, *Inventing American Broadcasting, 1899–1922* (Baltimore: Johns Hopkins University Press, 1987), xiv.

27 Frederic William Wile, *Emile Berliner: Maker of the Microphone* (Indianapolis: Bobbs-Merrill, 1926), 67–94, 183–93.

28 Charles Sumner Tainter and Chichester A. Bell to Alexander Graham Bell, 29 November 1881, AGB, box 25, folder "Bell, Chichester, Alexander."

29 Hugh G. J. Aitken, *Syntony and Spark: The Origins of Radio* (New York: Wiley, 1976), 159. Vacuum tubes are also central to understanding the aesthetics of recorded and amplified live music over the course of the twentieth century; this is a topic that I hope to pursue elsewhere.

30 Emily Thompson, "Mysteries of the Acoustic: Architectural Acoustics in America, 1800–1932" (Ph.D. diss., University of Pennsylvania, 1992), 267–76; Susan Smulyan, *Selling Radio: The Commercialization of American Broadcasting, 1920–1934* (Washington, D.C.: Smithsonian Institution Press, 1994), 37–64.

31 On the history of Muzak, see Simon Jones and Thomas Schumacher, "Muzak: On Functional Music and Power," *Critical Studies in Mass Communication* 9 (1992): 156–63; Joseph Lanza, *Elevator Music: A Surreal History of Muzak, Easy-Listening, and Other Moodsong* (New York: St. Martin's, 1994), 22–31. I discuss programmed music in a more contemporary context in my "Sounds Like the Mall of America: Programmed Music and the Architectonics of Commercial Space," *Ethnomusicology* 41, no. 1 (winter 1997): 22–50.

32 Winslow A. Duerr, "Will Radio Replace the Phonograph?" *Radio Broadcast*, November 1922, 52–54.

33 James P. Kraft, *Stage to Studio: Musicians and the Sound Revolution, 1890–1950* (Baltimore: Johns Hopkins University Press, 1996), esp. 59–88, 162–93.

34 Kraft, *Stage to Studio*, 66; Sanjek, *American Popular Music and Its Business*, 392–420.

35 Kraft, *Stage to Studio*, 47–58, 61, 64–66.

36 David Nasaw, *Going Out: The Rise and Fall of Public Amusements* (New York: Basic, 1993), 241–42.

37 Arnold Pacey, *The Culture of Technology* (Cambridge, Mass.: MIT Press, 1983).

38 Richard Ohmann, *Selling Culture: Magazines, Markets, and Class at the Turn of the Century* (New York: Verso, 1996), 171; Stuart Blumin, *The Emergence of the Middle Class: Social Experience in the American City, 1760–1900* (New York: Cambridge University Press, 1989), 11 and passim.

39 Reynold Weidenaar, *Magic Music from the Telharmonium* (Metuchen, N.J.: Scarecrow, 1995), 3–4, 17. See also Elliott Sivowitch, "Musical Broadcasting in the Nineteenth Century," *Audio* 51, no. 6 (June 1967): 19–23. On telephone broadcasting in Munich, see Margarete Rhem, "Information und Kommunikation in Geschichte und Gegenwart" (available on-line at http://www.ib.hu-berlin.de/~wumsta/rehm8.html; last accessed 18 January 2001).

40 See Carolyn Marvin, *When Old Technologies Were New: Thinking about Electric Communication in the Nineteenth Century* (New York: Oxford University Press, 1988), 223–28; Weidenaar, *Magic Music,* 16–17; and Sivowitch, "Musical Broadcasting in the Nineteenth Century."

41 Marvin, *When Old Technologies Were New,* 230–31. See also Michèle Martin, *"Hello, Central?": Gender, Technology, and Culture in the Formation of Telephone Systems* (Montreal: McGill-Queen's University Press, 1991), 14–27.

42 Chichester Alexander Bell, entry dated 27 February 1882, Home Notes, bk. 2, p. 14, AGB, box 25, folder "Bell, C. A. Scientific Experiments 1881–4," 2d folder.

43 "Phono Chat," *Phonogram I* 2, no. 3 (March 1892): 86. See also "Hears Sister's Voice on Phonograph," *Phonogram II* 1, no. 6 (October 1900): 178; and "Letters by Phonography," *Phonogram II* 1, no. 6 (October 1900): 181.

44 Alexander Graham Bell, Laboratory Notes, vol. 4, pp. 55–61, AGB, box 256, folder "Phonograph—Miscellany."

45 Alexander Graham Bell to Chichester A. Bell and Charles Sumner Tainter, 14 June 1885, AGB, box 256, folder "Charles Sumner Tainter."

46 Herbert Berliner, untitled speech (n.d.), EB, scrapbook "Phonograph, Graphophone, Gramophone: Historical Accounts."

47 George H. Clark to Lee DeForest, 1940, GHC, series 135, box 532. Clark spent a good deal of time working on a definition of broadcasting and finally decided that the for-profit model of RCA was the ideal-type of broadcasting: a large, centralized source broadcasting to individuals or small groups of listeners in their homes and selling that audience to advertisers for profit. One could easily object that this definition's triumphalism renders it useless for historical purposes, and that would be true if one sought the "origin" of broadcasting as such. My purpose here, however, is simply to contrast the commercial form that radio eventually took with earlier forms of radio

communication, including antecedents of the current "American-style" capitalist broadcasting system.

48 Susan Douglas provides the richest account of this period in *Inventing American Broadcasting;* Susan Smulyan's *Selling Radio* picks up where Douglas leaves off, in the early 1920s; and Robert McChesney's *Telecommunications, Mass Media, and Democracy: The Battle for Control of U.S. Broadcasting, 1928–1935* (New York: Oxford University Press, 1994) provides a history of several broadcast reform movements in the United States and a useful critique of triumphalist accounts of American-style capitalist broadcasting.

49 J. Andrew White, "Report on the Broadcasting of the Dempsey-Carpentier Fight by RCA" (typescript apparently copied by George H. Clark from the original report), GHC, series 135, box 532.

50 Ibid., 8. Additionally, KDKA of Pittsburgh planned to simulcast the fight to local theaters and to Forbes Field. In fact, it re-created the fight on the basis of telegraph transmissions from New Jersey. See Thomas White, "'Battle of the Century': The WJY Story" (available on-line at http://www.ipass.net/~whitetho/wjy.htm; last accessed 18 January 2001).

51 Marvin, *When Old Technologies Were New,* 230–31.

52 Kraft, *Stage to Studio,* 68–69. The alternative would have been to construct a state-based system of radio broadcasting and implement a tax on the ownership of radio sets to finance such broadcasting. Although this system eventually took hold in the United Kingdom and elsewhere, a variety of factors helped undermine the establishment of a similar system in the United States. See McChesney, *Telecommunications, Mass Media, and Democracy,* 100–101, 166–67.

53 Pierce B. Collison, "Shall We Have Music or Noise?" *Radio Broadcast,* September 1922, 434–36; Joseph H. Jackson, "Should Radio Be Used for Advertising?" *Radio Broadcast,* November 1922, 72–77.

54 Marvin, *When Old Technologies Were New,* 63–108.

55 Claude S. Fischer, *America Calling: A Social History of the Telephone to 1940* (Berkeley and Los Angeles: University of California Press, 1992), 40, 58–59, 78–79; Martin, "Hello, Central?" 140–67.

56 "Indeed, it was women's extensive use of the telephone that eventually forced the industry to change some of its plans" (Martin, "Hello, Central?" 5).

57 Fischer, *America Calling,* 231.

58 Martin, "Hello, Central?" 148, 150.

59 Lana F. Rakow, *Gender on the Line: Women, the Telephone, and Community Life* (Urbana: University of Illinois Press, 1992), passim.

60 Marvin, *When Old Technologies Were New,* 102–8; John Brooks, *Telephone: The First Hundred Years* (New York: Harper and Row, 1976), 83–85.

61 Fischer, *America Calling,* 256.

62 McChesney, *Telecommunications, Mass Media, and Democracy*, 18, 25–28.
63 Raymond Williams's often-cited phrase, *mobile privatization,* would also be apropos here. Williams (*Television,* 20) references the increasing dependence on transportation and communication technologies among the emergent consumerist middle class. Clearly, this marks a change in the geography of public life.
64 "An Important Suggestion," *Phonogram I* 1, no. 1 (January 1891): 6.
65 Columbia Phonograph Co., *List of Users of the Phonograph and Phonograph-Graphophone in the District of Columbia, Maryland, and Delaware* (Washington, D.C.: Terry Bros., March 1891). Note that counts cannot be exact since phonographs in residences may be used for both work and leisure activities.
66 Fischer, *America Calling,* 40; Read and Welch, *From Tin Foil to Stereo,* 39; Millard, *America on Record,* 40–41.
67 Read and Welch, *From Tin Foil to Stereo,* 40–41.
68 Fischer, *America Calling,* 41.
69 Unfortunately, I know of no other existing phonograph directories from this period for comparison. On the uses of the phonograph by the federal government, see V. H. McRae, "The Present Position of the Phonograph and a Résumé of Its Merits," *Phonogram I* 1, no. 4 (April 1891): 83–84; and the collection of letters making the same point as McRae, *Phonogram I* 1, no. 4 (April 1891): 85–86.
70 See Nasaw, *Going Out,* 120–34.
71 Read and Welch, *From Tin Foil to Stereo,* 269.
72 Paraphrasing Edison as quoted in Roland Gelatt, *The Fabulous Phonograph, 1877–1977* (New York: Appleton-Century, 1977), 29; Attali, *Noise,* 93; and Chanan, *Repeated Takes,* 4.
73 "*The Phonogram* . . . has suggested to the manufacturers that albums be constructed, varying in size to suit purchasers, so that they may hold two, four, six, eight or even a hundred cylinders, and that these be prepared artistically, to resemble, as much as possible, in form, a photograph album, yet possessing the conveniences for holding the wax phonograms and keeping them intact" ("'Being Dead, He Yet Speaketh,'" *Phonogram I* 2, no. 11 [November 1892]: 249).
74 Cylinders, too, could be mass produced (at least in theory), but no such scheme caught on between 1888 and 1895, when the gramophone was first being marketed.
75 Ohmann, *Selling Culture,* 140–49.
76 Emile Berliner, "The Gramophone: Etching the Human Voice," *Journal of the Franklin Institute* 75, no. 6 (June 1888): 445–46.
77 Nasaw, *Going Out.*
78 Jacques Derrida, Briankle Chang, and John Peters all argue in their own

way that this potential for dissemination is, in fact, the defining characteristic of all communication. While this may be the case, modern sound culture explicitly "problematized" (i.e., made a theoretical and practical issue of) both the sound event itself and the conditions under which it could become mobile. Dissemination became an explicitly social, economic, and cultural problem. See chapter 1 above; Jacques Derrida, *The Postcard: From Socrates to Freud and Beyond,* trans. Alan Bass (Chicago: University of Chicago Press, 1987); Briankle Chang, *Deconstructing Communication: Representation, Subject, and Economies of Discourse* (Minneapolis: University of Minnesota Press, 1996), esp. 171–221; and John Durham Peters, *Speaking into the Air: A History of the Idea of Communication* (Chicago: University of Chicago Press, 1999), 33–62.

79 Michael Warner, "The Mass Public and the Mass Subject," in *Habermas and the Public Sphere,* ed. Craig Calhoun (Cambridge, Mass.: MIT Press, 1994), 387–91; Carey, *Communication as Culture,* 1–36. Interestingly, Warner's analysis can be read as a psychological restatement of the often-cited refeudalization thesis in Jürgen Habermas's *Structural Transformation of the Public Sphere: An Inquiry into a Category of Bourgeois Society* (trans. Thomas Burger [Cambridge, Mass.: MIT Press, 1989]).

80 The United States Gramophone Co., "E. Berliner's Gramophone: Directions for Users of the Seven-Inch American Hand Machine," EBM, folder "Printed Matter."

81 *Edison Phonograph Monthly* 1, no. 1 (March 1903): 1.

82 *The 1897 Sears Roebuck Catalogue,* ed. Fred L. Israel, with introductions by S. J. Perelman and Richard Rovere (New York: Chelsea House, 1968), 485.

83 *The 1902 Edition of the Sears Roebuck Catalogue,* with an introduction by Cleveland Amory (New York: Bounty, 1969), 156–64. In its copious and tiny copy, the catalog does not once mention the business uses of sound recording. Given that it offered other kinds of business supplies, the decision to market the graphophone primarily as an entertainment device was clearly a deliberate choice.

84 *Appointment by Telephone,* Library of Congress, Division of Motion Pictures, Broadcasting, and Recorded Sound, Paper Prints Collection, Edison 1902, LC 1460/FLA 4474.

85 Martin, *"Hello, Central?"* 146.

86 That said, residential telephone subscription was still relatively confined to the middle class and the rich. Telephones would become affordable to working people only after World War II. See Fischer, *America Calling,* 236–42.

87 Douglas, *Inventing American Broadcasting,* passim; Smulyan, *Selling Radio,*

37–64. For a discussion of listening for distance, or DXing, see the beginning of chapter 2.

88 The popularity of "race" and "ethnic" records suggests that a similar kind of cosmopolitanism was at work in phonographic listening practices during the same period. On race records, see William Barlow, "The Music Industry: Cashing In: 1900–1939," in *Split Image: African Americans in the Mass Media,* ed. Jannette L. Dates and William Barlow (Washington, D.C.: Howard University Press, 1990), 25–56.

89 AT&T, "Multiplying Man-Power," NWA, series 2, box 21, folder 3.

90 The Central District and Printing Telegraph Co., "The Telephone as Employe [*sic*]," NWA, box 21, folder 1.

91 AT&T, "Marshaling the Telephone Forces," NWA, series 2, box 21, folder 3.

92 AT&T, "Highways of Speech," NWA, series 2, box 21, folder 3, "The Clear Track," NWA, series 2, box 21, folder 3, and "A Highway of Communication," NWA, series 2, box 21, folder 1.

93 AT&T, "The Center of Population: A Title That Fits Every Bell Telephone," WBA, telephone box, folder "AT&T #3."

94 AT&T, "The Implement of the Nation," NWA, series 2, box 21, folder 1, "A United Nation," NWA, series 2, box 21, folder 1, and "The U.S. Is Only a Few Minutes Wide," NWA, series 2, box 21, folder 3.

95 Two short films, both entitled *The Telephone,* Library of Congress, Division of Motion Pictures, Broadcasting, and Recorded Sound, Paper Prints Collection, Edison 1898m, LC 1075/FLA 4089 and LC 1076/FLA 4090.

96 "Here Is a Thrifty Habit for You," ad for the Central District and Printing Telegraph Co., NWA, series 2, box 21, folder 1. The ad reads in part: "You can keep in personal contact with more people; you can be active in more affairs; you can make more money. Cultivate the telephone habit. It develops quick thinking and decisive action. It gives you mastery of self and surroundings. It is a liberal education."

97 "The Imaginary and the Real Phonograph," *Phonogram I* 2, no. 10 (October 1892): 205.

98 "Facts about the Phonograph When You Want a Stenographer," *Phonogram I* 3, no. 2 (February 1893): viii.

99 Lubar, *Infoculture,* 171.

100 *The Stenographer's Friend; or, What Was Accomplished by an Edison Business Phonograph* (a reissue of a 1904 film), Library of Congress, Division of Motion Pictures, Broadcasting, and Recorded Sound, Edison 1910.

101 Lizabeth Cohen, *Making a New Deal: Industrial Workers in Chicago, 1919–1939* (New York: Cambridge University Press, 1990), 99–158. See also "The Phonograph Uniting the Nation," *Phonogram I* 3, nos. 3–4 (March and April 1893): 351; and Attali, *Noise,* 92.

5. The Social Genesis of Sound Fidelity

1. Victor Talking Machine Co., "Which Is Which?" WBA, phonographs box 1, folder 33, "Victor Talking Machine Co."
2. Victor Talking Machine Co., "Both Are Caruso," *Ladies Home Journal,* October 1913, 64.
3. Hillel Schwartz, *Culture of the Copy: Striking Likenesses, Unreasonable Facsimiles* (New York: Zone, 1996).
4. These are the two most common instances of the comparison of live and reproduced sound, although their frequency should not lead us to assume that they are transparently and transportably universal examples of "live" sound. Other examples could yield very different results.
5. I'm borrowing the phrase *vanishing mediator* from Slavoj Žižek (who borrows it from Fredric Jameson), but to argue against a philosophy of mediation (whereas Žižek uses the term to extend a philosophy of mediation). For Žižek, a vanishing mediator is the fourth movement in Hegel's dialectic, where the mediating term disappears, leaving its structure and effect but not its form. His reading of Weber's account of the capitalist ethos posits Protestantism as the vanishing mediator since Weber argues that, by the twentieth century, capitalism no longer needed Protestantism for an ethos. While this proves a useful argument for Žižek's purposes, I am using the term to slightly different ends. When technologies of reproduction are idealized as vanishing mediators, they are alternately fetishized as historical agents in and of themselves and instrumentalized as merely a means to an end. In fact, they do not mediate between originals and copies at all. For Žižek's discussion, see his *For They Know Not What They Do: Enjoyment as a Political Factor* (New York: Verso, 1991), 179–88. For Jameson's original use, see his "The Vanishing Mediator; or, Max Weber as Storyteller," in *Ideologies of Theory* (Minneapolis: University of Minnesota Press, 1988), 2:3–34.
6. James Lastra, *Sound Technology and American Cinema: Perception, Representation, Modernity* (New York: Columbia University Press, 2000), 123–53.
7. Ibid.
8. Eric W. Rothenbuhler and John Durham Peters, "Defining Phonography: An Experiment in Theory," *Musical Quarterly* 81, no. 2 (summer 1997): 246 (first quotation), 252 (second quotation), 260.
9. The argument is fully developed in Rick Altman, "The Material Heterogeneity of Recorded Sound," in *Sound Theory, Sound Practice,* ed. Rick Altman (New York: Routledge, 1992), 15–31.
10. Raymond Williams, "Base and Superstructure in Marxist Cultural Theory," in *Problems in Materialism and Culture* (London: Verso, 1980), 47.

11 Walter Benjamin, "The Work of Art in the Age of Mechanical Reproduction," in *Illuminations,* trans. Harry Zohn (New York: Schocken, 1968), 221.
12 Gilles Deleuze, *The Logic of Sense,* trans. Mark Lester (New York: Columbia University Press, 1990), 259.
13 Benjamin, "The Work of Art," 243, 233, 232.
14 See, e.g., Kristin Thompson's well-known work on early cinema as a studio art in David Bordwell, Janet Staiger, and Kristin Thompson, *The Classical Hollywood Cinema: Film Style and Mode of Production to 1960* (New York: Columbia University Press, 1985), 155–240.
15 Sarah Thornton, *Club Cultures: Music, Media, and Subcultural Capital* (Hanover, N.H.: Wesleyan University Press, 1996), 42–43.
16 Lastra, *Sound Technology and American Cinema,* 131.
17 Deleuze, *The Logic of Sense,* 257, 262.
18 Barry Truax, *Acoustic Communication* (Norwood, N.J.: Ablex, 1984), 8–9.
19 For a discussion of Western Electric's development of electric recording, see Andre Millard, *American on Record: A History of Recorded Sound* (New York: Cambridge University Press, 1995), 140–44. For an account of the development of electroacoustic measurements, see Emily Thompson, "Mysteries of the Acoustic: Architectural Acoustics in America, 1800–1932" (Ph.D. diss., University of Pennsylvania, 1992), 265–76.
20 "The Human Voice *Is* Human on the New Orthophonic Victrola," advertisement proof, NWA, series I, box 294, folder 1.
21 This is Latour and Woolgar's often-cited point that representations of processes are attempts to delimit and order them. While Latour and Woolgar's analysis was based on laboratory notes (as is some of mine), it can also be generalized to descriptions of mechanical processes outside strictly "scientific" contexts. See Bruno Latour and Steve Woolgar, *Laboratory Life: The Construction of Scientific Facts* (Princeton, N.J.: Princeton University Press, 1986), 45–53, 244–52.
22 Andreas Huyssen, "Mass Culture as Woman: Modernism's Other," *Studies in Entertainment,* ed. Tania Modleski (Bloomington: Indiana University Press, 1986), 188–207. See also Lynn Spigel, *Make Room for TV: Television and the Family Ideal in Postwar America* (Chicago: University of Chicago Press, 1992), 11–35; Richard Leppert, *The Sight of Sound: Music, Representation, and the History of the Body* (Berkeley and Los Angeles: University of California Press, 1993); David Nasaw, *Going Out: The Rise and Fall of Public Amusements* (New York: Basic, 1993); and Richard Ohmann, *Selling Culture: Magazines, Markets, and Class at the Turn of the Century* (New York: Verso, 1996). On women and the telephone, see Michèle Martin, *"Hello, Central?": Gender, Technology, and Culture in the Formation of Telephone Systems* (Montreal:

McGill-Queen's University Press, 1991); and Lana F. Rakow, *Gender on the Line: Women, the Telephone, and Community Life* (Urbana: University of Illinois Press, 1992).

23 Elisha Gray's patent caveats for the telephone employ similar artistic conventions.

24 For a discussion of the ear phonautograph and the tympanic in general, see chapter 1.

25 Bugs and other devices for recording sonic events that were not specifically being produced for reproduction were developed over the course of the 1920s and 1930s. A cultural history of electronic eavesdropping has yet to be written, although Clinton Heylin's *Bootleg: The Secret History of the Other Recording Industry* (New York: St. Martin's, 1995) offers some interesting tidbits on the early history of secret recordings.

26 My argument here echoes that of Jürgen Habermas in *Structural Transformation of the Public Sphere* (Cambridge, Mass.: MIT Press, 1989), 31–43, 159–74. In order for a cultural practice to cohere and operate effectively, it needs spaces. Thus, for Habermas, the shift from salons to suburbs creates a physical crisis in the possibility of a public sphere. I am arguing here that the spatiality of reproduction is also essential to the very existence of reproducibility itself, not in some abstract sense, but in the real existence of phone lines and booths, recording and radio studios, and parlors and homes for listeners.

27 Chichester Alexander Bell, entry dated 1 March 1882, Home Notes, bk. 2, p. 15, AGB, box 25, folder "Bell, C. A. Scientific Experiments 1881–4," 2d folder.

28 Eldridge Johnson, quoted in *Talking Machine World,* September 1910, 47.

29 Steven Jones, "A Sense of Space: Virtual Reality, Authenticity, and the Aural," *Critical Studies in Mass Communication* 10 (1993): 238–52.

30 C. E. Le Massena, "How Opera Is Broadcasted: Difficulties That Must Be Overcome in Order to Obtain the Best Results; How Singers Must Be Especially Drilled and Grouped, and How the Opera Must Be Revised, Interpreted, and Visualized to Make Up for the Lack of Action, Costumes, and Scenery; Artists Are Put in a Musical Straitjacket; Moving, Whispering, Even Deep Breathing a Crime," *Radio Broadcast,* August 1922, 286. The lengthy subtitle suggests the degree to which studio workers understood music's abstraction from event to sound.

31 For an interesting contemporary case, see the discussion of hip hop and recording in Greg Dimitriadis, *Performing Identity/Performing Culture: Hip Hop as Text, Pedagogy, and Lived Practice* (New York: Peter Lang, 2001), 15–34. Dimitriadis's argument parallels mine: as hip hop moved from a live to

a studio performance setting, the form, sound, content, and meaning of the music changed significantly.

32 Thomas A. Watson, *The Birth and Babyhood of the Telephone* (reprint, n.p.: American Telephone and Telegraph Co., Information Department, 1934), 28.

33 "Phonograph Singers," *Phonogram II* 2, no. 1 (November 1900): 7.

34 Early studio manager quoted in E. W. Mayo, "A Phonographic Studio," *Antique Phonograph Monthly* 6, no. 6 (June 1980): 7. Mayo speculates that the studio in which this manager worked was Bettini's.

35 Herbert A. Shattuck, "The Making of a Record," *Phonogram II* 2, no. 5 (March 1901): 183–84.

36 Leon Alfred Duthernoy, "Singing to Tens of Thousands: Impressions of an Artist during His First Radio Concert," *Radio Broadcast*, November 1922, 49.

37 Ibid., 50–51.

38 Ibid., 51.

39 Max Weber, *The Protestant Ethic and the Spirit of Capitalism*, trans. Talcott Parsons (New York: Scribner's, 1958), 181.

40 Watson, *Birth and Babyhood of the Telephone*, 27.

41 This account is taken from Tim Brooks, "The Last Words of Harry Hayward: A True Record Mystery," *Antique Phonograph Monthly* 1, no. 6 (June–July 1973): 1, 3–9.

42 Jacques Attali, *Noise: The Political Economy of Music*, trans. Brian Massumi (Minneapolis: University of Minnesota Press, 1985), 101.

43 William Howland Kenney, *Recorded Music in American Life: The Phonograph and Popular Memory, 1890–1930* (New York: Oxford University Press, 1999), 11; Ruth Cowan, *More Work for Mother: The Ironies of Household Technology from the Open Hearth to the Microwave* (New York: Basic, 1983); J. Macgregor Wise, "Community, Affect, and the Virtual: The Politics of Cyberspace," in *Virtual Publics: Policy and Community in an Electronic Age*, ed. Beth Kolko (New York: Columbia University Press, 2003). See also Miles Orvell, *The Real Thing: Imitation and Authenticity in American Culture, 1880–1940* (Chapel Hill: University of North Carolina Press, 1989).

44 *Columbia Disc Records Catalogue, 1904*, personal collection of George Kimball. Thanks to George Kimball for sharing this document with me.

45 Based on a listening survey of descriptive specialties on file at the Library of Congress Recorded Sound Reference Center. With the exception of an industry magazine reporting that a young woman had accidentally come across the testimonials of a murderer and fainted, I have yet to discover an account of someone listening to this kind of recording. Listener response is,

thus, difficult to gauge except to note that the genre died out in the late 1910s.

46 Rudolf Arnheim, *Radio,* trans. Margaret Ludwig and Herbert Read (London: Faber and Faber, 1936), 42.

47 Kristin Thompson, "The Formulation of the Classical Style, 1909–28," in *The Classical Hollywood Cinema: Film Style and Production to 1960,* by David Bordwell, Janet Staiger, and Kristin Thompson (New York: Columbia University Press, 1985), 157–73.

48 J. Andrew White, "Report on the Broadcasting of the Dempsey-Carpentier Fight by RCA" (typescript apparently copied by George H. Clark from the original report), GHC, series 135, box 532. Pittsburgh's KDKA did one better by broadcasting a voice description of the fight based on telegraph transmissions from Hoboken, N.J., where operators listened to the RCA transmission. See Thomas White, "'Battle of the Century': The WJY Story" (available on-line at http://www.ipass.net/~whitetho/wjy.htm; last accessed 18 January 2001).

49 Roland Barthes argues that there are two levels of signification: denotation, which is the propositional content of the message, and connotation, which is the effective content of the message. For instance, a photograph of a black Algerian soldier saluting the French flag denotes the event but connotes a much more insidious colonial message. See Roland Barthes, "The Photographic Message," in *A Barthes Reader,* ed. Susan Sontag (New York: Hill and Wang, 1982), 194–210.

50 Bruno Latour, "Mixing Humans and Nonhumans Together: The Sociology of a Door-Closer," *Social Problems* 35, no. 1 (June 1988): 298–310.

51 Robert V. Bruce, *Bell: Alexander Graham Bell and the Conquest of Solitude* (Boston: Little, Brown, 1973), 181.

52 Jacques Attali (*Noise,* 87) writes that, in antiquity, the power to record sound was reserved for the gods. Perhaps the hope was to extend this power to rulers.

53 Bruce, *Bell,* 197. Bruce recounts the entire episode on pp. 193–97.

54 "Telephony: Audible Speech by Telegraph," *Scientific American,* suppl. 48 (25 November 1876): 765. For a discussion of visual verification of earlier audio technologies, see chapters 2 and 3.

55 Millard, *America on Record,* 24–25.

56 Watson, *Birth and Babyhood of the Telephone,* 25.

57 Claude S. Fischer, *America Calling: A Social History of the Telephone to 1940* (Berkeley and Los Angeles: University of California Press, 1992), 60.

58 "The Telephone: Professor Bell's Lecture in Music Hall Last Evening: A Novel Entertainment—History of the Telephone Invention—Music and

Speech from Somerville and Providence," unattributed newspaper clipping, EG, box 2, folder 3.

59 John Durham Peters, *Speaking into the Air: A History of the Idea of Communication* (Chicago: University of Chicago Press, 1999), 180.

60 Latour and Woolgar, *Laboratory Life,* 154–68, 236–44.

61 Charles Sumner Tainter, "The Talking Machine and Some Little Known Facts in Connection with Its Early Development," 7, CST, series 1, box 1, folder 5, Home Notes, 1881, 3:22–23, CST, series II, box 2, folder 4, and Home Notes, 1881, 3:25, CST, series II, box 2, folder 4. See also Charles Sumner Tainter, Home Notes, 1881, 2:67, CST, series II, box 2, folder 3, Home Notes, 1881, 3:5, CST, series II, box 2, folder 4, Home Notes, 1882, 8:99, CST, series II, box 2, folder 9.

62 "A Red Letter Day for Photophony!" (anonymous typescript copy of A. G. Bell's "Work-Room Notes. Vol. 1," 158), AGB, box 256, folder "Photophone—Miscellany."

63 Thomas A. Edison, "The Phonograph and Its Future," *Telegraphic Journal,* 15 June 1878, 250. Edison's article also marks one of the earliest uses of fidelity, in this case referring to the ability of the reproducing needle and embossed tinfoil fully to reverse the recording process. The question of permanence as it is dealt with here and elsewhere is the subject of chapter 6.

64 Charles Sumner Tainter, Home Notes, 1881, 1:19, 39, 41, 65, 87, 95, CST, series II, box 2, folder 2, Home Notes, 1881, 3:7, 19, 23, 73, 83, CST, series II, box 2, folder 4, Home Notes, 1881, 4:5–7, CST, series II, box 2, folder 5, Homes Notes, 1882, 8:57, CST, series II, box 2, folder 9, Home Notes, 1883, 11:83–93, CST, series II, box 2, folder 10, and Home Notes, 1883, 12:85, CST, series II, box 2, folder 11.

65 Chichester Alexander Bell, entry dated 11–13 February 1882, Home Notes, 2:5–8, AGB, box 25, folder "Bell, C. A. Scientific Experiments 1881–4," 2d folder.

66 Charles Sumner Tainter, Home Notes, 1882, 8:101 (see also 99–101), CST, series II, box 2, folder 9.

67 Charles Sumner Tainter, Home Notes, 1881, 2:9, CST, series II, box 2, folder 3. See also Charles Sumner Tainter, Homes Notes, 1881, 3:25, CST, series II, box 2, folder 4.

68 Chichester Alexander Bell, entry dated 17 February 1882, Home Notes, 2:11, AGB, box 25, folder "Bell, C. A. Scientific Experiments 1881–4," 2d folder.

69 For example, Charles Sumner Tainter, Home Notes, 1882, 8:87, CST, series II, box 2, folder 9, and Home Notes, 1883, 12:25, CST, series II, box 2, folder 13.

70 John Corbett, *Extended Play: Sounding Off: From John Cage to Dr. Funkenstein* (Durham, N.C.: Duke University Press, 1994), 36.

71 Jean-Louis Baudry, "Ideological Effects of the Basic Cinematographic Apparatus" and "The Apparatus: Metapsychological Approaches to the Impression of Reality in Cinema," in *Narrative, Apparatus, Ideology: A Film Theory Reader,* ed. Philip Rosen (New York: Columbia University Press, 1986), 286–98, 299–318; Laura Mulvey, *Visual and Other Pleasures* (Bloomington: Indiana University Press, 1989).

72 Alexander Graham Bell to Chichester A. Bell and Charles Sumner Tainter, 14 June 1885, AGB, box 256, folder "Charles Sumner Tainter." Bell's letter was also probably an artifact of the Volta Laboratory's running out of money and needing to seek new sources of income. See Tainter, "The Talking Machine," 73–75, 96–99.

73 No known recordings of Alexander Graham Bell's voice survive from this period.

74 See David Morton, *Off the Record: The Technology and Culture of Sound Recording in America* (New Brunswick, N.J.: Rutgers University Press, 2000), 76–86.

75 In "Machines, Music, and the Quest for Fidelity: Marketing the Edison Phonograph in America, 1877–1925," *Musical Quarterly* 79 (spring 1995): 131–75, Emily Thompson distinguishes between two types of tone tests, one featuring major artists in large halls of major cities, the other using less-renowned artists in small towns (the latter was far more common).

76 Ibid., 149.

77 "Two Good Ones from Wichita, Kansas," *Phonogram II* 3, no. 5 (September 1901): 77.

78 The reviews are discussed in Thompson, "Machines, Music, and the Quest for Fidelity," 157. Although Thompson wants to privilege consumption at the abstract level (in the persona of the consumer), this privilege disappears in her analysis; the Edison Company, advertisers, promoters, performers, machines, and listeners are all actors in her history—a history that appears to be driven at least as much by capitalization and management planning as by consumption.

79 Victor Talking Machine Co., "No Need to Wait for Hours in the Rain," WBA, phonographs box 1, folder 33.

80 "The Acme of Realism," advertisement in *Phonogram II* 2, no. 5 (March 1901): 187.

81 Letter to the Editor, *Phonogram II* 3, no. 2 (June 1901): 29.

82 Stuart Hall, "The Narrative Construction of Reality: An Interview with Stuart Hall," *Southern Review* 17 (March 1984): 8.

83 "Dr. Jekyll and Mr. Hyde," clipped, undated AT&T advertisement from *Town and Country,* NWA, telephone box 1, folder "AT&T #4."

84 "Telephone Etiquette," NWA, telephone box 1, folder "AT&T #4"; New York Telephone Co., "Courtesy between Telephony Users," WBA, telephone box 3, folder "NY Telephone #2," and "The Bell Telephone: For Cheap and Instantaneous Communication by Direct Sound," WBA, telephone box 2, folder "NY Telephone #2."

85 "Courtesy between Telephony Users."

86 Undated letter on U.S. House of Representatives letterhead, AGB, box 255, folder "Phonograph Correspondence."

87 Charles Sumner Tainter, "The Graphophone," *Electrical World,* 14 July 1888, 16, and "The Talking Machine," 85.

88 Sales tags for gramophone, dated 1894, EB, scrapbook 3.

89 Owner's manual for seven-inch American Hand Gramophone, dated 1894, EB, scrapbook 3.

90 The lack of standardized pitch and rotation speed has proved to be a significant problem when early ethnographic recordings are replayed today. Although some ethnographers had the foresight to use a pitch pipe or some other method by which pitch could be standardized, even these markers can be unclear. In some cases, recording engineers have not been able to discern whether the pitch pipe was tuned to A or C.

91 Edison Phonograph Works, *Inspector's Handbook of the Phonograph* (Orange, N.J., 1889), passim, National Museum of American History, Division of Mechanisms.

92 "The Victor System of Changeable Needles Gives You Complete Musical Control," *Ladies Home Journal,* June 1913, 68.

93 E.G., "The Brunswick Method of Reproduction: Certainly Different! Certainly Better!" *Ladies Home Journal,* August 1920, 179. Brunswick offered a built-in tone arm with multiple needles. This tradition continues—in other forms, e.g., equalization—with consumer electronics down to the present day.

94 "Another Famous Tower," clipping from *Wireless Age,* May 1925, GHC, series 60, box 331; "Loud Speaker Qualities: A Most Important Feature in Listening: Results of a Test," GHC, series 60, box 331; "The End of a Perfect Howl," *Radio Broadcast,* July 1922, n.p. (advertising section); "Stop Buzzing and Sizzling," *Radio Broadcast,* July 1922, n.p. (advertising section).

95 "Ye Telephonists of 1877: Harmonious Internal Working," in *Lightning Flashes and Electric Dashes: A Volume of Choice Telegraphic Literature, Humor, Fun, Wit, and Wisdom, Contributed to by All of Principal Writers in the Ranks*

of Telegraphic Literature as Well as Some Well-Known Outsiders, with Numerous Wood-Cut Illustrations* (New York: W. J. Johnston, 1877).

96 Geo. S. Beetle to Elisha Gray, 13 August 1877, EG, box 2, folder 4.
97 On the quality of telephone sound, see also Fischer, *America Calling,* 166–67.
98 Barry Pain, "Diary of a Baby," *The Reader,* October 1906, 566. Thanks to Adrienne Berney for sending me this source.
99 "We Thought It Was a Bunch of Tin Cans," spoof advertisement clipped from *Judge* magazine, n.d., GHC, series 169, box 579a.
100 "A Piercing Shriek . . . ," GHC, series 169, box 572.
101 "Broadcasting Close to Nature," clipping from the *Boston Post,* Tuesday, 13 May 1924, GHC, series 169, box 572.
102 "Somebody's . . . All for $7.35," GHC, series 169, box 572.
103 "Our Tattler," *Phonogram II* 4, no. 5 (March 1902): 67.
104 Carolyn Marvin, *When Old Technologies Were New: Thinking about Electric Communication in the Late Nineteenth Century* (New York: Oxford University Press, 1988), 191–231.
105 Typescript dated 4 April 1877, AGB, box 386, folder "Telephone: Series of Lectures, Boston, April, May 1877."
106 "A True Mirror of Sound: Bettini Micro-Phonograph: Micro-Diaphragms for Phonograph and Graphophone," National Museum of American History, Library.
107 "Columbia Grafonola De Luxe," WBA, phonographs series 3, box 5, folder 7.
108 "The Trill of Galli-Curci's Voice as It Rises and Soars," advertisement, *Radio Broadcast,* June 1922, 4.
109 "Stupendous Advance over Former Recording Instruments," NWA, series I, box 294, folder 1.
110 "All the Roundness and Warmth of the Original," advertisement proof, NWA, series I, box 294, folder 1.
111 "My Voice Being Reflected Back to Me," advertisement proof, NWA, series I, box 293, folder 1.
112 "Its Performance Is Gorgeous, Amazing," advertisement proof, NWA, series I, box 294, folder 1.
113 "Reproduces the Spirit of the Interpreter," advertisement proof, NWA, series I, box 294, folder 1; "Glittering Symphony Music—Colorful Fabrics of Sound for Your Music Library," advertisement proof, NWA, series I, box 294, folder 1.
114 Millard, *America on Record,* 142–47.
115 John Mowitt, "The Sound of Music in the Era of Its Electronic Reproducibility," in *Music and Society: The Politics of Composition, Performance, and*

Reception, ed. Richard Leppert and Susan McClary (New York: Cambridge University Press, 1987), 194; Peter Manuel, *Cassette Culture: Popular Music and Technology in North India* (Chicago: University of Chicago Press, 1993). To Manuel's examples, we could add the popularity of "lo-fi" recording techniques in North American genres like hardcore punk and techno.

116 Oliver Read and Walter L. Welch, *From Tin Foil to Stereo: Evolution of the Phonograph* (Indianapolis: H. W. Sams, 1976), 237–75. See also Roland Gelatt, *The Fabulous Phonograph: From Edison to Stereo* (New York: Appleton-Century, 1965), 219–28.

117 Tony Faulkner, "FM: Frequency Modulation or Fallen Man?" in *Radiotext(e)*, ed. Neil Strauss (New York: Semiotext[e], 1993), 61–65; Lawrence Lessig, *Man of High Fidelity: Edwin Howard Armstrong* (New York: Bantam, 1956). Lessig's biography is very much a great-man narrative, but it does provide some useful historical information on FM and Armstrong.

118 Tainter, "The Talking Machine," 34. Similarly, "The Telephone" (article clipping dated March 1878, AGB, box 305, folder "Telephone, Printed Matter 1877–1925") lamented the "feebleness" of telephonic speech while praising the invention of the telephone itself.

119 Incomplete clipping from the *"Cam. (N.Y.) Advertiser,"* 8 December 1888, EB, Berliner scrapbook 2.

120 *Electrical World,* 29 June, 375.

121 "E. Berliner's Multiphone (Multiplex Gramophone)," program for an undated public performance, EB, Berliner scrapbook 2.

122 "Fake Records," *Phonoscope* 2, no. 11 (November 1898): 10. The *Phonoscope* was essentially an extension of the Edison empire, so criticism of the gramophone cannot be taken to be impartial.

123 "Gramophone Is Suppressed: It Took Martial Law to Do It in St. Petersburg," *New York Times,* Sunday, 17 October 1909, C2.

124 "May Be Heard Ten Miles Away," *Phonogram II* 2, no. 2 (December 1900): 71. The article goes on to argue that this (likely entirely fictional) invention was in fact accomplished by Edison at the unlikely date of 1879, with a machine called the "aerophone . . . for projecting the human voice an indefinite distance" (73–75).

125 "An Instrument of Satan," *Phonogram II* 2, no. 3 (January 1901): 99; "January Notes," *Phonogram II* 2, no. 3 (January 1901): 127.

126 Anne McClintock, *Imperial Leather: Race, Gender, and Sexuality in the Colonial Contest* (New York: Routledge, 1995), 223–26; and Ohmann, *Selling Culture,* 203–5.

127 Sigmund Spaeth, *Listening* (New York: Federal-Brandes, 1927), 4, 5, NWA, radio box 1, Kolster folder.

128 Ibid., 7, 9, 19, 20, 28.

129 This is Lastra's argument concerning the "identity" theorists whom he criticizes. His point, albeit argued on different grounds, is also that copies are not necessarily best understood as debased versions of an original. See Lastra, *Sound Technology and American Cinema,* 126–27.

130 It is also possible to celebrate this irrevocable difference between real and copy: for John Mowitt, the advent of the recording studio testifies "to the technological advances that made the present priority of cultural consumption over cultural production possible" ("The Sound of Music," 175). But the recording studio ties those possibilities for listening intimately to the possibilities for production. The relative weight of production or consumption is not the question here.

6. A Resonant Tomb

1 *"Fewkes Collection,* 39 ('40') taken from the Passamaquoddy Indians, 10 taken from the Zuñi Indians—includes non-field recordings," transcription made ca. 1980 by a Library of Congress sound engineer of cylinders (now barely audible) recorded by the anthropologist Jesse Walter Fewkes in the 1890s, Library of Congress, American Folklife Center, folder "Fewkes, Jesse Walter." The epigraph is a transcription of a test recording of Fewkes's voice.

2 C. B. Lewis, "Mr. Bowser's Tribulations," *Phonogram II* 2, no. 4 (February 1901): 151 (reprinted from an unspecified issue of *McClure's*).

3 D. L. LeMahieu, *A Culture for Democracy: Mass Communication and the Cultivated Mind in Britain between the Wars* (Oxford: Clarendon Press, 1988), 89.

4 Avital Ronell, *The Telephone Book: Technology—Schizophrenia—Electric Speech* (Lincoln: University of Nebraska Press, 1989), 438.

5 "Microphone Inventor Is Resident of the Capital," *Washington Post,* Sunday, 21 June 1925, EB, Berliner scrapbook 1.

6 Susan Douglas argues that spiritualism in fact helped shape the radio craze of the 1920s. See her *Listening In: Radio and the American Imagination from Amos 'n Andy and Edward R. Murrow to Wolfman Jack and Howard Stern* (New York: Random House, 1999), 40–48. Ronell (*The Telephone Book,* 245–50) makes a similar point about telephony, focusing on Thomas Watson's interest in spiritualism. See also Jeffrey Sconce, *Haunted Media: Electronic Presence from Telegraphy to Television* (Durham, N.C.: Duke University Press, 2000)—which arrived in my hands too late to be seriously addressed in this book.

7 I discuss the exteriority of sound in the introduction and first three chapters.

8 John Durham Peters, *Speaking into the Air: A History of the Idea of Communication* (Chicago: University of Chicago Press, 1999), 147.

9 In addition to the account in Douglas's *Listening In,* see Peters, *Speaking into the Air,* 188–94; and John Morley, *Death, Heaven, and the Victorians* (Pittsburgh: University of Pittsburgh Press, 1971), esp. 19–31, 102–11.
10 Friedrich Kittler, *Gramophone, Film, Typewriter,* trans. Geoffrey Winthrop-Young and Michael Wutz (Stanford, Calif.: Stanford University Press, 1999), 5–13 (quotation from 13). See also Peters, *Speaking into the Air,* 137–76; Peters's provocative observation of "the unity of communication at a distance and communication with the dead" (248) captures this transhistorical impulse quite well.
11 John Philip Sousa, "The Menace of Mechanical Music," *Appleton's Magazine* 8 (September 1906): 278–84, 279. See also William Howland Kenney, *Recorded Music in American Life: The Phonograph and Popular Memory, 1890–1945* (New York: Oxford University Press, 1999), 31.
12 Ruth Schwartz Cowan, *More Work for Mother: The Ironies of Household Technology from the Open Hearth to the Microwave* (New York: Basic, 1983), 73. See also Kenneth Kiple and Kreimhild Conee Ornelas, eds., *The Cambridge World History of Food* (New York: Cambridge University Press, 2000), 2:1314; and Waverly Lewis Root and Richard de Rochemont, *Eating in America: A History* (New York: William Morrow, 1976).
13 Sousa's aesthetic criticisms of recorded music were of the zero-sum variety—listening to recorded music would discourage the production of "live" music and lower Americans' musical tastes. See Kenney's discussion of Sousa in *Recorded Music in American Life,* 31. And see also Sousa, "The Menace of Mechanical Music."
14 Amy Lawrence, *Echo and Narcissus: Women's Voice in Classical Hollywood Cinema* (Berkeley and Los Angeles: University of California Press, 1991), 13–14; James Lastra, *Sound Technology and American Cinema: Perception, Representation, Modernity* (New York: Columbia University Press, 2000), 61–91; Theodor Adorno, "The Curves of the Needle," trans. Thomas Levin, *October,* no. 55 (winter 1990): 54.
15 On the history of the body, see also my discussion in the introduction. The reference to biopower is taken from the last section of Michel Foucault, *The History of Sexuality,* vol. 1, *An Introduction,* trans. Robert Hurley (New York: Vintage, 1978), 133–59.
16 Christine Quigley, *The Corpse: A History* (Jefferson, N.C.: McFarland, 1996), 62.
17 Christine Quigley, *Modern Mummies: The Preservation of the Human Body in the Twentieth Century* (Jefferson, N.C.: McFarland, 1998), 5; Robert G. Mayer, *Embalming: History, Theory, and Practice,* 2d ed. (Stamford, Conn.: Appleton and Lange, 1996), 113.

18   Jean-Nicolas Gannal, *History of Embalming, and of Preparations in Anatomy, Pathology, and Natural History, Including an Account of a New Process for Embalming,* trans. Richard Harlan (Philadelphia: J. Dobson, 1840).
19   Quigley, *The Corpse,* 55.
20   Robert W. Habenstein and William M. Lamers, *The History of American Funeral Directing* (Milwaukee: Bulfin Printers, 1955), 328.
21   "A complaint was made to General Grant by the railroad companies of offensive odor from the bodies on trains going North. An order was issued to Jacob Weaver, Undertaker at Baltimore to board every train and remove all bodies that were offensive, put them into a vault, or enclose them in tight zinc boxes and notify their friends, and every embalmer at City Point was ordered to leave the army. Large numbers of bodies which had been dead only two or three days were taken from the trains" (*The Casket,* May 1892, quoted in Habenstein and Lamers, *The History of American Funeral Directing,* 334–35).
22   Robert Kastenbaum and Beatrice Kastenbaum, eds., *The Encyclopedia of Death* (Phoenix: Oryx, 1989), 109.
23   Habenstein and Lamers, *The History of American Funeral Directing,* 333.
24   Ibid., 448.
25   Quigley, *The Corpse,* 56; Habenstein and Lamers, *The History of American Funeral Directing,* 335.
26   Kastenbaum and Kastenbaum, eds., *The Encyclopedia of Death,* 110.
27   E. W. Mayo, "A Phonographic Studio," *Antique Phonograph Monthly* 6, no. 6 (June 1980): 7 (reprinted from the July 1899 issue of *Quaker Magazine*).
28   "A Wonderful Invention—Speech Capable of Indefinite Repetition from Automatic Records," *Scientific American,* 17 November 1877, 304.
29   Jacques Derrida, *Of Grammatology,* trans. Gayatri Chakravorty Spivak (Baltimore: Johns Hopkins University Press, 1976), 4, 7.
30   "An Imperishable Phonograph Record," *Phonogram II* 4, no. 6 (April 1902): 93 (reprinted from an unspecified edition of the *Kansas City Star*).
31   "The Indestructible Records," NWA, phonograph box 1, folder 18, "Indestructible Records Company"; "Indestructible Phonographic Records Do Not Wear Out," ibid.; Brian Philpot to Dealers, 30 September 1908, ibid.; Indestructible Record Co. to Clayton P. Olin, 1 April 1908, 2 April 1908, 21 July 1908, and 8 August 1908, ibid.; "Special" to Dealers, n.d., ibid.; Oliver Read and Walter L. Welch, *From Tin Foil to Stereo: Evolution of the Phonograph* (New York: Bobbs-Merrill, 1976), 96, 100, 103, 193, 197. Interestingly, some listeners and collectors have noted that the original celluloid Indestructible records did hold up well over repeated playings and long-term aging.
32   Leonard Petts, *The Story of "Nipper" and the "His Master's Voice" Picture*

Painted by Francis Barraud (Bournemouth: Ernie Bayly for the Talking Machine Review International, 1983). See also Lawrence, *Echo and Narcissus*, 14.

33 Kittler, *Gramophone, Film, Typewriter*, 69.
34 Morley, *Death, Heaven, and the Victorians*, 201 (Morley's pl. 1 is an image of a dog attending a cradle); Peters, *Speaking into the Air*, 161.
35 See the images of attentive dogs in Richard Leppert, *The Sight of Sound: Music, Representation, and the History of the Body* (Berkeley and Los Angeles: University of California Press, 1993), 78, 167.
36 Robert Feinstein, "His Master's Casket: Notes on Some Phonographic Undertakings," *Antique Phonograph Monthly* 6, no. 7 (July 1980): 1, 3–6.
37 "The Strangest Funeral Ever Heard," *Phonogram I* 2, no. 11 (November 1892): 246–47.
38 "Preached His Own Funeral Sermon by Phonograph," *Edison Phonograph Monthly* 3, no. 3 (May 1905): 12. For singers, see "A Phonograph at a Funeral," *Edison Phonograph Monthly* 3, no. 2 (April 1905): 10.
39 Feinstein, "His Master's Casket," 6.
40 Duplex Phonograph Co., "Let Us Send You This Two-Horn Duplex Phonograph on Trial Direct from Our Factory to Your Own Home," WBA, phonograph box 1, folder "Duplex Phonograph."
41 "Form B-94," gramophone advertisement handbill, probably ca. 1894, EB, scrapbook 2.
42 Quoted in Roland Gelatt, *The Fabulous Phonograph, 1877–1977* (New York: Appleton-Century, 1977), 29. The Victorian family album is discussed in chapter 4.
43 "The Voice of the Late William J. Florence Is Always with Us, Thanks to Mr. Edison's Phonograph," *Phonogram I* 1, nos. 11–12 (November–December 1891): 253.
44 "The Voice of the Dead," *Phonogram I* 2, no. 1 (January 1892): 8.
45 "A Quartette of Australian Good Ones," *Phonogram II* 3, no. 6 (October 1901): 86.
46 "August Notes," *Phonogram II* 3, no. 4 (August 1901): 63.
47 For a discussion of delegation, see the previous chapter and also Bruno Latour, "Mixing Humans and Nonhumans Together: The Sociology of a Door-Closer," *Social Problems* 35, no. 1 (June 1988): 298–310.
48 Adorno, "The Curves of the Needle," 54.
49 "Voices of the Dead," *Phonoscope* 1, no. 1 (15 November 1896): 1.
50 Leon Alfred Duthernoy, "Singing to Tens of Thousands: Impressions of an Artist during His First Radio Concert," *Radio Broadcast*, November 1922, 49–51. See the discussion in chapter 5.
51 Kittler, *Gramophone, Film, Typewriter*, 55.

52 Alexander Graham Bell, typed transcription of an excerpt from *Laboratory Note Books,* Sunday, 22 January 1888, AGB, box 256, folder "Phonograph-Smithsonian."

53 Library of Congress, Recorded Sound Reference Center, early recording optical disk.

54 Gary Gumpert, *Talking Tombstones and Other Tales of the Media Age* (New York: Oxford University Press, 1987), 5.

55 Mattei Calinescu, *Five Faces of Modernity: Modernity, Avant-Garde, Decadence, Kitch, Postmodernism* (Durham, N.C.: Duke University Press, 1987), 13.

56 Janet Lyon, *Manifestoes: Provocations of the Modern* (Ithaca, N.Y.: Cornell University Press, 1999), 203.

57 Jacques Attali, *Noise: The Political Economy of Music,* trans. Brian Massumi (Minneapolis: University of Minnesota Press, 1985), 101.

58 See, e.g., Kittler, *Gramophone, Film, Typewriter,* 3.

59 Johannes Fabian, *Time and the Other: How Anthropology Makes Its Object* (New York: Columbia University Press, 1982), 17.

60 Philip J. Deloria, *Playing Indian* (New Haven, Conn.: Yale University Press, 1998), 106. Both racial categories, *Native American* and *white,* are, of course, fraught with internal contradictions and complexities. I am *not* claiming a single, unified purpose or experience for either group (since there were many important ethnic and political divisions among whites and Native Americans), although I do think that this binary division is heuristically useful because it a founding dichotomy for the thinking that went into federal policy and for Native American responses to federal policy—as well as for ethnology and ethnography.

61 Francis Paul Prucha, *The Great Father: The United States Government and the American Indians,* abridged ed. (Lincoln: University of Nebraska Press, 1986), 153.

62 Peter Nabokov, ed., *Native American Testimony: A Chronicle of Indian-White Relations from Prophesy to the Present, 1492–1992* (New York: Penguin, 1991), 146, 237, 258, 259; Frederick E. Hoxie, "Exploring a Cultural Borderland: Native American Journeys of Discovery in the Early Twentieth Century," *Journal of American History* 79, no. 3 (December 1992): 970–71.

63 Lewis Henry Morgan, *League of the Ho-Dé-No-Sau-Nee, or Iroquois* (Rochester, N.Y.: Sage and Bros., 1851), ix, and *Ancient Society,* ed. Leslie White (Cambridge, Mass.: Harvard University Press, 1964), 40; Robert F. Berkhofer Jr., *The White Man's Indian: Images from Columbus to the Present* (New York: Vintage, 1978), 54. On the influence of Boas, see Curtis M. Hinsley Jr., *Savages and Scientists: The Smithsonian Institution and the Development of American Anthropology, 1846–1910* (Washington, D.C.: Smithsonian Institution Press, 1981).

64 Fabian, *Time and the Other*, 21.

65 Hoxie, "Exploring a Cultural Borderland," 973.

66 John Peabody Harrington, lines composed in 1922, quoted in Erika Brady, *A Spiral Way: How the Phonograph Changed Ethnography* (Jackson: University Press of Mississippi, 1999), 52.

67 Hoxie, "Exploring a Cultural Borderland," 993. See also Russel Thornton, *American Indian Holocaust and Survival: A Population History since 1492* (Norman: University of Oklahoma Press, 1987), passim.

68 Brady, *A Spiral Way*, 52–59; "Past Is Present: Women's Money and the 'Study of Man': The Hemenway Expeditions, Part II," *Anthropology Newsletter*, April 1988, 12. Hemenway may, in fact, have been a major motivating (as well as economic) force in Fewkes's use of recording in the field.

69 Joseph Charles Hickerson, "Annotated Bibliography of North American Indian Music North of Mexico" (M.A. thesis, Indiana University, 1961), 10–12.

70 "I believe that the memory of Indians for the details of a story is often better than that of white men" (Jesse Walter Fewkes, "A Contribution of Passamaquoddy Folk-Lore," *Journal of American Folk-Lore* 3, no. 9 [October–December 1890]: 258).

71 Ibid., 277. Fewkes supports this claim with a list of words that, he said, he learned to pronounce by listening to a phonograph and then having an Indian offer the correct English translation.

72 Benjamin Ives Gilman, "Zuni Melodies," *Journal of American Ethnology and Archaeology* 1 (1891): 63–91; John Comfort Fillmore, "The Zuni Music as Translated by Mr. Benjamin Ives Gilman," *Music* 5 (1893–94): 49–56; "Professor Stumpf on Mr. Gilman's Transcription of the Zuni Songs," *Music* 5 (1894): 649–52. See Hickerson's narrative of the debate in "Annotated Bibliography, 20–22. Interestingly, some years later, Theodor Adorno would come to see the virtue of sound recording to be its affinity for listening as examination, for attention to technical and melodic detail. See Thomas Y. Levin, "For the Record: Adorno on Music in the Age of Its Technological Reproducibility," *October*, no. 55 (winter 1990): 45.

73 Brady, *A Spiral Way*, 63; Michael Yates, "Percy Grainger and the Impact of the Phonograph," *Folk Music Journal* 4 (1982): 265–75.

74 See Franz Boas, "The Limitations of the Comparative Method in Anthropology," *Science* 4, no. 103 (18 December 1896): 901–8.

75 Jesse Walter Fewkes, "Additional Studies of Zuni Songs and Rituals with the Phonograph," *American Naturalist*, November 1890, 1095, 1098.

76 Fewkes, "Contribution," e.g., 259, 272–75.

77 "The Phonograph and the Mojave," *Antique Phonograph Monthly* 1, no. 1 (January 1973): 2.

78. Brady, *A Spiral Way,* 59; Curtis Hinsley Jr., "Ethnographic Charisma and Scientific Routine: Cushing and Fewkes in the American Southwest, 1879–1893," in *Observers Observed: Essays on Ethnographic Fieldwork,* ed. George Stocking (Madison: University of Wisconsin Press, 1983).
79. Richard K. Spottwood, ed., *Religious Music, Solo and Performance* (LP available from the Library of Congress Music Division, Recorded Sound Section, Washington, D.C., 20540), liner notes for track B9.
80. Description paraphrased from Fewkes, "Contribution," 262.
81. For a fuller elaboration of this argument, see chapter 5.
82. Edmund Nequtewa, "Dr. Fewkes Plays Like a Child," in Nabokov, ed., *Native American Testimony,* 228, 229. For the full account, see ibid., 227–29.
83. Brady, *A Spiral Way,* 31.
84. Joan Mark, *A Stranger in Her Native Land: Alice Fletcher and the American Indians* (Lincoln: University of Nebraska Press, 1988), 223, 231–32, 254.
85. Ibid., 254–55, 305.
86. Brady, *A Spiral Way,* 99.
87. "Perpetuating the Beauty of the Indian," *Ladies Home Journal* 24 (April 1907): 26.
88. Brady, *A Spiral Way,* 78–80.
89. Frances Densmore, "The Study of American Indian Music," in *Frances Densmore and American Indian Music: A Memorial Volume,* ed. Charles Hoffman (New York: Museum of the American Indian Heye Foundation, 1968), 102, 105.
90. Ibid., 105 (re Henry Thunder), 106 (quotation).
91. "The transcribing of records is seldom done in the field" (ibid., 109).
92. Ibid., 103 (quotation), 109 (on problems determining pitch).
93. D. K. Wilgus, *Anglo-American Folksong Scholarship since 1898* (New Brunswick, N.J.: Rutgers University Press, 1959), 232 (argument about transcription vs. preservation), 160 (quotation). See also Brady, *A Spiral Way,* 62. I write of Wilgus's point as a hypothesis because it is not universally true; the American Museum of Natural History and other institutionally backed and organized traditions quickly developed an ethos of preservation that has resulted in a collection of very usable cylinders from the 1910s and 1920s (Marilyn Graf, archivist, Indiana University, Archives of Traditional Music, telephone interview by author, 25 September 1998). My point here is simply that this ethos of preservation had to be learned; permanence was not inherent in the medium or in the machine.
94. "*Fewkes Collection,* 39 ('40') taken from. . . ."
95. Fewkes, "Contribution," 258–59.
96. Anthony Seeger and Louise Spear, eds., *Early Field Recordings: A Catalogue*

of *Cylinder Collections at the Indiana University Archives of Traditional Music* (Bloomington: Indiana University Press, 1976), 11.

97 Read and Welch, *From Tin Foil to Stereo,* 40–41; Kay Kaufman Shulemay, "Recording Technology, the Record Industry, and Ethnomusicological Scholarship," in *Comparative Musicology and the Anthropology of Music,* ed. Bruno Nettl and Philip V. Bohlman (Chicago: University of Chicago Press, 1991), 282.

98 For instance, Read and Welch (*From Tin Foil to Stereo,* 76) write of a particularly "promising find of Bettini cylinders [discovered] near Syracuse, N.Y. in 1952, some of them in a barn." It included records by various opera artists famous at the turn of the twentieth century, and some were "excellent recordings, others poor." The Bettini example illustrates the haphazard nature of archiving practices from the early history of recording. In this particular case, the preservation of the recordings (and their discovery) appears to have been more or less accidental.

99 "'Being Dead, He Yet Speaketh,'" *Phonogram I* 2, no. 11 (November 1892): 249.

100 J. Mount Beyer, "Living Autograms," *Phonogram I* 3, no. 1 (January 1893): 298.

101 "Notes," *Phonogram II* 4, no. 4 (February 1902): 51.

102 Philip Mauro to Alexander Graham Bell, 11 March 1904, AGB, box 255, folder "Phono-Correspondence."

103 George List, foreword to *A Catalog of Phonorecordings of Music and Oral Data Held by the Archives of Traditional Music* (Boston: G. K. Hall, 1975), iv; Erich Von Hornbostel, "The Problems of Comparative Musicology," in *Hornbostel Opera Omnia I,* ed. Klaus P. Wachmann, Dieter Christensen, and Hans-Peter Reinecke (The Hague: Martinus Nijhoff, 1975), 252.

104 Dieter Christensen, "Erich M. von Hornbostel, Carl Stumpf, and the Institutionalization of Comparative Musicology," in *Comparative Musicology and the Anthropology of Music,* ed. Bruno Nettl and Philip V. Bohlman (Chicago: University of Chicago Press, 1991), 204–5.

105 Von Hornbostel, "The Problems of Comparative Musicology," 252. Von Hornbostel also notes the possibility afforded by synchronized films and recordings for the study of work songs, "music of secular feasts, theater and in particular, dance music" (ibid., 268).

106 Von Hornbostel quoted in Kittler, *Gramophone, Film, Typewriter,* 3–4.

107 Von Hornbostel, "The Problems of Comparative Musicology," 270.

108 "Library of Congress Plans to Preserve American Folksongs in National Collection," *U.S. Daily* (Washington, D.C.), 21 April 1928, C2. For a fuller discussion of the history of folk-song collections at the Library of

Congress, see Peter Bartis, "A History of the Archive of Folk Song at the Library of Congress: The First Fifty Years" (Ph.D. diss., University of Pennsylvania, 1982).

109 "Preserving the Songs of a Dying Nation," *Phonogram II* 1, no. 3 (July 1900): 80.

110 Marilyn Graf, archivist, Indiana University, Archives of Traditional Music, telephone interview by author, 25 September 1998.

111 Dennis Hastings, "Reflections on the Omaha Cylinder Recordings," in *Omaha Indian Music: Historical Recordings from the Fletcher LaFlesche Collection* (Library of Congress catalog no. 84-743254), ed. Dorothy Sara Lee and Maria La Vigna (Washington, D.C.: Library of Congress, 1985), liner notes.

112 Brady, *A Spiral Way*, 108–9.

113 Ibid., 124.

Conclusion: Audible Futures

1 Karl Marx and Friedrich Engels, "The Communist Manifesto," in *The Marx-Engels Reader*, 2d ed., ed. Robert C. Tucker (New York: Norton, 1978), 476; Jean-François Lyotard, *The Postmodern Condition: A Report on Knowledge,* trans. Geoff Bennington and Brian Massumi (Minneapolis: University of Minnesota Press, 1984), 3. See also Perry Anderson, *Origins of Postmodernity* (New York: Verso, 1988).

2 Some conclusions are scholarly summations. Others retrace the territories of their books as if from above. Still others veer widely off on tangents. On the argumentative difficulty of the conclusion, see Bruce Lincoln, *Discourse and the Construction of Society: Comparative Studies of Myth, Ritual, and Classification* (New York: Oxford University Press, 1989), 171–72.

3 See Pierre Bourdieu, *Outline of a Theory of Practice,* trans. Richard Nice (New York: Cambridge University Press, 1977), 164–71. The term *doxa* has a history that can be traced back through twentieth-century sociology, medieval Christian theology, and Platonic thought, but, here, I am using it in Bourdieu's sense.

4 See Michael Bull, *Sounding Out the City: Personal Stereos and Everyday Life* (New York: New York University Press, 2000).

5 On the home studio, see Paul Théberge, *Any Sound You Can Imagine: Making Music/Consuming Technology* (Hanover, N.H.: Wesleyan University Press, 1997). On virtual reality, see Steven Jones, "A Sense of Space: Virtual Reality, Authenticity, and the Aural," *Critical Studies in Mass Communication* 10 (1993): 238–52.

6 George Wilhelm Friedrich Hegel, *Philosophy of Right,* trans. T. M. Knox (Oxford: Clarendon, 1952), 13. Clearly, this gestures toward a whole debate in the human sciences about scholars' distance from their objects of study. Without rehearsing that whole debate here, let me simply acknowledge that there are both advantages and disadvantages to maintaining scholarly distance from one's object of study.

7 Ibid., 12–13.

8 To be fair to Hegel, he was writing in a monarchy. To disown one's own purchase on a society may be a rhetorical maneuver meant to avoid possible persecution.

9 Lawrence Grossberg, *We Gotta Get Out of This Place: Popular Conservatism and Postmodern Politics* (New York: Routledge, 1992), 1.

10 Walter Ong, *The Presence of the Word: Some Prolegomena for Cultural and Religious History* (Minneapolis: University of Minnesota Press, 1981), 311–24.

11 Plato, *Phaedrus,* in *Collected Dialogues of Plato,* ed. Edith Hamilton and Huntington Cairns, trans. R Hackforth (Princeton, N.J.: Princeton University Press, 1961), 475–525; Jürgen Habermas, *The Structural Transformation of the Public Sphere: An Inquiry into a Category of Bourgeois Society,* trans. Thomas Burger (Cambridge, Mass.: MIT Press, 1989).

12 R. Murray Schafer, *The Soundscape: Our Sonic Environment and the Turning of the World* (Rochester, N.Y.: Destiny, 1994), 215–16.

13 As Steven Feld has shown with respect to the Kaluli, a busy and cacophonous soundscape can be a part of a rich cultural life, even in less-populous cultures. See Steven Feld, "Aesthetics as Iconicity of Style (Uptown Title); or (Downtown Title) 'Lift-up-over Sounding': Getting into the Kaluli Groove," in *Music Grooves,* Charles Keil and Steven Feld (Chicago: University of Chicago Press, 1994), 115–31.

14 Oskar Negt and Alexander Kluge, *The Public Sphere and Experience: Toward an Analysis of the Bourgeois and Proletarian Public Sphere,* trans. Peter Labanji, Jamie Owen Daniel, and Assenka Oksiloff (Minneapolis: University of Minnesota Press, 1993); Nancy Frasier, *Justice Interruptus: Critical Reflections on the "Postsocialist" Condition* (New York: Routledge, 1997); Dan Schiller, *Theorizing Communication: A History* (New York: Oxford University Press, 1996).

15 On the necessity of reconsidering the question of social form in social theory, see, e.g., Roberto Unger, *Social Theory: Its Situation and Its Task: A Critical Introduction to "Politics," a Work in Constructive Theory* (New York: Cambridge University Press, 1987); Iris Marion Young, *Justice and the Politics of Difference* (Princeton, N.J.: Princeton University Press, 1990); and David Harvey, *Spaces of Hope* (Berkeley and Los Angeles: University of California Press, 2000).

16 Jacques Derrida, *Speech and Phenomena and Other Essays on Husserl's Theory of Signs,* trans. David B. Allison (Evanston, Ill.: Northwestern University Press, 1973), 70–87; Gayatri Chakravorty Spivak, "Can the Subaltern Speak?" in *Marxism and the Interpretation of Culture,* ed. Cary Nelson and Lawrence Grossberg (Urbana: University of Illinois Press, 1988); Paul Carter, *The Sound In-Between: Voice, Space, Performance* (Kensington: New South Wales University Press, 1992); Briankle Chang, *Deconstructing Communication* (Minneapolis: University of Minnesota Press, 1996), 187–220; John Durham Peters, "The Gaps of Which Communication Is Made," *Critical Studies in Mass Communication* 11, no. 2 (1994): 117–40, and *Speaking into the Air: A History of the Idea of Communication* (Chicago: University of Chicago Press, 1999), 33–108.

17 Raymond Cohen, *Culture and Conflict in Egyptian-Israeli Relations: Dialogue of the Deaf* (Bloomington: Indiana University Press, 1990); D. D. Khanna and Kishmore Kumar, *Dialogue of the Deaf: The India-Pakistan Divide* (Delhi: Konark, Advent, 1992); Lord Beloff, *Britain and the European Union: Dialogue of the Deaf* (London: Palgrave, 1996).

18 Susan Plann, *A Silent Minority: Deaf Education in Spain, 1550–1835* (Berkeley and Los Angeles: University of California Press, 1997), 13–15; James Woodward, *How You Gonna Get to Heaven If You Can't Talk with Jesus? On Depathologizing Deafness* (Silver Spring, Md.: T.J., 1982); Brenda Jo Brueggemann, *Lend Me Your Ear: Rhetorical Constructions of Deafness* (Washington, D.C.: Gallaudet University Press, 1999), 30–33; Horst Biesold, *Crying Hands: Eugenics and Deaf People in Nazi Germany,* trans. Henry Friedlander (Washington, D.C.: Gallaudet University Press, 1999), 13–27; Joachim-Ernst Berendt, *The Third Ear: On Listening to the World,* trans. Tim Nevill (New York: Owl, 1988), 11–13.

19 Oliver Sacks, *Seeing Voices: A Journey into the World of the Deaf* (Berkeley and Los Angeles: University of California Press, 1989), xi. See also Lennard Jeffries, *Enforcing Normalcy: Disability, Deafness, and the Body* (New York: Verso, 1995); Plann, *A Silent Minority,* 1–9; Harlan Lane, *When the Mind Hears: A History of the Deaf* (New York: Random House, 1984); Richard Winefield, *Never the Twain Shall Meet: Bell, Gallaudet, and the Communications Debate* (Washington, D.C.: Gallaudet University Press, 1987); and Douglas Baynton, *Forbidden Signs: American Culture and the Campaign against Sign Language* (Chicago: University of Chicago Press, 1996).

20 Walter Benjamin, *Illuminations,* trans. Harry Zohn (New York: Schocken, 1968), 222.

21 C. Wright Mills, *The Sociological Imagination* (New York: Oxford University Press, 1959), 3–24; Karl Marx, *Economic and Philosophic Manuscripts of 1844,* trans. Martin Milligan (New York: International, 1968), 140–41; Malcolm

Bull, "Where Is the Anti-Nietzsche?" *New Left Review,* ser. 2, no. 3 (May/June 2000): 121–45; Unger, *Social Theory,* 18–25.

22 Jay Allison, "About Quest for Sound" and "Introduction to the Quest," both available on-line at http://www.npr.org/programs/lnfsound/quest/index.html (last accessed November 2001).

Bibliography

Adorno, Theodor. *Introduction to the Sociology of Music.* Translated by E. B. Ashton. New York: Seabury, 1976.
———. "On the Fetish-Character in Music and the Regression of Listening." In *The Essential Frankfurt School Reader,* ed. Andrew Arato and Eike Gebhardt. New York: Continuum, 1982.
———. "The Curves of the Needle." Translated by Thomas Y. Levin. *October,* no. 55 (winter 1990): 49–56.
———. "The Form of the Phonograph Record." Translated by Thomas Y. Levin. *October,* no. 55 (winter 1990): 57–62.
Aeschylus. *The Agamemnon.* Translated by Louis MacNeice. London: Faber and Faber, 1936.
Aitken, Hugh G. J. *Syntony and Spark: The Origins of Radio.* New York: Wiley, 1976.
———. *The Continuous Wave: Technology and American Radio, 1900–1932.* Princeton, N.J.: Princeton University Press, 1985.
Altman, Rick. "The Evolution of Sound Technology." In *Film Sound: Theory and Practice,* ed. Elisabeth Weis and John Belton. New York: Columbia University Press, 1985.
———, ed. *Sound Theory, Sound Practice.* New York: Routledge, 1992.
Anderson, Perry. *Origins of Postmodernity.* New York: Verso, 1988.
Aristotle. *Nicomachean Ethics.* Translated by Martin Ostwald. New York: Bobbs-Merrill, 1962.

Arnheim, Rudolf. *Radio.* Translated by Margaret Ludwig and Herbert Read. London: Faber and Faber, 1936.

Attali, Jacques. *Noise: The Political Economy of Music.* Translated by Brian Massumi. Minneapolis: University of Minnesota Press, 1985.

Auenbrugger, Leopold. "On the Percussion of the Chest." Translated by John Forbes. *Bulletin of the History of Medicine* 4 (1936): 373–403.

Bacon, Francis. *The Advancement of Learning and the New Atlantis.* Oxford: Clarendon, 1974.

Bacon, Gorham. *A Manual of Otology.* With an introduction by Clarence John Blake. New York: Lea Bros., 1898.

Barlow, William. "The Music Industry: Cashing In, 1900–1939." In *Split Image: African Americans in the Mass Media,* ed. Jannette L. Dates and William Barlow. Washington, D.C.: Howard University Press, 1990.

Barthes, Roland. "The Photographic Message." In *A Barthes Reader,* ed. Susan Sontag. New York: Hill and Wang, 1982.

Bartis, Peter. "A History of the Archive of Folk Song at the Library of Congress: The First Fifty Years." Ph.D. diss., University of Pennsylvania, 1982.

Baudrillard, Jean. *For a Critique of the Political Economy of the Sign.* Translated by Charles Levin. New York: Telos, 1981.

Baudry, Jean-Louis. "The Apparatus: Metapsychological Approaches to the Impression of Reality in Cinema." In *Narrative, Apparatus, Ideology: A Film Theory Reader,* ed. Philip Rosen. New York: Columbia University Press, 1986.

———. "Ideological Effects of the Basic Cinematographic Apparatus." In *Narrative, Apparatus, Ideology: A Film Theory Reader,* ed. Philip Rosen. New York: Columbia University Press, 1986.

Bauman, Zygmunt. *Modernity and Ambivalence.* Cambridge: Polity, 1991.

Baynton, Douglas C. "'Savages and Deaf-Mutes': Evolutionary Theory and the Campaign against Sign Language in the Nineteenth Century." In *Deaf History Unveiled: Interpretations from the New Scholarship,* ed. John Vickery Van Cleve. Washington, D.C.: Gallaudet University Press, 1993.

———. *Forbidden Signs: American Culture and the Campaign against Sign Language.* Chicago: University of Chicago Press, 1996.

Beck, Ulrich. *Risk Society: Towards a New Modernity.* Translated by Mark Ritter. Newbury Park, Calif.: Sage, 1992.

Békésy, Georg, and Walter Rosenblinth. "The Early History of Hearing—Observations and Theories." *Journal of the Acoustical Society of America* 20, no. 6 (November 1948): 727–48.

Bell, Charles. *Ideas of a New Anatomy of the Brain: A Facsimile of the Privately Printed Edition of 1811.* London: Dawsons of Pall Mall, 1966.

———. *The Nervous System of the Human Body: Embracing the Papers Delivered to the Royal Society on the Subject of Nerves.* Washington, D.C.: Stereotyped by D. Green, for the Register and Library of Medicine and Chirurgical Science, 1833.

Benjamin, Walter. *Illuminations.* Translated by Hannah Arendt. New York: Schocken, 1968.

Bennett, John Hughes. *Clinical Lectures on the Principles and Practice of Medicine.* New York: Samuel S. and William Wood, 1860.

Bennion, Elisabeth. *Antique Medical Instruments.* Berkeley and Los Angeles: University of California Press, 1979.

———. *Antique Hearing Devices.* London: Vernier, 1994.

Berendt, Joachim-Ernst. *The Third Ear: On Listening to the World.* Translated by Tim Nevill. New York: Owl, 1988.

Berger, Kenneth. *The Hearing Aid: Its Operation and Development.* Livonia, Mich.: National Hearing Aid Society, 1974.

Berkhofer, Robert F., Jr. *The White Man's Indian: Images from Columbus to the Present.* New York: Vintage, 1978.

Berland, Jody. "Cultural Technologies and the Production of Space." In *Cultural Studies,* ed. Lawrence Grossberg, Cary Nelson, and Paula Treichler. New York: Routledge, 1992.

Berman, Marshall. *All That Is Solid Melts into Air: The Experience of Modernity.* New York: Penguin, 1992.

Biesold, Horst. *Crying Hands: Eugenics and Deaf People in Nazi Germany.* Translated by Henry Friedlander. Washington, D.C.: Gallaudet University Press, 1999.

Bijker, Wiebe. *Of Bicycles, Bakelites, and Bulbs: Toward a Theory of Sociotechnical Change.* Cambridge, Mass.: MIT Press, 1995.

Blake, Clarence J. "Sound and the Telephone." Paper read before the British Society of Telegraph Engineers, London, 8 May 1878. An offprint of this talk can be found at the Library of Congress.

———. "The Use of the Membrana Tympani as a Phonautograph and Logograph." *Archives of Ophthalmology and Otology* 5 (1878): 108–13.

Blondheim, Menahem. *News over the Wires: The Telegraph and the Flow of Public Information in America, 1844–1897.* Cambridge, Mass.: Harvard University Press, 1994.

Blumin, Stuart. *The Emergence of the Middle Class: Social Experience in the American City, 1760–1900.* New York: Cambridge University Press, 1989.

Boas, Franz. "The Limitations of the Comparative Method in Anthropology." *Science* 4, no. 103 (18 December 1896): 901–8.

Bordwell, David, Janet Staiger, and Kristin Thompson. *The Classical Hollywood Cinema: Film Style and Production to 1960.* New York: Columbia University Press, 1985.

Bourdieu, Pierre. *Outline of a Theory of Practice.* Translated by Richard Nice. New York: Cambridge University Press, 1977.

———. *The Logic of Practice.* Translated by Richard Nice. Stanford, Calif.: Stanford University Press, 1980.

———. *Distinction: A Social Critique of the Judgment of Taste.* Translated by Richard Nice. Cambridge, Mass.: Harvard University Press, 1984.

———. *Homo Academicus.* Translated by Peter Collier. Stanford, Calif.: Stanford University Press, 1988.

Bourdieu, Pierre, and Loïc J. D. Wacquant, *An Invitation to Reflexive Sociology.* Chicago: University of Chicago Press, 1992.

Brady, Erika. "The Disembodied Voice: First Encounters with the Phonograph." Paper read before the seventy-ninth annual meeting of the American Anthropological Association, Washington, D.C., 7 December 1980.

———. *A Spiral Way: How the Phonograph Changed Ethnography.* Jackson: University Press of Mississippi, 1999.

Braudel, Fernand. "History and Social Science: The Longue Duree." In *Economy and Society in Early Modern Europe,* ed. Peter Burke. London: Routledge and Kegan Paul, 1972.

Breuer, Josef, and Sigmund Freud. *Studies in Hysteria.* Translated by A. A. Brill. New York: Nervous and Mental Disease Publishing Co., 1936.

Brooks, John. *Telephone: The First Hundred Years.* New York: Harper and Row, 1976.

Bruce, Robert V. *Bell: Alexander Graham Bell and the Conquest of Solitude.* Boston: Little, Brown, 1973.

Brueggemann, Brenda Jo. *Lend Me Your Ear: Rhetorical Constructions of Deafness.* Washington, D.C.: Gallaudet University Press, 1999.

Brühl, Gustav, and Adam Politzer. *Atlas and Epitome of Otology.* Translated and edited by S. MacCuen Smith. Philadelphia: W. B. Saunders, 1903.

Bull, Malcolm, "Where Is the Anti-Nietzsche?" *New Left Review,* ser. 2, no. 3 (May/June 2000): 121–45.

Bull, Michael. *Sounding Out the City: Personal Stereos and Everyday Life.* New York: New York University Press, 2000.

Burdick, Alan. "Now Hear This: Listening Back on a Century of Sound." *Harper's Magazine* 303, no. 1804 (July 2001): 70–77.

Burnett, Charles. "Sound and Its Perception in the Middle Ages." In *The Second Sense: Studies in Hearing and Musical Judgement from Antiquity to the Seventeenth Century,* ed. Charles Burnett, Michael Fend, and Penelope Gouk. London: Warburg Institute, 1991.

Calinescu, Matei. *Five Faces of Modernity: Modernity, Avant-Garde, Decadence, Kitch, Postmodernism.* Durham, N.C.: Duke University Press, 1987.

Candland, Douglas Keith. *Feral Children and Clever Animals: Reflections on Human Nature.* New York: Oxford University Press, 1993.

Cantril, Hadley, and Gordon Allport. *The Psychology of Radio.* New York: Harper and Bros., 1935.

Carey, James. *Communication as Culture: Essays on Media and Society.* Boston: Unwin Hyman, 1988.

Carrick, George L. "On the Differential Stethoscope and Its Value in the Diagnosis of Diseases of the Lungs and Heart." *Aberdeen Medical and Chirurgical Tracts* 12, no. 9 (1873): 894–916.

Carter, Paul. *The Sound in Between: Voice, Space, Performance.* Kensington: New South Wales University Press, 1992.

Cartwright, Lisa. *Screening the Body: Tracing Medicine's Visual Culture.* Minneapolis: University of Minnesota Press, 1995.

Chadarevian, Soyara de. "Graphical Method and Discipline: Self-Recording Instruments in Nineteenth-Century Physiology." *Studies in the History and Philosophy of Science* 24, no. 2 (1993): 267–91.

Chanan, Michael. *Musica Practica: The Social Practice of Music from Gregorian Chant to Postmodernism.* New York: Verso, 1994.

———. *Repeated Takes: A Short History of Recording and Its Effects on Music.* New York: Verso, 1995.

Chang, Briankle. *Deconstructing Communication: Representation, Subject, and Economies of Discourse.* Minneapolis: University of Minnesota Press, 1996.

Chernoff, John Miller. *African Rhythm and African Sensibility: Aesthetics and Social Action in African Musical Idioms.* Chicago: University of Chicago Press, 1979.

Chion, Michel. *Audio-Vision: Sound on Screen.* Translated by Claudia Gorbman. New York: Columbia University Press, 1994.

Christensen, Dieter. "Erich M. von Hornbostel, Carl Stumpf, and the Institutionalization of Comparative Musicology." In *Comparative Musicology and the Anthropology of Music,* ed. Bruno Nettl and Philip V. Bohlman. Chicago: University of Chicago Press, 1991.

Coe, Lewis. *The Telegraph: A History of Morse's Inventions and Its Predecessors in the United States.* Jefferson, N.C.: McFarland, 1993.

Cohen, Lizabeth. *Making a New Deal: Industrial Workers in Chicago, 1919–1939.* New York: Cambridge University Press, 1990.

Cooley, Charles Horton. *Social Organization: A Study of the Larger Mind.* Glencoe, Ill.: Free Press, 1909.

Corbett, John. "Free, Single, and Disengaged: Listening Pleasure and the Popular Music Object." *October,* no. 54 (fall 1990): 79–101.

———. *Extended Play: Sounding Off: From John Cage to Dr. Funkenstein.* Durham, N.C.: Duke University Press, 1994.

Corbin, Alain. *The Foul and the Fragrant: Odor and the French Social Imagination.* Cambridge, Mass.: Harvard University Press, 1986.

———. *Time, Desire, and Horror: Toward a History of the Senses.* Translated by Jean Birrell. Cambridge: Blackwell, 1995.

———. *Village Bells: Sound and Meaning in the Nineteenth-Century French Countryside.* New York: Columbia University Press, 1999.

Cowan, Ruth Schwartz. *More Work for Mother: The Ironies of Household Technology from the Open Hearth to the Microwave.* New York: Basic, 1983.

Crary, Jonathan. *Techniques of the Observer: On Vision and Modernity in the Nineteenth Century.* Cambridge, Mass.: MIT Press, 1990.

Culley, Richard Spellman. *A Handbook of Practical Telegraphy.* 8th ed. London: Longmans, Green, 1885.

Czitrom, Daniel. *Media and the American Mind: From Morse to McLuhan.* Chapel Hill: University of North Carolina Press, 1982.

Darwin, Erasmus. *The Letters of Erasmus Darwin.* Edited by Desmond King-Hele. Cambridge: Cambridge University Press, 1981.

Davis, Audrey B. *Medicine and Its Technology: An Introduction to the History of Medical Instrumentation.* Westport, Conn.: Greenwood, 1981.

Davis, Audrey B., and Uta C. Merzbach. *Early Auditory Studies: Activities in the Psychology Laboratories of American Universities.* Washington, D.C.: Smithsonian Institution Press, 1975.

Degravers, Peter. *A Complete Physico-Medical and Chirurgical Treatise on the Human Eye, Second Editions, Corrected and Considerably Enlarged, to Which Is Now Added a Treatise on the Human Ear.* Edinburgh: Printed for the Author, 1788.

Deleuze, Gilles. *Foucault.* Translated by Sean Hand. Minneapolis: University of Minnesota Press, 1988.

———. *Spinoza: Practical Philosophy.* San Francisco: City Lights, 1988.

———. *The Logic of Sense.* Translated by Mark Lester. New York: Columbia University Press, 1990.

Deleuze, Gilles, and Felix Guattari. *A Thousand Plateaus: Capitalism and Schizophrenia.* Translated by Brian Massumi. Minneapolis: University of Minnesota Press, 1987.

Deloria, Philip J. *Playing Indian.* New Haven, Conn.: Yale University Press, 1998.

Densmore, Frances. "The Study of American Indian Music." In *Frances Densmore and American Indian Music: A Memorial Volume,* ed. Charles Hoffman. New York: Museum of the American Indian Heye Foundation, 1968. Originally published in *Smithsonian Annual Report,* Publication no. 3651 (Washington, D.C.: Smithsonian Institution, 1941), 101–14.

Derrida, Jacques. *Speech and Phenomena and Other Essays on Husserl's Theory of Signs.*

Translated by David B. Allison. Evanston, Ill.: Northwestern University Press, 1973.

———. *Of Grammatology.* Translated by Gayatri Chakravorty Spivak. Baltimore: Johns Hopkins University Press, 1976.

———. *Dissemination.* Translated by Barbara Johnson. Chicago: University of Chicago Press, 1981.

———. *The Postcard: From Socrates to Freud and Beyond.* Translated by Alan Bass. Chicago: University of Chicago Press, 1987.

Descartes, René. *Treatise of Man.* Translated by Thomas Steele Hall. Cambridge, Mass.: Harvard University Press, 1972.

———. *Discourse on Method and Meditations on First Philosophy.* Translated by Donald A. Cress. 4th ed. Indianapolis: Hackett, 1998.

Dimitriadis, Greg. *Performing Identity/Performing Culture: Hip Hop as Text, Pedagogy, and Lived Practice.* New York: Peter Lang, 2001.

Doane, Mary Ann. "Ideology and the Practice of Sound Editing and Mixing." In *Film Sound: Theory and Practice,* ed. Elisabeth Weis and John Belton. New York: Columbia University Press, 1985.

Douglas, Susan. *Inventing American Broadcasting, 1899–1922.* Baltimore: Johns Hopkins University Press, 1987.

———. *Listening In: Radio and the American Imagination from Amos 'n Andy and Edward R. Murrow to Wolfman Jack and Howard Stern.* New York: Times Books/Random House, 1999.

Duffin, Jacalyn. *To See with a Better Eye: A Life of R. T. H. Laennec.* Princeton, N.J.: Princeton University Press, 1998.

Durant, Alan. *Conditions of Music.* Albany: State University of New York Press, 1984.

Ediger, Jeffrey. "A Phenomenology of the Listening Body." Ph.D. diss., University of Illinois, 1993.

Edwards, Linden F. *The History of Human Dissection.* Fort Wayne, Tex.: Fort Wayne and Allen County Public Libraries, 1955.

Eisenberg, Evan. *The Recording Angel: The Experience of Music from Aristotle to Zappa.* New York: Penguin, 1987.

Eisler, Hanns, and Theodor Adorno. *Composing for the Films.* New York: Oxford University Press, 1947.

Elias, Norbert. *The Civilizing Process: Sociogenetic and Psychogenetic Investigations.* Translated by Edmund Jephcott. Edited by Eric Dunning, Johan Goudsblom, and Stephen Mennell. Rev. ed. Malden: Blackwell, 2000.

Ellul, Jacques. *The Technological Society.* New York: Vintage, 1964.

Fabian, Johannes. *Time and the Other: How Anthropology Makes Its Object.* New York: Columbia University Press, 1983.

Faulkner, Tony. "FM: Frequency Modulation or Fallen Man?" In *Radiotext(e)*, ed. Neil Strauss. New York: Semiotext(e), 1993.

Feld, Steven. "Orality and Consciousness." In *The Oral and the Literate in Music*, ed. Yoshiko Tokumaru and Osamu Yamaguti. Tokyo: Academia Music, 1986.

———. "Aesthetics as Iconicity of Style (Uptown Title); or (Downtown Title) 'Lift-up-over Sounding': Getting into the Kaluli Groove." In *Music Grooves*, Charles Keil and Steven Feld. Chicago: University of Chicago Press, 1994.

Felski, Rita. *The Gender of Modernity*. Cambridge, Mass.: Harvard University Press, 1995.

Fewkes, Jesse Walter. "Additional Studies of Zuni Songs and Rituals with the Phonograph." *American Naturalist*, November 1890, 1094–98.

———. "A Contribution to Passamaquoddy Folk-Lore." *Journal of American Folk-Lore* 3, no. 9 (October–December 1890): 257–80.

Fillmore, John Comfort. "The Zuni Music as Translated by Mr. Benjamin Ives Gilman." *Music* 5 (1893–94): 39–46.

———. "Professor Stumpf on Mr. Gilman's Transcription of the Zuni Songs." *Music* 5 (1894): 649–52.

Finkelstein, Sidney. *Composer and Nation: The Folk Heritage in Music*. New York: International, 1989.

Fischer, Claude S. *America Calling: A Social History of the Telephone to 1940*. Berkeley and Los Angeles: University of California Press, 1992.

Flinn, Caryl. *Strains of Utopia: Gender, Nostalgia, and Hollywood Film Music*. Princeton, N.J.: Princeton University Press, 1992.

Flint, Austin. *A Manual of Percussion and Auscultation; of the Physical Diagnosis of Diseases of the Lungs and Heart, and of Thoracic Aneurism*. Philadelphia: Henry C. Lea, 1876.

Forbes, John. *Original Cases with Descriptions and Observations Illustrating the Use of the Stethoscope and Percussion in the Diagnosis of Diseases of the Chest; Also Commentaries on the Same Subjects Selected and Translated from Auenbrugger, Corvisart, Laennec, and Others*. London: Printed for T. and G. Underwood, Fleet Street, 1824.

Foucault, Michel. *The Archaeology of Knowledge and the Discourse on Language*. Translated by A. M. Sheridan Smith. New York: Pantheon, 1972.

———. *The Birth of the Clinic: An Archaeology of Medical Perception*. Translated by A. M. Sheridan Smith. New York: Pantheon, 1973.

———. *Discipline and Punish: The Birth of the Prison*. Translated by Alan Sheridan. New York: Vintage, 1977.

———. "Nietzsche, Genealogy, History." In *Language, Counter-Memory, Practice: Selected Essays and Interviews*, ed. Donald Bouchard, trans. Donald Bouchard and Sheery Simon. Ithaca, N.Y.: Cornell University Press, 1977.

———. *The History of Sexuality.* Vol. 1, *An Introduction.* Translated by Robert Hurley. New York: Vintage, 1978.

———. "Questions of Method: An Interview with Michel Foucault." In *The Foucault Effect: Studies in Governmentality,* ed. Graham Burchell, Colin Gordon, and Peter Miller. Chicago: University of Chicago Press, 1991.

Frank, Felicia. *The Mechanical Song: Women, Voice, and the Artificial in Nineteenth-Century French Narrative.* Stanford, Calif.: Stanford University Press, 1995.

Frasier, Nancy. *Justice Interruptus: Critical Reflections on the "Postsocialist" Condition.* New York: Routledge, 1997.

French, Roger. "The Anatomical Tradition." In *Companion Encyclopedia to the History of Medicine,* ed. W. F. Bynum and Roy Porter. London: Routledge, 1993.

Frith, Simon. *Performing Rites: On the Value of Popular Music.* Cambridge, Mass.: Harvard University Press, 1996.

Gannal, Jean-Nicolas. *History of Embalming, and of Preparations in Anatomy, Pathology, and Natural History, Including an Account of a New Process for Embalming.* Translated by Richard Harlan. Philadelphia: J. Dobson, 1840.

Gannon, Jack R. *Deaf Heritage: A Narrative History of Deaf America.* Silver Spring, Md.: National Association of the Deaf, 1981.

Gelatt, Roland. *The Fabulous Phonograph: From Edison to Stereo.* New York: Appleton-Century, 1954.

Giddens, Anthony. *The Constitution of Society: Outline of the Theory of Structuration.* Berkeley and Los Angeles: University of California Press, 1984.

Gilman, Benjamin Ives. "Zuni Melodies." *Journal of American Ethnology and Archaeology* 1 (1891): 63–91.

Goffman, Erving. *Presentation of Self in Everyday Life.* Garden City, N.Y.: Doubleday Anchor, 1959.

Gomery, Douglas. "The Coming of Sound: Technological Change in the American Film Industry." In *Film Sound: Theory and Practice,* ed. Elisabeth Weis and John Belton. New York: Columbia University Press, 1985.

———. "Economic Struggle and Hollywood Imperialism: Europe Converts to Sound." In *Film Sound: Theory and Practice,* ed. Elisabeth Weis and John Belton. New York: Columbia University Press, 1985.

Goodwin, Andrew. *Dancing in the Distraction Factory: Music, Television, and Popular Culture.* Minneapolis: University of Minnesota Press, 1992.

Gorbman, Claudia. *Unheard Melodies: Narrative Film Music.* Bloomington: Indiana University Press, 1987.

Gorman, Michael. *Simulating Science: Heuristics, Mental Models, and Technoscientific Thinking.* Bloomington: Indiana University Press, 1992.

Gouk, Penelope. "Some English Theories of Hearing in the Seventeenth Century: Before and After Descartes." In *The Second Sense: Studies in Hearing and*

Musical Judgement from Antiquity to the Seventeenth Century, ed. Charles Burnett, Michael Fend, and Penelope Gouk. London: Warburg Institute, 1991.

Gracyk, Theodore. *Rhythm and Noise: An Aesthetics of Rock.* Durham, N.C.: Duke University Press, 1996.

Grossberg, Lawrence. *We Gotta Get Out of This Place: Popular Conservatism and Postmodern Culture.* New York: Routledge, 1992.

———. *Dancing in Spite of Myself: Essays on Popular Culture.* Durham, N.C.: Duke University Press, 1997.

Gumpert, Gary. *Talking Tombstones and Other Tales of the Media Age.* New York: Oxford University Press, 1987.

Habenstein, Robert W., and William M. Lamers. *The History of American Funeral Directing.* Milwaukee: Bulfin Printers, 1955.

Habermas, Jürgen. *The Structural Transformation of the Public Sphere: An Inquiry into a Category of Bourgeois Society.* Translated by Thomas Burger with Frederick Lawrence. Cambridge, Mass.: MIT Press, 1991.

Hacking, Ian. *The Social Construction of What?* Cambridge, Mass.: Harvard University Press, 1999.

Hall, Edward. *The Hidden Dimension.* London: Bodley Head, 1966.

Hall, Stuart. "The Narrative Construction of Reality: An Interview with Stuart Hall." *Southern Review* 17 (March 1984): 3–17.

———. "On Postmodernism and Articulation." *Journal of Communication Inquiry* 10, no. 2 (1986): 45–60.

Handel, Stephen. *Listening: An Introduction to the Perception of Auditory Events.* Cambridge, Mass.: MIT Press, 1989.

Handelman, Susan. *Slayers of Moses: The Emergence of Rabbinic Interpretation in Modern Literary Theory.* Albany: State University of New York Press, 1982.

Hankins, Thomas L., and Robert J. Silverman. *Instruments and the Imagination.* Princeton, N.J.: Princeton University Press, 1995.

Hansen, Miriam. *Babel and Babylon: Spectatorship in American Silent Film.* Cambridge, Mass.: Harvard University Press, 1991.

Harvey, David. *Spaces of Hope.* Berkeley and Los Angeles: University of California Press, 2000.

Havelock, Eric. *Preface to Plato.* Cambridge, Mass.: Harvard University Press, 1963.

———. *The Muse Learns to Write: Reflections on Orality and Literacy from Antiquity to the Present.* New Haven, Conn.: Yale University Press, 1986.

Hegel, George Wilhelm Friedrich. *Philosophy of Right.* Translated by T. M. Knox. Oxford: Clarendon, 1952.

Heidegger, Martin. *The Question concerning Technology and Other Essays.* Translated by William Lovitt. New York: Harper Torchbooks, 1977.

Helmholtz, Hermann. *The Mechanism of the Ossicles of the Ear and the Membrana*

Tympani. Translated by Albert H. Buck and Normand Smith. New York: William Wood, 1873.

———. "On the Interaction of Natural Forces." In *Popular Lectures on Scientific Subjects,* trans. E. Atkinson. New York: Appleton, 1885.

———. *On the Sensations of Tone as a Physiological Basis for the Theory of Music.* 2d English ed. Translated from the 4th German ed. by Alexander J. Ellis. New York: Dover, 1954.

Heylin, Clinton. *Bootleg: The Secret History of the Other Recording Industry.* New York: St. Martin's, 1995.

Hickerson, Joseph Charles. "Annotated Bibliography of North American Indian Music North of Mexico." M.A. thesis, Indiana University, 1961.

Hinsley, Curtis M., Jr. *Savages and Scientists: The Smithsonian Institution and the Development of American Anthropology, 1846–1910.* Washington, D.C.: Smithsonian Institution Press, 1981.

———. "Ethnographic Charisma and Scientific Routine: Cushing and Fewkes in the American Southwest, 1879–1893." In *Observers Observed: Essays on Ethnographic Fieldwork,* ed. George Stocking. Madison: University of Wisconsin Press, 1983.

Horkheimer, Max, and Theodor W. Adorno. *Dialectic of Enlightenment.* New York: Continuum, 1944.

Hornbostel, Erich Von. "The Problems of Comparative Musicology." In *Hornbostel Opera Omnia I,* ed. Klaus P. Wachmann, Dieter Christensen, and Hans-Peter Reinecke. The Hague: Martinus Nijhoff, 1975.

Howes, David. *The Varieties of Sensory Experience: A Sourcebook in the Anthropology of the Senses.* Toronto: University of Toronto Press, 1991.

Hoxie, Frederick E. "Exploring a Cultural Borderland: Native American Journeys of Discovery in the Early Twentieth Century." *Journal of American History* 79, no. 3 (December 1992): 969–96.

Huxley, T. H. *On the Physical Basis of Life.* New Haven, Conn.: Charles C. Chatfield, 1870.

Huyssen, Andreas. "Mass Culture as Woman: Modernism's Other." In *Studies in Entertainment,* ed. Tania Modleski. Bloomington: Indiana University Press, 1986.

Iggers, Georg C. *Historiography in the Twentieth Century: From Scientific Objectivity to the Postmodern Challenge.* Hanover, N.H.: Wesleyan University Press, 1997.

Ihde, Don. *Listening and Voice: A Phenomenology of Sound.* Athens: University of Ohio Press, 1978.

Innis, Harold. *The Bias of Communication.* Toronto: University of Toronto Press, 1951.

Itard, Jean-Marc Gaspard. *The Wild Boy of Aveyron.* Translated by George Humphrey and Muriel Humphrey. New York: Meredith, 1962.

Jameson, Fredric. *Ideologies of Theory.* Vol. 2. Minneapolis: University of Minnesota Press, 1988.

Jay, Martin. *Downcast Eyes: The Denigration of Vision in Twentieth Century French Thought.* Berkeley and Los Angeles: University of California Press, 1993.

Jeffries, Lennard. *Enforcing Normalcy: Disability, Deafness, and the Body.* New York: Verso, 1995.

John, Richard R. *Spreading the News: The American Postal System from Franklin to Morse.* Cambridge, Mass.: Harvard University Press, 1995.

Johnson, James H. *Listening in Paris: A Cultural History.* Berkeley and Los Angeles: University of California Press, 1995.

Jones, Simon, and Thomas Schumacher. "Muzak: On Functional Music and Power." *Critical Studies in Mass Communication* 9, no. 2 (June 1992): 156–69.

Jones, Steven. "A Sense of Space: Virtual Reality, Authenticity, and the Aural." *Critical Studies in Mass Communication* 10 (1993): 238–52.

Kahn, Douglas. "Histories of Sound Once Removed." In *Wireless Imagination: Sound, Radio, and the Avant-Garde,* ed. Douglas Kahn and Gregory Whitehead. Cambridge, Mass.: MIT Press, 1992.

———. *Noise, Water, Meat: A History of Sound in the Arts.* Cambridge, Mass.: MIT Press, 1999.

Kassabian, Anahid. *Hearing Film: Tracking Identifications in Contemporary Hollywood Film Music.* New York: Routledge, 2001.

Kassler, Jamie C. "Man—a Musical Instrument: Models of the Brain and Mental Functioning before the Computer." *History of Science* 22 (1984): 59–92.

Kastenbaum, Robert, and Beatrice Kastenbaum, eds. *The Encyclopedia of Death.* Phoenix: Oryx, 1989.

Kennedy, P. *Ophthalmographia; or, A Treatise of the Eye in Two Parts.* London: Bernard Lintott, 1713.

Kenney, William Howland. *Recorded Music in American Life: The Phonograph and Popular Memory, 1890–1945.* New York: Oxford University Press, 1999.

Kern, Stephen. *The Culture of Time and Space, 1800–1918.* Cambridge, Mass.: Harvard University Press, 1983.

Kiple, Kenneth, and Kreimhild Conee Ornelas, eds. *The Cambridge World History of Food.* Vol. 2. New York: Cambridge University Press, 2000.

Kittler, Friedrich. *Gramophone, Film, Typewriter.* Translated by Geoffrey Winthrop-Young and Michael Wutz. Stanford, Calif.: Stanford University Press, 1999.

Kraft, James P. *Stage to Studio: Musicians and the Sound Revolution, 1890–1950.* Baltimore: Johns Hopkins University Press, 1996.

Laennec, R. T. H. *A Treatise on the Diseases of the Chest and on Mediate Auscultation.* 3d ed. Translated by John Forbes. New York: Samuel Wood; Collins and Hannay, 1830.

Lane, Harlan. *When the Mind Hears: A History of the Deaf.* New York: Random House, 1984.

Lanza, Joseph. *Elevator Music: A Surreal History of Muzak, Easy-Listening, and Other Moodsong.* New York: St. Martin's, 1994.

Lastra, James. *Sound Technology and American Cinema: Perception, Representation, Modernity.* New York: Columbia University Press, 2000.

Latour, Bruno. "Mixing Humans and Nonhumans Together: The Sociology of a Door-Closer." *Social Problems* 35, no. 1 (June 1988): 298–310.

———. *We Have Never Been Modern.* Translated by Catherine Porter. Cambridge, Mass.: Harvard University Press, 1993.

Latour, Bruno, and Steve Woolgar. *Laboratory Life: The Construction of Scientific Facts.* Princeton, N.J.: Princeton University Press, 1986.

Lawrence, Amy. *Echo and Narcissus: Women's Voice in Classical Hollywood Cinema.* Berkeley and Los Angeles: University of California Press, 1991.

Lefebvre, Henri. *Introduction to Modernity.* Translated by John Moore. New York: Verso, 1995.

LeMahieu, D. L. *A Culture for Democracy: Mass Communication and the Cultivated Mind in Britain between the Wars.* Oxford: Clarendon, 1988.

Leppert, Richard. *The Sight of Sound: Music, Representation, and the History of the Body.* Berkeley and Los Angeles: University of California Press, 1993.

Lessig, Lawrence. *Man of High Fidelity: Edwin Howard Armstrong.* New York: Bantam, 1956.

Levin, David Michael. *The Listening Self: Personal Growth, Social Change, and the Closure of Metaphysics.* New York: Routledge, 1989.

———, ed. *Modernity and the Hegemony of Vision.* Berkeley: University of California Press, 1993.

Levin, Thomas Y. "For the Record: Adorno on Music in the Age of Its Technological Reproducibility." *October,* no. 55 (winter 1990): 23–48.

Lincoln, Bruce. *Discourse and the Construction of Society: Comparative Studies of Myth, Ritual, and Classification.* New York: Oxford University Press, 1989.

List, George. Foreword to *A Catalog of Phonorecordings of Music and Oral Data Held by the Archives of Traditional Music.* Boston: G. K. Hall, 1975.

Locke, John. *An Essay concerning Human Understanding.* 1689. New York: Meridian, 1964.

Lomax, John. "Field Experiences with Recording Machines." *Southern Folklore Quarterly* 1, no. 2 (June 1937): 57–60.

Lowe, Donald M. *History of Bourgeois Perception.* Chicago: University of Chicago Press, 1982.

Lubar, Steven. *Infoculture.* Boston: Houghton Mifflin, 1993.

Lyon, Janet. *Manifestoes: Provocations of the Modern.* Ithaca, N.Y.: Cornell University Press, 1999.

Lyotard, Jean-François. *The Postmodern Condition: A Report on Knowledge.* Translated by Geoff Bennington and Brian Massumi. Minneapolis: University of Minnesota Press, 1984.

———. *The Postmodern Explained: Correspondence, 1982-1985.* Translated by Don Barry et al. Minneapolis: University of Minnesota Press, 1993.

Mackenzie, Catherine F. *Alexander Graham Bell: The Man Who Contracted Space.* Boston: Houghton-Mifflin, 1928.

Manuel, Peter. *Cassette Culture: Popular Music and Technology in North India.* Chicago: University of Chicago Press, 1993.

Mark, Joan. *A Stranger in Her Native Land: Alice Fletcher and the American Indians.* Lincoln: University of Nebraska Press, 1988.

Marks, Laura. *The Skin of the Film: The Senses in Intercultural Cinema.* Durham, N.C.: Duke University Press, 1999.

Marland, E. A. *Early Electrical Communication.* New York: Abelard-Schuman, 1964.

Marsh, R. O. *White Indians of the Darian.* New York: Putnam's, 1934.

Martin, Michèle. *"Hello, Central?": Gender, Technology, and Culture in the Formation of Telephone Systems.* Montreal: McGill-Queen's University Press, 1991.

Marvin, Carolyn. *When Old Technologies Were New: Thinking about Electrical Communication in the Late Nineteenth Century.* New York: Oxford University Press, 1988.

Marx, Karl. *Capital: A Critique of Political Economy.* Vol. 1, *The Process of Capitalist Production.* Translated by Samuel Moore and Edward Aveling. Edited by Frederick Engels. New York: International, 1967.

———. *Economic and Philosophic Manuscripts of 1844.* Translated by Martin Milligan. New York: International, 1968.

Marx, Karl, and Friedrich Engels. "The Communist Manifesto." In *The Marx-Engels Reader,* 2d ed., ed. Robert C. Tucker. New York: Norton, 1978.

Mauss, Marcel. "Body Techniques." In *Sociology and Psychology: Essays,* trans. Ben Brewster. Boston: Routledge and Kegan Paul, 1979.

Mayer, Robert G. *Embalming: History, Theory, and Practice.* 2d ed. Stamford, Conn.: Appleton and Lange, 1996.

McChesney, Robert. *Telecommunications, Mass Media, and Democracy: The Battle for Control of U.S. Broadcasting, 1928–1935.* New York: Oxford University Press, 1994.

McClintock, Anne. *Imperial Leather: Race, Gender, and Sexuality in the Colonial Contest.* New York: Routledge, 1995.

McLuhan, Marshall. "Five Sovereign Fingers Taxed the Breath." In *Explorations in Communication: An Anthology,* ed. Edmund Carpenter and Marshall McLuhan. Boston: Beacon, 1960.

———. *The Gutenberg Galaxy: The Making of Typographic Man.* Toronto: University of Toronto Press, 1962.

———. *Understanding Media: The Extensions of Man.* New York: McGraw-Hill, 1964.

McLuhan, Marshall, and Edmund Carpenter. "Acoustic Space." In *Explorations in Communication: An Anthology,* ed. Edmund Carpenter and Marshall McLuhan. Boston: Beacon, 1960.

Merleau-Ponty, Maurice. *The Primacy of Perception and Other Essays on Phenomenological Psychology, the Philosophy of Art, History, and Politics.* Edited by James M. Edie. Evanston, Ill.: Northwestern University Press, 1964.

Millard, Andre. *Edison and the Business of Innovation.* Baltimore: Johns Hopkins University Press, 1990.

———. *America on Record: A History of Recorded Sound.* New York: Cambridge University Press, 1995.

Mills, C. Wright. *The Sociological Imagination.* New York: Oxford University Press, 1959.

Morgan, Lewis Henry. *League of the Ho-Dé-No-Sau-Nee, or Iroquois.* Rochester, N.Y.: Sage and Bros., 1851.

———. *Ancient Society.* Edited by Leslie White. Cambridge, Mass.: Harvard University Press, 1964.

Morland, S. *Tuba Stentoro-Phonica: An Instrument of Excellent Use, as Well at Sea, as at Land; Invented and Variously Experimented in the Year 1670 and Humbly Presented to the Kings Most Excellent Majesty Charles II in the Year 1671.* London: Printed by W. Godbid and Sold by M. Pitt, 1672. A copy is housed in the Library of Congress Performing Arts Collection.

Morley, John. *Death, Heaven, and the Victorians.* Pittsburgh: University of Pittsburgh Press, 1971.

Morton, David. *Off the Record: The Technology and Culture of Sound Recording in America.* New Brunswick, N.J.: Rutgers University Press, 2000.

Mowitt, John. "The Sound of Music in the Era of Its Electronic Reproducibility." In *Music and Society: The Politics of Composition, Performance, and Reception,* ed. Richard Leppert and Susan McClary. New York: Cambridge University Press, 1987.

Müller, Johannes. *Elements of Physiology.* Translated by William Baly. Arranged from the 2d London ed. by John Bell. Philadelphia: Lea and Blanchard, 1843.

Mulvey, Laura. "Visual Pleasure and Narrative Cinema." In *Narrative, Apparatus, Ideology: A Film Theory Reader,* ed. Philip Rosen. New York: Columbia University Press, 1986.

———. *Visual and Other Pleasures.* Bloomington: Indiana University Press, 1989.

Munro, Alexander. *Observations on the Organ of Hearing in Man and Other Animals.* Edinburgh: Adam Neill, 1797.

Nabokov, Peter, ed. *Native American Testimony: A Chronicle of Indian-White Relations from Prophesy to the Present, 1492–1992.* New York: Penguin, 1991.

Nasaw, David. *Going Out: The Rise and Fall of Public Amusements.* New York: Basic, 1993.

Neale, Stephen. *Cinema and Technology: Image, Sound, Colour.* Bloomington: Indiana University Press, 1985.

Negt, Oskar, and Alexander Kluge. *The Public Sphere and Experience: Toward an Analysis of the Bourgeois and Proletarian Public Sphere.* Translated by Peter Labanji, Jamie Owen Daniel, and Assenka Oksiloff. Minneapolis: University of Minnesota Press, 1993.

Nerone, John. *Violence against the Press: Policing the Public Sphere in U.S. History.* New York: Oxford University Press, 1994.

Nicolson, Malcolm. "The Introduction of Percussion and Stethoscopy to Early Nineteenth-Century Edinburgh." In *Medicine and the Five Senses,* ed. W. F. Bynum and Roy Porter. New York: Cambridge University Press, 1993.

Nietzsche, Friedrich. *On the Genealogy of Morals and Ecce Homo.* Translated by Walter Kauffman. New York: Vintage, 1967.

Novick, Peter. *That Noble Dream: The "Objectivity Question" and the American Historical Profession.* New York: Cambridge University Press, 1988.

Ohmann, Richard. *Selling Culture: Magazines, Markets, and Class at the Turn of the Century.* New York: Verso, 1996.

Ong, Walter. *The Presence of the Word: Some Prolegomena for Cultural and Religious History.* New Haven, Conn.: Yale University Press, 1967.

———. *Orality and Literacy: The Technologizing of the Word.* New York: Routledge, 1982.

Ord-Hume, Arthur W. J. G. *Clockwork Music: An Illustrated History of Mechanical Musical Instruments from the Musical Box to the Pianola, from Automaton Lady Virginal Players to Orchestrion.* New York: Crown, 1973.

Orvell, Miles. *The Real Thing: Imitation and Authenticity in American Culture, 1880–1940.* Chapel Hill: University of North Carolina Press, 1989.

Ott, Katherine. *Fevered Lives: Tuberculosis in American Culture since 1870.* Cambridge, Mass.: Harvard University Press, 1996.

Pacey, Arnold. *The Culture of Technology.* Cambridge, Mass.: MIT Press, 1983.

Pappas, Dennis. "Anton Friedrich von Tröltsch (1829–1890): The Beginning of Otology in Germany." *Ear, Nose, and Throat Journal* 75, no. 10 (October 1996): 50–51.

Peirce, Charles Sanders. *Philosophical Writings of Peirce.* Selected and edited and with an introduction by Justus Buchler. New York: Dover, 1955.

Peters, John Durham. "The Gaps of Which Communication Is Made." *Critical Studies in Mass Communication* 11, no. 2 (June 1994): 117–40.

———. *Speaking into the Air: A History of the Idea of Communication.* Chicago: University of Chicago Press, 1999.

———. "Helmholtz, Edison, and Sound History." In *Memory Bytes: History, Technology, and Digital Culture,* ed. Lauren Rabinovitz. Forthcoming. A translation and revision of "Helmholtz und Edison: Zur Endlichkeit der Stimme." In *Zur kulturellen und Medien-geschichte der Stimme,* ed. Sigrid Weigel, Friedrich A. Kittler, and Thomas Macho, trans. Antje Pfannkuchen. Berlin: Akademie, in press.

Petts, Leonard. *The Story of "Nipper" and the "His Master's Voice" Picture Painted by Francis Barraud.* Bournemouth: Ernie Bayly for the Talking Machine Review International, 1983.

Plann, Susan. *A Silent Minority: Deaf Education in Spain, 1550–1835.* Berkeley and Los Angeles: University of California Press, 1997.

Plato. *Collected Dialogues.* Edited by Edith Hamilton and Huntington Cairns. Princeton, N.J.: Princeton University Press, 1961.

Politzer, Adam. *History of Otology.* Translated by Stanley Milstein, Collice Portnoff, and Antje Coleman. Phoenix: Columella, 1981.

Preston, William, Jr., Edward S. Herman, and Herbert I. Schiller. *Hope and Folly: The United States and UNESCO, 1945–1985.* Minneapolis: University of Minnesota Press, 1989.

Price, Derek J. de Solla. "Automata and the Origins of Mechanism and Mechanistic Philosophy." *Technology and Culture* 5 (1964): 9–23.

Prucha, Francis Paul. *The Great Father: The United States Government and the American Indians.* Abridged ed. Lincoln: University of Nebraska Press, 1986.

Quigley, Christine. *The Corpse: A History.* Jefferson, N.C.: McFarland, 1996.

———. *Modern Mummies: The Preservation of the Human Body in the Twentieth Century.* Jefferson, N.C.: McFarland, 1998.

Rakow, Lana F. *Gender on the Line: Women, the Telephone, and Community Life.* Urbana: University of Illinois Press, 1992.

Read, Oliver, and Walter L. Welch. *From Tin Foil to Stereo: Evolution of the Phonograph.* New York: Herbert W. Sams, 1976.

Reiser, Stanley Joel. *Medicine and the Reign of Technology.* Cambridge: Cambridge University Press, 1978.

Richardson, Ruth. *Death, Dissection, and the Destitute.* New York: Routledge and Kegan Paul, 1987.

Rodman, Gilbert. *Elvis after Elvis: The Posthumous Career of a Living Legend.* New York: Routledge, 1996.

Ronnel, Avital. *The Telephone Book: Technology, Schizophrenia, Electric Speech.* Lincoln: University Press of Nebraska, 1989.

Root, Waverly Lewis, and Richard de Rochemont. *Eating in America: A History.* New York: William Morrow, 1976.

Rothenbuhler, Eric W., and John Durham Peters. "Defining Phonography: An Experiment in Theory." *Musical Quarterly* 81, no. 2 (summer 1997): 242–64.

Rüdinger, N. *Rüdinger Atlas of the Osseous Anatomy of the Human Ear, Comprising a Portion of the Atlas of the Human Ear.* Translated and with notes and an additional plate by Clarence Blake. Boston: A. Williams, 1874.

Russell, R. W. *History of the Invention of the Electric Telegraph, Abridged from the Works of Lawrence Turnbull, M.D., and Edward Highton, C.E. with Remarks on Royal E. House's American Printing Telegraph and the Claims of Samuel F. B. Morse as an Inventor.* New York: Wm. C. Bryant, 1853.

Sabine, Robert, C.E. *The History and Progress of the Electric Telegraph with Description of Some of the Apparatus.* 2d ed. New York: D. Van Nostrund, 1869.

Sacks, Oliver. *Seeing Voices: A Journey into the World of the Deaf.* Berkeley and Los Angeles: University of California Press, 1989.

Said, Edward. *Culture and Imperialism.* New York: Vintage, 1993.

Sanjek, Russell. *American Popular Music and Its Business: The First Four Hundred Years.* Vol. 2, *From 1790 to 1909.* New York: Oxford University Press, 1988.

Saxton, Alexander. *The Rise and Fall of the White Republic: Class Politics and Mass Culture in Nineteenth Century America.* New York: Verso, 1990.

Schafer, R. Murray. *The New Soundscape.* Vienna: Universal Edition, 1969.

———. *The Soundscape: Our Sonic Environment and the Tuning of the World.* Rochester, N.Y.: Destiny, 1994.

Schiller, Dan. *Theorizing Communication: A History.* New York: Oxford University Press, 1996.

Schwartz, David. *Listening Subjects: Music, Psychoanalysis, Culture.* Durham, N.C.: Duke University Press, 1997.

Schwartz, Hillel. *Culture of the Copy: Striking Likenesses, Unreasonable Facsimiles.* New York: Zone, 1996.

Sconce, Jeffrey. *Haunted Media: Electronic Presence from Telegraphy to Television.* Durham, N.C.: Duke University Press, 2000.

Scott, Joan. *Gender and the Politics of History.* New York: Columbia University Press, 1988.

Scripture, Edward Wheeler. *The Elements of Experimental Phonetics.* New York: Scribner's, 1902.

———. *Researches in Experimental Phonetics: The Study of Speech Curves.* Washington, D.C.: Carnegie Institution of Washington, 1906.

Seeger, Anthony, and Louise Spear, eds. *Early Field Recordings: A Catalogue of Cylinder Collections at the Indiana University Archives of Traditional Music.* Bloomington: Indiana University Press, 1976.

Shaw, George Bernard. Preface to *Pygmalion*. In *Bernard Shaw: Collected Plays with Their Prefaces*, vol. 4. New York: Dodd, Mead, 1972.

———. *Pygmalion*. In *Bernard Shaw: Collected Plays with Their Prefaces*, vol. 4. New York: Dodd, Mead, 1972.

Shelemay, Kay Kaufman. "Recording Technology, the Record Industry, and Ethnomusicological Scholarship." In *Comparative Musicology and the Anthropology of Music*, ed. Bruno Nettl and Philip V. Bohlman. Chicago: University of Chicago Press, 1991.

Shepherd, John. *Music as a Social Text*. Cambridge: Polity, 1991.

Shiers, George. *The Telegraph: An Historical Anthology*. New York: Arno, 1977.

Shultz, Suzanne. *Body Snatching: The Robbing of Graves for the Education of Physicians*. Jefferson, N.C.: McFarland, 1993.

Silverman, Kaja. *The Acoustic Mirror: The Female Voice in Psychoanalysis and Cinema*. Bloomington: Indiana University Press, 1988.

Silverman, Robert. "Instrumentation, Representation, and Perception in Modern Science: Imitating Human Function in the Nineteenth Century." Ph.D. diss., University of Washington, 1992.

Simpson, Christopher. *Science of Coercion: Communication Research and Psychological Warfare, 1945–1960*. New York: Oxford University Press, 1994.

Sivowitch, Elliott. "Musical Broadcasting in the Nineteenth Century." *Audio* 51, no. 6 (June 1967): 19–23.

Skoda, Josef. *Auscultation and Percussion*. Translated by W. O. Markham. Philadelphia: Lindsay and Blakiston, 1854.

Slack, Jennifer Daryl. *Communication Technologies and Society: Conceptions of Causality and the Politics of Technological Intervention*. Norwood, N.J.: Ablex, 1984.

Small, Christopher. *Music-Society-Education*. London: John Calder, 1977.

Smith, Bonnie. *The Gender of History*. Cambridge, Mass.: Harvard University Press, 1998.

Smith, Bruce R. *The Acoustic Culture of Early-Modern England: Attending to the O-Factor*. Chicago: University of Chicago Press, 1999.

Smulyan, Susan. *Selling Radio: The Commercialization of American Broadcasting, 1920–1934*. Washington, D.C.: Smithsonian Institution Press, 1994.

Snyder, Charles. "Clarence John Blake and Alexander Graham Bell: Otology and the Telephone." *Annals of Otology, Rhinology, and Laryngology*, vol. 83, no. 4, pt. 2, suppl. 13 (July–August 1974).

Solymar, Lazlo. *Getting the Message: A History of Communications*. New York: Oxford University Press, 1999.

Sousa, John Philip. "The Menace of Mechanical Music." *Appleton's Magazine* 8 (September 1906): 278–84.

Spigel, Lynn. *Make Room for TV: Television and the Family Ideal in Postwar America*. Chicago: University of Chicago Press, 1992.

Spivak, Gayatri Chakravorty. "Can the Subaltern Speak?" In *Marxism and the Interpretation of Culture,* ed. Cary Nelson and Lawrence Grossberg. Urbana: University of Illinois Press, 1988.

Stabile, Carol. *Feminism and the Technological Fix.* New York: St. Martin's, 1994.

Starr, Paul. *The Social Transformation of American Medicine.* New York: Basic, 1982.

Sterne, Jonathan. "Sounds Like the Mall of America: Programmed Music and the Architectonics of Commercial Space." *Ethnomusicology* 41, no. 1 (winter 1997): 22–50.

———. "Television under Construction: American Television and the Problem of Distribution, 1926–62." *Media, Culture, and Society* 21, no. 4 (July 1999): 503–50.

———. "Sound Out of Time/Modernity's Echo." In *Turning the Century,* ed. Carol Stabile. Boulder, Colo.: Westview, 2000.

Stool, Sylvan, Marlyn Kemper, and Bennett Kemper. "Adam Politzer, Otology, and the Centennial Exhibition of 1876." *Laryngoscope* 85, no. 11, pt. 1 (November 1975): 1898–1904.

Streeter, Thomas. *Selling the Air: A Critique of the Policy of Commercial Broadcasting in the United States.* Chicago: University of Chicago Press, 1996.

Taussig, Michael. *Mimesis and Alterity: A Particular History of the Senses.* New York: Routledge, 1993.

Théberge, Paul. *Any Sound You Can Imagine: Making Music/Consuming Technology.* Hanover, N.H.: Wesleyan University Press, 1997.

Thompson, Emily. "Mysteries of the Acoustic: Architectural Acoustics in America, 1800–1932." Ph.D. diss., University of Pennsylvania, 1992.

———. "Machines, Music, and the Quest for Fidelity: Marketing the Edison Phonograph in America, 1877–1925." *Musical Quarterly* 79 (spring 1995): 131–73.

Thompson, John. *The Media and Modernity: A Social Theory of the Media.* Stanford, Calif.: Stanford University Press, 1995.

Thornton, Russel. *American Indian Holocaust and Survival: A Population History since 1492.* Norman: University of Oklahoma Press, 1987.

Thornton, Sarah. *Club Cultures: Music, Media, and Subcultural Capital.* Hanover, N.H.: Wesleyan University Press, 1996.

Townsend, John Wilson. *The Life of James Francis Leonard, the First Practical Sound-Reader of the Morse Alphabet.* Louisville: John P. Morton, 1909.

Toynbee, Joseph. *The Diseases of the Ear: Their Nature, Diagnosis, and Treatment.* London: John Churchill, 1860.

Tröltsch, Anton von. *Treatise on the Diseases of the Ear, Including the Anatomy of the Organ.* Translated by D. B. St. John Roosa. 2d American ed., from the 4th German ed. New York: William Wood, 1869.

Truax, Barry. *Acoustic Communication.* Norwood, N.J.: Ablex, 1984.

Turino, Thomas. "Signs of Imagination, Identity, and Experience: A Peircian Semiotic Theory for Music." *Ethnomusicology* 43, no. 2 (1999): 221–55.

Unger, Roberto. *Social Theory: Its Situation and Its Task: A Critical Introduction to "Politics," a Work in Constructive Theory.* New York: Cambridge University Press, 1987.

Vail, Alfred. *The American Electric Magnetic Telegraph: With the Reports of Congress and a Description of All Telegraphs Known, Employing Electricity or Galvanism.* Philadelphia: Lea and Blanchard, 1845.

Vico, Giambattista. *The New Science of Giambattista Vico.* Translated by Thomas Bergin and Max Fisch. Ithaca, N.Y.: Cornell University Press, 1984.

Vogel, Stephan. "Sensation of Tone, Perception of Sound, and Empiricism: Helmholtz's Physiological Acoustics." In *Hermann von Helmholtz and the Foundations of Nineteenth-Century Science,* ed. David Cahan. Berkeley and Los Angeles: University of California Press, 1993.

Waksman, Steve. *Instruments of Desire: The Electric Guitar and the Shaping of Musical Experience.* Cambridge, Mass.: Harvard University Press, 1999.

Wallerstein, Immanuel. "Fernand Braudel, Historian, 'Homme de la Conjoncture.'" In *Unthinking Social Science: The Limits of Nineteenth Century Paradigms.* Cambridge: Polity Press, 1991.

Wallis, Roger, and Krister Malm. *Big Sounds from Small Peoples: The Music Industry in Small Countries.* New York: Pendragon, 1984.

Walser, Robert. *Running with the Devil: Power, Gender, and Madness in Heavy Metal Music.* Hanover, N.H.: Wesleyan University Press, 1993.

Warner, Michael. *The Letters of the Republic: Publication and the Public Sphere in Eighteenth-Century America.* Cambridge, Mass.: Harvard University Press, 1990.

———. "The Mass Public and the Mass Subject." In *Habermas and the Public Sphere,* ed. Craig Calhoun. Cambridge, Mass.: MIT Press, 1994.

Watson, Thomas A. *The Birth and Babyhood of the Telephone.* N.p.: American Telephone and Telegraph Co., Information Department, 1934. A reprint of "an address delivered before the Third Annual Convention of the Telephone Pioneers of America at Chicago, October 17, 1913."

Weidenaar, Reynold. *Magic Music from the Telharmonium.* Metuchen, N.J.: Scarecrow, 1995.

Weir, Neil. *Otolaryngology: An Illustrated History.* Boston: Butterworths, 1990.

———. "Adam Politzer's Influence on the Development of International Otology." *Journal of Laryngology and Otology* 110, no. 9 (September 1996): 824–28.

Wheatstone, Sir Charles. *The Papers of Sir Charles Wheatstone.* London: Taylor and Francis, 1879.

White, Hayden. *Tropics of Discourse: Essays in Cultural Criticism.* Baltimore: Johns Hopkins University Press, 1978.

Wiener, Norbert. *The Human Use of Human Beings: Cybernetics and Society.* New York: Doubleday, 1954.

Wilde, William R. *Practical Observations on Aural Surgery and the Nature and Treatment of Diseases of the Ear with Illustrations.* Philadelphia: Blanchard and Lea, 1853.

Wile, Frederic William. *Emile Berliner: Maker of the Microphone.* Indianapolis: Bobbs-Merrill, 1926.

Wilgus, D. K. *Anglo-American Folksong Scholarship since 1898.* New Brunswick, N.J.: Rutgers University Press, 1959.

Williams, Raymond. *The Long Revolution.* Westport, Conn.: Greenwood, 1975.

———. *Marxism and Literature.* New York: Oxford University Press, 1977.

———. *Problems in Materialism and Culture.* London: Verso, 1980.

———. *Television: Technology and Cultural Form.* Middletown, Conn.: Wesleyan University Press, 1992.

Winefield, Richard. *Never the Twain Shall Meet: Bell, Gallaudet, and the Communications Debate.* Washington, D.C.: Gallaudet University Press, 1987.

Winner, Langdon. *Autonomous Technology: Technics-out-of-Control as a Theme in Political Thought.* Cambridge, Mass.: MIT Press, 1977.

Winston, Brian. *Media Technology and Society: A History: From the Telegraph to the Internet.* New York: Routledge, 1998.

Wise, J. Macgregor. *Exploring Technology and Social Space.* Thousand Oaks, Calif.: Sage, 1997.

———. "Community, Affect, and the Virtual: The Politics of Cyberspace." In *Virtual Publics: Policy and Community in an Electronic Age,* ed. Beth Kolko. New York: Columbia University Press, 2003.

Wittgenstein, Ludwig. *Philosophical Investigations.* New York: Macmillan, 1958.

Wolff, Janet. "The Ideology of Autonomous Art." In *Music and Society: The Politics of Composition, Performance, and Reception,* ed. Richard Leppert and Susan McClary. New York: Cambridge University Press, 1987.

Woodward, James. *How You Gonna Get to Heaven If You Can't Talk with Jesus? On Depathologizing Deafness.* Silver Spring, Md.: T.J., 1982.

Yates, Michael. "Percy Grainger and the Impact of the Phonograph." *Folk Music Journal* 4 (1982): 265–75.

Young, Iris Marion. *Justice and the Politics of Difference.* Princeton, N.J.: Princeton University Press, 1990.

Young, Thomas. *A Course of Lectures on Natural Philosophy and the Mechanical Arts.* London: J. Johnson, 1807.

Žižek, Slavoj. *For They Know Not What They Do: Enjoyment as a Political Factor.* New York: Verso, 1991.

Index

Abraham, Otto, 329
Acousmatic sound. *See* Sound: acousmatic
Acoustics, 42–45, 51, 60, 62–67, 130, 268, 373–74 n.44, 382 n.51; role of instrumentation in, 58
Acoustic space. *See* Space: acoustic
Adorno, Theodor, 6, 31, 49–50, 157–58, 307, 381 n.39, 407 n.72
Aeschylus, 141
Aesthetics, 218–24, 236–42, 244–46, 266–74, 293, 295–97, 299; debates over, 274–82; of functionality, 246–56; of transparency, 218, 225, 256–61, 269, 286, 319–20, 326. *See also* Copy; Realism
Affect, 241–43, 251, 289, 305–6
Agency, 206–7, 209–13, 259, 344–45
Allison, Jay, 349–50
Altman, Rick, 18, 219
American Society of Composers, Authors and Performers (ASCAP), 190
American Telephone and Telegraph (AT&T, formerly Bell Telephone), 168–71, 188, 189, 196, 197–200, 208, 210–11, 214, 234–35, 265–66
American Telephone Booth Company, 158–59
Anatomy, 58–59, 120, 126–27, 130. *See also* Medicine; Otology; Physiology.
Anatomy acts, 68–69
Anthropology, 27, 311–26, 330–32
Apache, 315
Appointment by Telephone (film), 207–8
Arcade, 201. *See also* Phonograph: parlors
Archive. *See* Sound recording: archives
Archives of Traditional Music, 320
Armstrong, Edwin, 278
Arnheim, Rudolf, 245

Articulation: as connection, 24–26, 183, 185, 359 n.56. *See also* Sound reproduction: fidelity of; Speech
Attali, Jacques, 242–43, 310, 396 n.52
Audience, 160–67, 196, 225–29, 238–40, 245–46, 250–51, 261–65, 276, 308, 382 n.46, 395–96 n.45
Audile, 96, 372 n.17. *See also* Technique, audile
Audile technique. *See* Technique, audile
Audion. *See* Vacuum tube
Audiovisual litany, 15–19, 95, 127, 342, 346
Auenbrugger, Leopold, 119–122, 290
Augustine, Saint, 16, 23, 347
Aura. *See* Authenticity
Aural, 10–11
Auricular, 10–11
Auriscope, 57
Auscultation, mediate, 90, 102–3, 109–17, 120, 126–28; defined, 99–100, 372 n.22; vs. immediate auscultation, 100, 106–8, 110–111, 113–14; pedagogy of, 112, 115, 129–136, 161; and professionalization of medicine, 101; publicity of, 116; vs. sound telegraphy, 137, 150, 152; as way of knowing, 108, 110
Authenticity, 21, 218–21, 226–28, 261; as artifice, 241–46, 264, 285. *See also* Aesthetics; Copy; Realism
Authoritarianism, 206, 248, 280, 343
Automata, 70–77, 80–81, 367 n.86, 382 n.48
Autopsy, 121, 123–24. *See also* Dissection

Ayer, N. W. (agency), 168–71, 177, 213

Bacon, Francis, 72
Baker, Theodor, 316
Barnum, P. T., 148, 309
Barraud, Francis, 302–3
Barthes, Roland, 245, 396 n.49
Bauman, Zygmunt, 357 n.33
Baynton, Douglas, 40
Beck, Ulrich, 69
Békésy, George, and Walter Rosenblinth, 57–58
Bell, Alexander Graham, 28, 44, 66, 74, 328; and automata, 75–76; and deaf education, 36–41, 46, 385 n.21; and ear phonautograph, 31–34, 52–53, 72; funding of, 188, 398 nn.72–73; and invention, 1, 180–81, 195–89, 353 n.1; and photophone, 255–56; and sound recording, 80, 179–80, 195, 257–60, 309; and stereo telephony, 156–57; and telephone, 184–85, 234–35, 247–51, 274
Bell, Charles, 59–60, 62–63
Bell, Chichester Alexander, 186–87, 194, 236
Bell, Melville, Sr., 37, 76
Benjamin, Walter, 6, 220–21, 348
Berkhofer, Robert, 313
Berliner, Emile, 28, 46–48, 66, 77–80, 189, 194, 289, 303; on possible uses of gramophone, 204–7
Berman, Marshall, 9
Bijker, Wiebe, 370 n.115
Birth metaphor, 180–81, 213–14
Blake, Clarence, 31–32, 34, 36, 52–56, 66, 68–69, 84–85
Bleyer, J. Mount, 328
Blondheim, Menachem, 140
Boas, Franz, 313, 317

Body, 153, 167–68, 216, 228–40, 262–63, 269, 289–90, 292–97, 306, 327, 332–33, 343, 345–46, 349, 368 n.93; and diseases, 118–21; history of, 12, 20–21, 50–51; percussion of, 119–21; skull, 75; sounds of, 109; and technique, 91–92. *See also* Auscultation, mediate; Corpse; Dissection; Ear; Elias, Norbert; Mauss, Marcel; Technique, audile

Bourdieu, Pierre, 92–93, 336, 360 n.1

Brady, Erika, 317–19

Broadcasting, 192–96, 204–7, 381 n.43, 387–88 n.47, 388 n.52. *See also* Radio; Sound recording; Telephone

Brooks, Tim, 242

Buisson, Matthieu-François-Régis, 100, 122, 128

Burdick, Alan, 2

Bureaucracy, 147, 200, 209–13, 328

Bureau of American Ethnology, 314, 322–23

Cahill, Thaddeus, 366 n.70

Calinescu, Matei, 310

Cammann, George, 112, 155

Canning. *See* Preservation: of food

Capital, 149, 168, 210, 212, 310

Capitalism, 3, 147, 183–84, 327, 336; industrial, 183, 292–93, 388 nn.47, 52. *See also* Class; Modernity; Music: political economy of; Sound: political economy of

Carey, James, 140–41, 206

Carrick, George, 156

Caruso, Enrico, 215–16, 223, 264

Castel, Louis-Bertrand, 73

Chappe, Claude, 141

Chladni, Ernst Florens Friedrich, 43–44, 50, 58, 290

Christianity, 294, 303–4. *See also* Spiritualism: Christian

Cinema, 245, 259, 360–61 n.3

Civil War (U.S.), 53, 292–93, 295–96, 404 n.21

Clark, George H., 387–88 n.47

Class, 25, 28, 37, 69–70, 92, 93, 95, 98–99, 113–17, 118, 136, 153–54, 160–61, 172, 174, 177, 183–84, 185, 191–92, 196–214, 228, 243–44, 279–80, 291, 293–96, 325, 344, 350, 367 n.82, 390 n.86

Cohen, Lizabeth, 213

Columbia Phonograph Company, 199, 244, 300, 327

Commodity. *See* Culture: consumer; Reification; Sound: commodification of

Communication, 20–22, 25, 37, 49, 143, 151, 153, 158–59, 172, 192, 206, 209, 217–18, 220–21, 226–28, 234–35, 239, 261–66, 284, 335, 342–45, 389–90 n.78

Consumerism. *See* Culture: consumer

Cook, William. *See* Telegraph: Cook and Wheatstone

Copy, 21, 25–26, 219–23, 226, 241, 247, 257, 259, 261–65, 267, 277, 282–85, 402 n.129. *See also* Aesthetics; Authenticity

Corbett, John, 20, 258–59

Corbin, Alain, 360 n.58

Corporations, 209–13, 279, 327. *See also* American Telephone and Telegraph (AT&T); Radio Corporation of America (RCA); Western Union

Corpse, 68–70, 294–97, 332, 404 n.21. *See also* Body; Embalming

Correspondence. *See* Aesthetics: of transparency

Corvisart, J. N., 105

Cosmopolitanism, 208–9

Cowan, Ruth Schwartz, 243, 293

Crary, Jonathan, 60–61
Cros, Charles, 41, 77–80
Culture, 203, 291–92, 305, 307, 311–25, 329–32; consumer, 203–4, 209, 243, 269–70, 288, 292–93, 336; mass, 25, 157, 167, 209, 228, 239, 245–46, 288, 291–92, 383 n.54. *See also* Communication; Nature; Preservation; Sound culture; Sound reproduction: as cultural artifact; Visual culture
Cushing, Frank Hamilton, 315
Cybernetics, 143
Czitrom, Daniel, 140–41, 147

Darwin, Erasmus, 74
Dawes Act, 312–13, 322
Deaf, the, 36–41, 346–47, 361–62 n.14
Deafness: cultural status of, 106, 346–47, 361–62 n.14; quack cures for, 81–83
Death, 289–309, 318–19, 332
Deconstruction. *See* Derrida, Jacques
Decorum, 116, 172, 265–67, 279–80
DeForest, Lee, 189
Delegation. *See* Hearing: delegation of; Speech: delegation of
DeLoria, Philip, 312
Dempsey-Carpentier fight (1921), 195–96, 245–46, 396 n.48
Densmore, Frances, 316, 323–24
Derrida, Jacques, 17, 46, 49–50, 298
Descartes, René, 72–73, 368 n.93
Descriptive specialty, 243–45, 309
Detail, 115, 157–58
Diagnosis, 117–18, 127
Diaphragm. *See* Tympanic function: diaphragm in
Digital sound. *See* Sound reproduction: digital

Dimitriadis, Greg, 394–95 n.31
Disgust, 114–16
Dissection, 55–57, 67–70, 118, 365 n.51, 375 n.63. *See also* Autopsy; Body; Ear; Medicine: professionalization of
Distance. *See* Space
Domesticity, 208–9
Douglas, Susan, 188–89, 402 n.6
Doxa, 336, 348
Duffin, Jacalyn, 100–103
Duthernoy, Leon Alfred, 239–41, 308–9

Ear, 31–32; abstraction of, 52, 56–57, 64, 67, 70, 83–85, 111, 239; as model for sound reproduction, 33–34, 39, 70, 77–81, 337, 370 n.116; status in relation to eye, 53–54, 143–44; understandings of, 51, 54–57, 62, 63, 65
Ear phonautograph, 22, 31–35, 41, 51, 338, 364 n.40; human ear in, 52–53, 69–70, 72
Edison, Thomas, 28, 41, 46, 80, 179–81, 256, 383–84 n.1, 384 n.2, 397 n.63; funding of, 188, 200; and invention, 1, 185–89, 250, 353 n.1; and possible uses of the phonograph, 199, 201–2, 305
Edison Phonograph Company, 261–65, 269
Elias, Norbert, 114, 118, 172, 374 n.51
Embalming, 12, 26, 294–97, 306–7, 404 n.21. *See also* Corpse; Death; Preservation
Embodiment. *See* Body; Hearing; Listening; Sound; Space; Speech; Technique, audile; Vision; Voice
Empiricism, 99, 106, 118–19
Enlightenment, 2–3, 51, 93, 110, 183, 343

Epistemic individual, 360 n.1
Erotics. *See* Sexuality
Ethnocentrism: audism, 28, 345–47; Eurocentrism, 27–28
Ethnography. *See* Anthropology; Sound recording: ethnographic
Ethnology. *See* Anthropology
Etiquette. *See* Decorum
Eugenics, 39–41, 315
Experience, 13, 19, 28–29, 133–36, 153, 154, 161, 167, 183, 241–44, 259, 261–66, 343, 345, 350. *See also* Audiovisual litany; Body; Communication; Culture; Sound: exteriority of
Exteriority. *See* Sound: exteriority of

Faber, Johannes, 71, 76–77, 79
Fabian, Johannes, 9, 27, 311, 313
Family album, 199, 203, 389 n.73
Farrar, Geraldine, 215–16
Federal Communications Commission (FCC), 278
Federal Radio Commission (FRC), 199; "public interest, convenience, or necessity," 199
Feinstein, Robert, 303–4
Feld, Steven, 97, 358 n.48, 411 n.13
Fewkes, Jesse Walter, 87–88, 315–21, 324–26, 331
Fidelity, Sound. *See* Sound reproduction: fidelity of
Fillmore, John Comfort, 317
Film. *See* Cinema
Fletcher, Alice, 316, 321–22, 324, 331
Flinn, Caryl, 15
Flint, Austin, 109, 112, 132–35
Foley effects, 245. *See also* Aesthetics; Descriptive specialty; Realism
Folklore. *See* Anthropology
Forbes, John, 126

Foucault, Michel, 12, 33, 91, 97, 99, 110, 116, 127, 294, 369–70 n.114
Freud, Sigmund, 375–76 n.75
Funeral, 291, 295–96, 303–4

Gallaudet, Edwin Miner, 39–40
Gender, 28, 107, 113–17, 152–54, 160, 166–74, 179–81, 197–98, 207–8, 210, 212–14, 226–30, 238, 272–73, 287, 359 n.55, 384 n.5, 388 n.56
Giddens, Anthony, 151, 153
Gilman, Benjamin Ives, 316–17, 321
Goffman, Erving, 151
Gramophone, 46–48, 77, 203–7, 236–37, 302–3, 305; compared with phonograph and graphophone, 203–4; instructions for using, 268–69; timbre of, 278–80; trade classification of, 195. *See also* Bell, Alexander Graham; Bell, Chichester Alexander; Berliner, Emile; Cros, Charles; Edison, Thomas; Graphophone; Phonograph; Sound recording; Sound reproduction; Tainter, Charles Sumner
Grant, Ulysses S., 312
Graphophone, 80, 179–80, 256–58, 260, 321, 323; depiction of, 230–34; timbre of, 278–80. *See also* Bell, Alexander Graham; Bell, Chichester Alexander; Berliner, Emile; Cros, Charles; Edison, Thomas; Gramophone; Phonograph; Sound recording; Sound reproduction; Tainter, Charles Sumner
Grave robbing, 67–68. *See also* Dissection

Gray, Elisha, 249, 270
Gumpert, Gary, 309–10

Habermas, Jürgen, 390 n.79, 394 n.26
Habitus, 92–93, 108, 113, 116, 172, 344. *See also* Bourdieu, Pierre; Class; Decorum; Elias, Norbert; Gender; Race
Hall, Edward, 151
Hall, Stuart, 20
Hankins, Thomas, and Robert Silverman, 58
Harrington, John Peabody, 314
Hastings, Dennis, 330–31
Hayward, Harry, 241–42
Headphones, 24, 87–89, 154–77, 319
Hearing: delegation of, 38–41, 83–84, 155, 233–35, 326, 331; history and theory of, 23, 34–35, 45, 51–70, 100, 154, 346–47, 371 n.15; as learned faculty, 12–13, 166; as "novel sense," 95, 117, 137, 281; supplementation of, 83–85, 106, 108, 110, 168–172, 257–58; words for, 9–10, 96. *See also* Audiovisual litany; Auscultation, mediate; Ear; Listening; Sound; Sound reproduction; Technique, audile
Hearing aids, 81–83, 105–6
Hegel, George Wilhelm Friedrich, 285, 339, 411 n.8
Heidegger, Martin, 355 n.19
Helmholtz, Hermann, 62–67, 77–81, 131–32, 142, 290, 366 n.70
Hemenway, Mary, 315, 321
Henry, O., 309
Hinsley, Curtis, 318–19
Hippocrates, 119, 128
"His Master's Voice," 301–3, 306–7

Historiography. *See* Sound: historiography of
Hooke, Robert, 141
Hopi, 317, 320–21
Horkheimer, Max, 157–58, 381 n.39. *See also* Adorno, Theodor
Hornbostel, Erich von, 329–30
Hoxie, Fred, 313–14
Hubbard, Gardiner, 188
Human telephone, 81–82. *See also* Deafness; Ear
Huyssen, Andreas, 228

Ideology, 20, 102, 168, 181, 263–64, 274, 285, 330. *See also* Audiovisual litany; Class; Enlightenment; Modernity; Technological determinism; Technology
Indestructible Phonograph Record Company, 299–300, 404 n.31. *See also* Preservation: of sound recordings
Industrialization. *See* Capitalism: industrial; Modernity
Innis, Harold, 140
Invention, history of, 185–86, 191. *See also* Bell, Alexander Graham; Edison, Thomas; Media: as industries; Research and development (R&D)
Itard, Jean-Marc Gaspard, 12, 19, 59

Jeffries, Lennard, 40
Johnson, Edward, 179–80, 188, 383–84 n.1, 384 n.2
Johnson, Eldridge, 236, 297
Johnson, James, 97–98, 167
Jukebox, 201

KDKA (Pittsburgh), 196, 388 n.50, 396 n.48
Kempelen, Wolfgang von, 74, 75, 76

Kenney, William, 163–65, 168, 174, 177, 243
Kittler, Friedrich, 41, 145, 302, 309
König, Rudolf, 77, 369 n.103
Kroeber, A. L., 318

Labor, 210, 212, 242–43, 309, 344
Laboratory. *See* Menlo Park; Volta Lab
Laennec, René-Théophile-Hyacinthe, 24, 90, 100, 104, 115, 155, 168, 269, 290, 377 n.99; body of, 114; criticisms of, 132–35; and invention of stethoscope, 101–3; on method, 120–23, 129–32; objections to immediate auscultation, 106–8, 110–11, 113–14; political belief of, 116; on sounds of the chest, 124–27
Language. *See* Sound reproduction: and language
Lastra, James, 358 n.47, 402 n.129
Latour, Bruno, 247, 254, 306, 393 n.21
LeMahieu, D. L., 6, 288
Leonard, James Francis, 147–48
Leppert, Richard, 160
Levin, Thomas, 363 n.28
Lévi-Strauss, Claude, 83
Library of Congress, 319, 325–26, 328, 330, 395–96 n.45
Lincoln, Abraham, 296
Lippincott, Jesse, 200
"Listeneritis," 272–73
Listening, 92, 96–99, 140, 167; active vs. passive, 100; "alone together," 138, 161–68; collective, 160–62, 167–68; as connoisseurship, 267–82; experience of, 241–43; vs. hearing, 19, 96; vs. looking, 167, 363–64 n.37; mediation of, 100–101; "structure of listening," 371 n.13; symbolic currency of, 94, 127–28, 137, 143–47, 174, 226–28; as test of sound reproduction, 246–56; virtuosity in, 106, 137–38, 149, 154. *See also* Auscultation, mediate; Hearing; Modernity; Sound; Technique, audile
Liveness, 221, 237, 239–40, 245–46, 262–64, 275–77, 284, 332, 392 n.4
"Lo-fi," 400–401 n.115
Lomax, John, 325
Looking. *See* Vision
Lost and Found Sound (radio show), 349–50
Lukács, Georg, 182
Lyon, Janet, 310
Lyotard, Jean-François, 335

Machine. *See* Technology
Maillardet, Henri, 77
Manometric flame, 369 n.103
Manuel, Peter, 277
Marconi, Gulielmo, 1, 48
Martin, Michèle, 198, 208
Marvin, Carolyn, 151–52, 194, 197, 273
Marx, Karl, 5, 12–13, 242–43, 335, 348
Mass culture. *See* Culture: mass
Mauro, Philip, 328
Mauss, Marcel, 13, 91–92
McKinley, William, 309
McLuhan, Marshall, 151, 251. *See also* Media: as extensions of senses
Media: as crystallized sets of social practices and relations, 160, 182–85, 191–96, 202, 223, 308–9, 315; and death, 291; as extensions of senses, 38, 65, 92, 154, 244, 289; as industries, 177, 185–91; as mediation, 20–22, 168,

Media (*continued*)
218–19, 222; media tourism, 242–44; as networks, 25, 150–54, 158, 177, 210, 213–14, 225–40, 246–56; plasticity of, 181–82, 184; senses as, 61, 72; sound, 25, 84, 179–214. *See also* Death; Sound reproduction; Technology

Mediation, 138, 163, 218, 222, 326–27; as epistemology, 106–8, 110, 128; vanishing, 147, 218, 225, 256–59, 261–66, 269, 282–86, 392 n.5. *See also* Listening: mediation of; Media: as mediation; Technology: as mediation

Medicine, 24, 373 n.37; orientations toward listening in, 93, 95, 99–136; professionalization of, 67–70, 101, 113, 115, 374 n.60, 375 n.69. *See also* Diagnosis; Ear: understandings of; Otology; Physiology; Stethoscope

Menlo Park, 186, 250

Merleau-Ponty, Maurice, 14

Metaphysics, 285, 289; of presence, 17, 48–49, 298, 345

Middle class. *See* Class; Sound reproduction: and middle class

Millard, Andre, 250

Milwaukee Wisconsin Telephone Company, 194

Mimesis, 241, 245, 264, 283

Mobile privatization, 389 n.63

Modernity, 9–10, 95, 118, 136, 140, 308, 322, 335, 337, 340, 349–50, 356–57 n.33, 380 n.32; bourgeois, 310–11, 332; defined, 9; as disorienting, 21; and ideology of progress, 9, 27, 310, 312–14, 318, 329–30; and listening, 92–93, 99, 101, 127, 136, 174, 281; and modernization, 9, 285; and sound 2–5, 9, 181–82, 206, 333, 343, 350. *See also* Time

Mojave, 318

Morgan, Lewis Henry, 313

Morley, John, 302

Morse, Samuel, 144–46, 148, 180. *See also* Morse code; Telegraph: Morse; Vail, Alfred

Morse code, 48, 145–46, 152, 225, 379 n.14

Mouth, 236, 239; as model of sound reproduction, 33, 70, 73–78, 367 n.86

Mowitt, John, 277, 371 n.13, 402 n.130

Müller, Johannes, 11, 60–64, 66, 76, 366 n.64

Munro, Alexander, 59

Music, 23, 26, 63, 65, 157–58, 160–62, 168, 184, 196, 243–45, 259, 269–70, 288–89, 316–25, 328–32, 337–38, 350–51, 359 n.49, 372 n.19; "canned," 292, 403 n.13; as metaphor, 152, 258, 275–77, 280–82, 382 n.48; notation, 41; political economy of, 385 n.11; publishing, 190; romantic notions of, 15, 357 n.38; as writing, 49–50. *See also* Aesthetics; Musicians

Musicians, 190, 221, 225–26, 236–40, 262–64, 275, 289, 308, 315, 350–51

Muzak, 189, 337, 382 n.46

Nasaw, David, 191, 201, 205

Nation, 28, 206, 211, 213, 363 n.28

National Public Radio, 349–51

Native Americans: and sound recording, 311–25, 330–32, 407 n.70; and U.S. policy, 312–15, 322,

331. *See also* Apache; Hopi; Mojave; Navajo; Omaha; Passamaquoddy; Race; Winnebago; Zuni
Nature, 10–19, 220, 313, 340–42; as concept, 11; vs. culture, 7, 10; historicity of, 12. *See also* Culture; Ear; Sound
Navajo, 315
New Jersey Telephone Herald, 193–94
Nicolson, Malcolm, 108
Nietzsche, Friedrich, 13, 308, 356–57 n.33
Noise, 166–67, 184, 279–81. *See also* Sound
Nostalgia, 220, 315, 321–22, 344, 350. *See also* Sound: historiography of; Time

Omaha, 330–31
Ong, Walter, 16–19, 342, 345, 347. *See also* Orality; Sound: historiography of
Ontology, 218–20
Orality, 16–19, 24, 45, 61, 342, 358 n.48. *See also* Feld, Steven; Nature; Ong, Walter; Sound: historiography of
Oratory, 303–6, 308–9, 327–28. *See also* Speech
Original. *See* Copy
Otology, 51, 53–57, 60, 64, 67–70, 365 n.51
Otoscope. *See* Auriscope

Pacey, Arnold, 191, 363 n.30
Parlor, 204. *See also* Phonograph: parlors; Space
Passamaquoddy, 315, 318–20
Patents, 181, 186–87
Peirce, Charles Sanders, 42, 130
Perception. *See* Ear; Hearing; Helmholtz, Hermann; Müller, Johannes; Senses; Tympanic function; Vision
Percussion. *See* Auenbrugger, Leopold; Body
Performance, 225–40, 249–51, 255, 261–65, 304–9, 317, 319–24. *See also* Copy; Musicians; Oratory; Speech
Periodization, 339–40. *See also* Sound: historiography of
Peters, John Durham, 37, 65, 218–19, 251, 291, 302
Phenomenology, 13–19, 21
Philadelphia International Exposition (1876), 57, 248–49
Phonautograph, 31, 35–36, 38, 41, 45–46, 51–52, 64, 71–72, 74, 80. *See also* Ear phonautograph; Scott de Martinville, Edouard-Léon; Sound reproduction
Phonogram, The (periodical), 199, 201–2
Phonogramm-Archiv, 328–29
Phonograph, 6, 22, 46, 70–71, 79, 81, 163–65, 168, 179–80, 187–88, 243, 261–65, 303–6, 319–20; business, 211–13; coin-in-the-slot, 201; compared with gramophone, 203–4; directory, 199–200, 389 n.69; inspection of, 269; parlors, 162; postal, 194–95; timbre of, 280, 401 n.124; tinfoil, 298. *See also* Bell, Alexander Graham; Bell, Chichester Alexander; Berliner, Emile; Cros, Charles; Edison, Thomas; Gramophone; Graphophone; Sound recording; Sound reproduction; Tainter, Charles Sumner
Phonography. *See* Sound recording
Phonoscope, The (periodical), 201–2
Photography, 57, 291, 293, 366 n.64

Photophone, 180–81, 187, 230–34
Physiology, 51, 57–67, 103, 117, 120, 126–27, 130. *See also* Anatomy; Medicine; Otology
Plato, 16
Politzer, Adam, 56–57, 64, 66, 365 n.51
Power. *See* Agency; Authoritarianism; Class; Deaf, the; Gender; Race; Sexuality
Presence. *See* Authenticity; Body; Communication; Copy; Experience; Liveness; Metaphysics: of presence
Preservation: of culture, 311–25, 330–31; of food, 12, 26, 292–93; of sound recordings, 24, 299, 309, 325–33, 397 n.63, 399 n.90, 404 n.31, 408 n.93, 409 n.98; of the voice, 12, 49, 287–90, 293–94, 297–99, 303–7, 311. *See also* Embalming; Sound recording: archives
Privacy, 138, 158–63, 167, 173–74, 199, 207–8, 295, 381 n.42, 382 n.46
Progress. *See* Modernity: and ideology of progress; Technology: and ideology of progress
Property, 95, 138, 159–63, 167, 173–74, 310, 312, 381 n.43
Prucha, Francis, 312
Psychoanalysis, 375 n.75
Public, 199, 207–8, 250–54, 295. *See also* Habermas, Jürgen; Radio: public vs. private; Sound recording: public vs. private; Telephone: public vs. private; Warner, Michael
Puskás, Tivador, 192–93
Pygmalion, 37–38

Quigley, Christine, 294

Race, 27–28, 158, 191, 228, 311–26, 329–30, 391 n.88, 406 n.60
Radio, 6, 22, 183, 237–40, 245–46, 309, 402 n.6; as broadcasting, 195–96, 387–88 n.47, 388 n.52; FM, 278; invention of, 1; and listening, 165–68, 172–74; names for, 50; as point-to-point medium, 195; public vs. private, 208–9; regulation of, 199; relation to sound recording and telephone, 189–90; vs. sound recording, 276–77; timbre of, 270–73, 276–77, 280–82. *See also* Federal Communications Commission; Federal Radio Commission; Sound reproduction
Radio Corporation of America (RCA), 195–96, 225–26, 235, 245–46, 278, 301–3
Rakow, Lana, 198
Rationality, 93, 94, 101, 114, 127–28, 136, 138, 356–57 n.33, 374 n.51. *See also* Audiovisual litany; Enlightenment; Modernity; Sound; Technique, audile; Vision
Read, Oliver, and Walter L. Welch, 199, 277–78, 409 n.98
Realism, 241–46, 258, 264, 285, 297, 324. *See also* Aesthetics; Authenticity; Copy
Reason, 99, 127–28, 136, 138, 183. *See also* Enlightenment; Modernity; Rationality
Recording. *See* Sound recording
Reification, 112–13, 182, 210, 310
Reis, Philip, 78
Reis telephone, 78
Representation, 215–16
Research and development (R&D), 185–86, 189
Richardson, Ruth, 68–69

Romanticism, 15
Rothenbuhler, Eric, 218–19
Rousseau, Jean-Jacques, 382 n.47

Sacks, Oliver, 347
Schaeffer, Pierre, 20
Schafer, R. Murray, 342–43, 345
Schizophonic sound. *See* Sound: schizophonic
Scientific American, 249, 297–99, 307
Scott de Martinville, Edouard-Léon (Leon Scott), 31, 35–36, 45–51, 78, 363 n.28. *See also* Phonautograph
Scripture, Edward Wheeler, 48–49
Seeger, Anthony, and Louise Spear, 326
Semiotics, 94, 103, 118, 112, 123–36, 150
Senses: history of, 5, 16, 18, 50, 61, 72, 142–43, 335, 340; separation of, 59–63, 66, 93, 110–112, 150, 155, 157, 167. *See also* Body; Ear; Hearing; Listening; Media; Sound; Sound reproduction; Vision
Sensorium, 16–17, 153. *See also* Senses
Sexuality, 73, 172–73, 226–28, 273
Shaw, George Bernard, 37–38, 46
Sight. *See* Vision; Visual culture
Sign language, 40
Sirens, 171–72
Skoda, Josef, 132, 135, 377 n.99
Smithsonian, 309, 328
Snake dance, 319–20. *See also* Aesthetics: of transparency; Fewkes, Jesse Walter; Trace
Snyder, Charles, 69
Social construction, 5, 7, 18. *See also* Nature
Social contract theory, 382 n.47
Solla Price, Derek J. de, 72

Sound: acousmatic, 20–21, 25, 215–86; commodification of, 159–60, 162–65, 167–68, 183–84; defined, 11–12; as "effect," 60; exteriority of, 13, 19, 24, 37, 61, 111–12, 128, 138, 150, 259–60, 283, 289, 298, 306–7, 320, 322–24, 332–33, 343–44, 380 n.34; historiography of, 28–29, 33, 91, 142, 339–51, 360 n.1, 411 n.6; history of, 10, 13, 20, 138, 347; metalanguage of, 94, 103, 128–36, 377 n.100; natural dimensions of, 11–12, 14–19; objectification of, 117; philosophy/theory of, 14–19, 23, 33, 39, 42–44, 60, 62, 71–85, 358 n.47; political economy of, 385 n.11; rationalization of, 131; schizophonic, 20–21, 25, 215–86; as sign, 118, 122–36; as "special case" in cultural theory, 14; visible, 38, 41, 44–50; writing of, 36–37, 41, 45–50. *See also* Ear; Hearing; Listening; Media; Modernity; Sound culture; Sound recording; Sound reproduction; Upper partials
Sound culture, 206, 214, 219, 285, 288, 292, 299, 305, 207, 336–39, 342–46, 349–51, 390 n.78; study of, 3–5, 27–29, 348
Sound recording, 183, 187, 190–91, 218–19, 235–37, 261–65, 289, 306, 323, 396 n.52; and agency, 211–12; archives, 288, 327–32, 337, 349–51; as broadcasting, 204–7; business vs. entertainment uses, 199–204, 207, 390 n.83; and death, 290–94, 297–309, 311–25; and eavesdropping, 273, 294 n.25; electric vs. acoustic, 276–78; ephemerality of, 287–89, 298–300, 310–11, 324–26, 350;

Sound recording (*continued*)
ethnographic, 313–33; and the future, 307–11; invention of, 1, 250; mass production vs. artesinal, 207; multitrack, 50; names for, 50; privilege of, 50; public vs. private, 207; relation to radio and telephone, 189–90; speed in playback of, 267–68; timbre of, 272; vs. transmission, 197. *See also* Bell, Alexander Graham; Bell, Chichester Alexander; Berliner, Emile; Cros, Charles; Edison, Thomas; Gramophone; Graphophone; Phonograph; Preservation; Sound; Sound reproduction; Tainter, Charles Sumner

Sound reproduction, 183; banality of, 6; as cultural artifact, 8, 290, 299, 338–39, 341, 344, 381 n.43; defined, 19–22, 34, 38, 51; digital, 22, 218–19, 277, 284–85, 335–39, 363–64 n.37; fidelity of, 25–25, 215–86, 397 n.63, 399 n.90; geographic development of, 192, 380 n.33; historical significance of, 6, 50, 84, 350; and history of invention, 185–91; and language, 48–49, 248–56, 283, 363 n.28, 370 n.1, 377 n.100, 407 n.71; and listening, 87–90, 93, 157–77; as mediation, 218–19, 222; and middle class, 201–14; models of, 33, 70–85; as network, 225–40, 246–56, 343; as object of study, 7, 19, 28, 33, 84, 223, 348; plasticity of, 181–82, 184, 191, 206, 214; possibility of, 2, 21–23, 33, 65, 99, 217, 220, 222, 241, 247–51, 255–56, 282–85, 332–33, 341; social character of, 185–91, 191–96, 219, 221, 223, 236, 239–41, 259, 274–83;

timbre of, 256–83. *See also* Hearing: delegation of; Media; Sound; Sound recording; Technique, audile; Tympanic function

Sousa, John Philip, 292, 403 n.13

Space, 9, 27, 192, 206, 210, 211, 244, 283, 311–12, 394 n.26; acoustic, 24, 50, 93; 103, 113–17, 138, 151–77, 308, 381 n.42, 382 n.46; analogy between physical and social, 107, 109, 113–17; of studio, 236–40

Spaeth, Sigmund, 281–82

Speech, 36–38, 117, 123, 126, 168, 261, 290, 298–99, 314, 327–28, 342–43, 375 nn.63, 75, 376 n.80; as defining human characteristic, 40; delegation of, 233–35, 263–64, 306–11; modeling of, 73–77; speech science, 49; as test of sound reproduction, 247–55; as type of sound, 43, 46; visible, 37–38

Spiritualism, 289, 291, 402 n.6; Christian, 14, 16–18, 23, 296

Squier, George O., 189

Squirrels, 11–12

Starr, Paul, 67–68, 101

Stenographers, 45–46, 49, 200–201, 212, 305

Stereo, 156–57, 193. *See also* Sound; Sound recording; Sound reproduction; Stethoscope: binaural

Stethoscope, 24, 90, 98, 99, 123, 125–26, 129, 136, 155, 225, 373–74 n.44, 376 n.76; binaural, 111–13; differential, 155–56; as innovation of ear trumpet, 105–6; instructional, 161–62; invention of, 101–3; monaural, 104–106; names for, 104. *See also* Auscultation, mediate; Laennec, René-Théophile-Hyacinthe

Studio, 205, 219–21, 223, 225–40,

297, 309, 321–24, 402 n.130
Stumpf, Carl, 317, 329
Subjectivity. *See* Body; Class; Experience; Gender; Hearing; Listening; Race; Senses; Sexuality; Speech; Vision
Synaesthesia, 39, 50, 56, 141–43
Synthesizer, 366 n.70

Tainter, Charles Sumner, 179, 186–87, 188, 228–30, 235, 256–61, 383–84 n.1
Technique, audile, 23–25, 92–95, 98–136, 142, 157–58, 407 n.72; limits of, 174; and sound reproduction, 161–77, 223–24, 256–65, 317; and telegraphy, 146–54; and timbre, 267–74. *See also* Body: and technique; Bourdieu, Pierre; Elias, Norbert; Habitus; Listening; Mauss, Marcel
Technological determinism, 7–8, 181, 355 n.19, 384 n.7
Technology, 180–83, 191, 203, 210, 217–19, 233–36, 244, 262–63, 269, 284, 288, 332, 336–39, 370 n.115, 374 n.47, 378 n.7; human assistance for, 246–56; and ideology of progress, 222, 274–82; and mechanistic philosophy, 72–73; as mediation, 107, 392 n.5; vs. medium, 206, 213–14, 251; and social relations, 225, 341; study of, 7–8, 348; and "virtuosity values," 363 n.30. *See also* Gramophone; Graphophone; Hearing: delegation of; Media; Mediation; Phonograph; Radio; Sound recording; Sound reproduction; Speech: delegation of; Stethoscope; Technique, audile; Technological determinism; Telegraph; Telephone
Telegraph: acoustic, 378–79 n.9; Cook and Wheatstone, 144, 378–79 n.9; electrical, 142; electrical vs. semaphoric, 141; historical significance of, 140–41, 146, 366 n.64; Morse, 142, 145–54, 180, 185; printing, 139
Telegraph operators, 24; and "crimes of confidence," 151–52, 173; and listening, 93, 137–40, 147–54, 158–59; professionalization of, 140, 154
Telegraphy, 76, 137, 168, 380 n.32; sound, 98, 139–40, 142, 146–54, 379 n.19, 380 n.23; wireless, 379 n.18. *See also* Auscultation, mediate; Synaesthesia
Telephone, 6, 22, 64–65, 70–71, 79, 81, 177, 183, 237–38, 265–66, 290, 337, 402 n.6; and agency, 210–11; booth for, 158–59; broadcasting, 192–94; business vs. entertainment uses, 197–98; as convenience, 198–99; depiction of, 230–34; directory, 200; exclusivity of, 198, 390 n.86; instructions for using, 266–67; invention of, 1, 41, 72, 247–48; and listening, 168–72; long distance, 208; as medium, 182; name of, 50; ear telephone, 84–85; public demonstrations of, 249, 250–51; public vs. private, 207–8; relation to radio and sound recording, 189–90; timbre of, 270–72, 401 n.118. *See also* Agency; Bell, Alexander Graham; Sound reproduction; Watson, Thomas
Telephon Hirmondó, 193
Telharmonium, 366 n.70
Théâtrophon (Paris), 192–93
Thompson, Emily, 263–64, 370 n.6, 398 nn.75, 77
Thornton, Sarah, 221

Thunder, Henry, 323
Time, 9, 27, 206, 242–43, 283, 294, 307–14, 317–19, 324–25, 328–32. *See also* Modernity; Nature: historicity of; Sound: historiography of; Space; Technology: and ideology of progress
Tone Tests, 261–65
Toynbee, Joseph, 55, 57
Trace, 153, 206, 309. *See also* Copy
Transducer, 22, 31, 33–35, 62, 277. *See also* Ear; Tympanic function
Tröltsch, Anton von, 54
Truax, Barry, 96–97
Tuba stentoro-phonica, 41–43, 53
Tympanic function, 22–23, 34–35, 38–39, 46, 50, 52, 59, 62, 64, 67, 71, 77–81, 83–85, 233–34, 337, 347, 360 n.2, 369 n.103; diaphragm in, 35–36, 56, 66, 72. *See also* Transducer

Upper partials, 64–65, 131, 157, 366 n.69

Vacuum tube, 189, 193, 276–77, 386 n.29
Vail, Alfred, 144–45, 147
Vanishing mediator. *See* Mediation: vanishing
Vaucanson, Jacques de, 73, 80–81
Vaudeville, 190–91, 243–44
Vico, Giambattista, 339–40
Victor Talking Machine Company, 215–17, 222–23, 269–70, 275–76, 302–3
Victrola, Orthophonic, 275–78
Visible speech. *See* Speech: visible
Vision, 5–6, 15–19, 363–64 n.37; prestige of, 50, 104, 122, 127; and proof, 121–22, 143–49, 353 n.5. *See also* Audiovisual litany; Sound: visible; Visual culture
Visual culture, 5–6, 20, 353 n.6; study of, 3–4, 10, 101, 143–47. *See also* Audiovisual litany; Sound: visible; Vision
Voice, 23, 76, 275, 319, 344–45, 359 n.55; disembodiment of, 1, 290, 349; as sound, 122–26, 128, 281, 337, 343; "voices of the dead," 26–27, 288–90, 293–94, 297–99, 301–10, 314, 317–19, 322–23, 328, 330, 332. *See also* Body; Canning; Embalming; Mouth; Speech
Volta Lab, 179–80, 195, 254–60, 278, 383–84 n.1; economics of, 186–89, 398 n.72

Warner, Michael, 106, 378 n.3, 390 n.79
Watson, Thomas, 1, 84–85, 186, 237–38, 248–51
Wave siren, 77
Western Union, 185, 197, 200
Wheatstone, Charles, 44, 75, 76, 144
White, Hayden, 29
Wilgus, D. K., 325, 408 n.93
Williams, Raymond, 219, 398 n.63
Winnebago, 323
Winner, Langdon, 374 n.47
Wise, J. MacGregor, 243
Woolgar, Steve, 254, 393 n.21

Young, Thomas, 44, 102

Zuni, 315, 317

Jonathan Sterne teaches in the Department of Communication and the Program for Cultural Studies at the University of Pittsburgh. He writes about media, technology, and the politics of culture, and is codirector of the on-line magazine *Bad Subjects: Political Education for Everyday Life.*

Library of Congress Cataloging-in-Publication Data
Sterne, Jonathan.
The audible past : cultural origins of sound reproduction / Jonathan Sterne.
p. cm.
Includes bibliographical references and index.
ISBN 0-8223-3004-0 (cloth : alk. paper)
ISBN 0-8223-3013-X (pbk. : alk. paper)
1. Sound—Recording and reproducing—History. 2. Sound recording industry—Social aspects. 3. Sound in mass media. 4. Sound recordings—Social aspects. 5. Popular culture. I. Title.
TK7881.4 .S733 2002
621.389'3'09—dc21   2002009196